**Sexual Selection
and the
Barn Swallow**

Oxford Series in Ecology and Evolution
Edited by Robert M. May and Paul H. Harvey

Sexual Selection and the Barn Swallow

Anders Pape Møller

Department of Zoology
Uppsala University, Sweden

with drawings by
Jens Gregersen

Oxford   New York   Tokyo
OXFORD UNIVERSITY PRESS
1994

Oxford University Press, Walton Street, Oxford OX2 6DP

Oxford New York Toronto
Delhi Bombay Calcutta Madras Karachi
Kuala Lumpur Singapore Hong Kong Tokyo
Nairobi Dar es Salaam Cape Town
Melbourne Auckland Madrid
and associated companies in
Berlin Ibadan

Oxford is a trade mark of Oxford University Press

Published in the United States
by Oxford University Press Inc., New York

A catalogue record for this book is available from the British Library

Library of Congress Cataloging in Publication Data
Møller, A. P. (Anders Pape)
Sexual selection and the barn swallow/Anders Pape Møller.
—(Oxford series in ecology and evolution)
Includes bibliographical references and indexes.
1. Barn swallow—Evolution. 2. Barn swallow—Behavior.
3. Sexual selection in animals. I. Title. II. Series
QL696.P247M64 1994   598.8′13—dc20   93-37324

ISBN 0 19 854029 9 (hbk)
ISBN 0 19 854028 0 (pbk)

Typeset by AMA Graphics Ltd., Preston, Lancs
Printed in Great Britain by
St. Edmundsbury Press, Bury St. Edmunds, Suffolk

Preface

The study of sexual selection is currently one of the hottest and most fascinating topics in evolutionary biology. My main aims with this book are twofold: first, to demonstrate the importance of sexual selection for almost every aspect of the life of animals, and second, to make a detailed comprehensive empirical study of all aspects of sexual selection. This has been done by presenting extensive reviews of sexual selection theory as a basis for the empirical study of sexual selection in a common bird species, the barn swallow. It is my hope that this approach will provide insights for both theoretically and empirically oriented evolutionary biologists, and that the book will help bridge the gap between theoreticians and empiricists. The level of argument should allow the readership to include advanced under-graduate or graduate students as well as those who have completed their studies.

My first encounters with barn swallows date back to my early childhood in the countryside in northern Denmark. I ringed my first barn swallow nestlings in 1970, unaware that this was later to become a long-term study, now in its twenty-third year. Local people in my study area have wondered about my many kinds of activities so much so that I have become known as the king of birds! Everyone has nicknames in the local communities of northern Denmark, but my name is one of the nicest!

This book is the result not only of many different people accepting my presence, but also of their direct help and support.

Several people read the manuscript in whole or in part and made constructive suggestions: T. R. Birkhead, P. H. Harvey, R. M. May, A. Pomiankowski, and S. Ulfstrand. They are all gratefully acknowledged for their help. The staff at Oxford University Press kindly provided efficient assistance throughout the process of producing this book.

The following farmers are gratefully acknowledged for allowing me to work on their properties at almost any hour of the day: P. Andersen, I. Bjørn Hansen, K. Bjørn Hansen, N. Bjørn Hansen, N. Bjørn Hansen, S. Bjørn Hansen, E. Christensen, O. Dam, S. Dam, E. Eriksen, H. Karlsen, E. Kristensen, J. Krog, P. Krog, N. Larsen, L. Melsen, P. Møller, A. Nielsen, K. Nørgaard, K. Olesen, J. L. Pedersen, L. Pedersen, K. Risager, S. Risager, S. Thomsen, and N. Zachariasen.

Several people helped with field work in Denmark and other places: P. Barnard, F. de Lope, E. Flensted-Jensen, K. Magnhagen, J. Moreno, A. M. Peklo, N. Saino, R. Simmons, N. Szczerbak, T. Szép, L. Szondi, A. A. Tokar, A. Ulfstrand, S. Ulfstrand, and W. C. Aarestrup. Their help is most appreciated.

Jens Gregersen kindly agreed to make the drawings for this book. His long-term knowledge of barn swallows allowed him to grasp details of the behaviour of live birds.

S. Nielsen at the Agricultural Research Station in Tylstrup kindly provided me with meteorological information for the period 1971–1992.

I highly appreciate access to the museum collections in the Zoological Museum, Helsinki; Natural History Museum, Stockholm; Zoological Museum, Copenhagen; Natural History Museum, Aarhus; British Museum (Natural History), Tring; Naturhistorisches Museum, Basel; Naturhistorisches Museum, Bern; Natural History Museum, Budapest; Zoological Museum, Kiev; Museo di Storia Naturale, Florence; Museo Nacional de Ciencias Naturales, Madrid; and Estación Biologica de Donaña, Seville. G. Jones, F. de Lope, T. Szép, L. Szondi, A. K. Turner, and M. Wellbourn kindly provided unpublished morphological data. G. F. Bennett kindly identified and scored blood smears for Haematozoa and answered questions about the life cycles and the fitness costs to hosts of avian blood parasites. He also provided excellent slides of several of the species of Haematozoa found in the barn swallow.

Financial support for parts of the project on which this book is based was provided by the Swedish Natural Science Research Council, the British Natural Environment Research Council, the Faculty of Natural Sciences at Uppsala University, and the Engkvistska Foundation. This book was written during a one-year stay at The Galton Laboratory, Department of Genetics and Biometry, University College London. I highly appreciated the stimulating company of J. S. Jones, J. Mallet, and particularly A. Pomiankowski.

Uppsala   A.P.M.
April 1993

Contents

1

Introduction

1.1 What is sexual selection?

When Darwin published *On the origin of species by means of natural selection* in 1859 he was basically able to account for the evolution of apparently adaptive characters and the process of speciation by means of natural selection. Variation in characters between individuals (which later became known as phenotypic variation), resource limitation, and the resulting non-random association between the expression of characters and a component of fitness were supposed to give rise to changes in the average character value. With the work of Mendel it was established that there was an association between phenotype and genotype, and differential fitness among phenotypes resulted in micro-evolutionary changes in gene frequencies.

However, Darwin was faced with the intriguing problem of a whole suite of apparently maladaptive characters which could not have evolved by natural selection. Extravagant traits such as the train of the peacock *Pavo cristatus* and the antlers of the red deer *Cervus elaphus* could not have evolved by natural selection. It is superficially a puzzle how peacocks with trains more than one metre long are able to survive in Indian forests with tigers *Panthera tigris* and other predators constantly on the look-out for a tasty meal. Similarly, red deer stags in some parts of the Holarctic have to survive with as much as twenty kilograms of monstrosities on their heads while fleeing from wolves *Canis lupus* or other agile predators. Such secondary sexual characters are unlikely to enhance the survival prospects of their possessors, since they are absent from females. They are also unlikely to increase the survival prospects of offspring in the same way as enlarged mammary glands which are restricted to a single sex.

The whole idea of sexual selection was laid out by Darwin in *The descent of man, and selection in relation to sex* in 1871. Extravagant secondary sexual characters were suggested to have evolved because of 'the advantage which

Peacock

Red deer

certain individuals have over other individuals of the same sex and species, in exclusive relation to reproduction' (Darwin 1871, Vol. 1, p. 256). Two different kinds of processes could account for the evolution of peacock trains and deer antlers. Competition for mates between members of the same sex, usually males, could result in the evolution of structures which served as weapons during fights. Males with the largest weapons were supposed to be able to win fights over males with smaller armaments. Male–male competition could therefore account for the evolution of such bizarre traits as antlers of deer, horns of antelopes and beetles, spurs of birds, and exaggerated canine teeth of several mammals.

The second process was active choice of particular individuals of one sex by individuals of the other, usually female choice of mates. Again, males with the largest secondary sexual characters were assumed to be at a selective advantage because they were more likely to be chosen by a female and they should therefore experience higher mating success. Mate choice could account for the evolution of song in insects, amphibians and birds, extravagant plumage ornaments in birds, and the bizarre display dances of several insects and birds. This original definition of sexual selection and the distinction between the processes of male–male competition and female choice have virtually remained unaltered until today.

Here I use the concept of sexual selection as defined by Darwin. Sexual selection occurs as a result of a non-random association between a (secondary sexual) character and a component of mating success. Mating success includes the direct acquisition of mates, but also other components of success such as sperm competition, differential abortion and infanticide, and differential parental effort. The two processes of sexual selection, male–male competition and female choice, are not necessarily easy to disentangle as we shall see in the next chapter. In Darwin's writings the process of mate choice was a bit more elusive than male–male combat. Here I shall assume that mate choice occurs when any display shown by individuals of one sex makes them more likely than others to mate with members of the opposite sex (Halliday 1983).

1.2 A brief history of sexual selection

Although extravagant secondary sexual characters had been described by scientists for centuries, it was not until Darwin addressed the evolution of sex traits by means of sexual selection that a coherent theory was put forward. Sexual selection was suggested to arise as a result of competition between conspecifics of one sex in relation to reproduction (Darwin 1859, 1871). The major findings were that (1) the evolution of apparently maladaptive secondary sexual characters could be explained by sexual selection, and (2) the two modes of sexual selection, male–male competition and female choice, were suggested to result in the evolution of different kinds of characters which were appropriate for fighting and subduing individuals of the same sex or charming and exciting individuals of the opposite sex.

Darwin's contemporaries readily accepted the idea that some secondary sexual characters had evolved as a consequence of competition between individuals of one sex for access to individuals of the opposite sex (for example, Wallace 1889). It was easy to understand how males with horns or other kinds of weaponry might achieve an advantage in fights with other males, and that males with the largest horns would be more intimidating or win fights more often than conspecifics with small horns. The major hurdle was the idea of female mate choice. Darwin (1871) stated that some secondary sexual characters such as the extravagant feathers of peacocks could not have evolved in the context of male combat, whereas a female mate preference for the most beautifully ornamented males may result in the evolutionary exaggeration process. Male displays had therefore evolved to excite and charm females, which were supposed to possess a sense of beauty. These claims were highly controversial and not acceptable to the scientific community.

This debate about female choice and the evolution of secondary sexual characters not used in the context of male–male competition continued well into the twentieth century (Belt 1874; Poulton 1890; Wallace 1891; Huxley 1938a, 1942; Cott 1940). Wallace (1891) continually attempted to find alternative explanations for extravagant displays and other secondary sexual characters, and many examples of what Darwin had claimed to be sexually selected colours could readily be explained as having a function in crypsis, mimicry, or as warning colours. Wallace's view was clearly that sexual selection could only favour characters as reinforcement of natural selection. Animal coloration was therefore also assumed to reflect health and vigour rather than being an arbitrary character used for attraction of mates (Wallace 1891). Huxley (1938b) also attempted to dismiss many of the examples of characters claimed by Darwin to have a role in female choice by assigning them different functions such as the deflection of predator attacks, warning coloration, recognition characters, and signals of threat. Huxley (1938b) even attributed the extreme degrees of display in many polygynous bird species to the 'advantage to the species in promoting more effective reproduction', which is a clear example of group selectionist thinking. It is clear from his writings that he consistently assumed that female preferences had evolved for male traits that were beneficial under natural selection. Huxley had not understood the concept of female choice and its role in generating maladaptive evolution in male secondary sexual characters. Empirical studies eventually demonstrated that female choice of many elaborate male displays was a fact, and the mechanisms of the process of mate choice were subsequently spelt out in detail (reviews in Searcy 1982; Bateson 1983; Bradbury and Andersson 1987).

The theory of sexual selection lay virtually dormant during most of the first half of the twentieth century, with a single important exception. Ronald Fisher elaborated one of the mechanisms of the evolution of female mate preferences in a short paper published in *The Eugenics Review* in 1915 and in his influential book, *The genetical theory of natural selection*, published in 1930. Fisher's twist to the theory of sexual selection was to demonstrate that the mate preference itself would evolve as a consequence of sexual selection. A female that prefers to mate with males of a particular phenotype because of a genetically determined mate preference would produce sons that carry genes for both the secondary sex trait and the female mate preference. By their mate choice females are therefore indirectly selecting for the mate preference that will become genetically coupled with the sex trait. The male secondary sexual character and the female mating preference may thus coevolve to ever more extreme expressions in an evolutionary runaway process. These arguments were not directly spelt out in a formal genetic model, although subsequent models have demonstrated that the mechanism is feasible (O'Donald 1962, 1967, 1980a; Lande 1981; Kirkpatrick 1982; Pomiankowski *et al.* 1991).

Fisher (1930) realized that the runaway process of sexual selection might result in sympatric speciation. The rapid divergence in secondary sexual characters along a cline caused by the runaway process would result in extreme exaggeration, even in the absence of physical barriers, provided that the female mate preference was sufficiently strong. The role of sexual selection by the runaway process in speciation has subsequently been confirmed in explicit genetic models (Lande 1981, 1982).

Fisher's final contribution to sexual selection theory concerns the evolution of mate choice in monogamous species (Fisher 1930). Darwin (1871) realized that extravagant sexual displays often occurred in monogamous organisms even though the variance in mating success was apparently much smaller than in polygynous species. If males arrived at the breeding sites earlier than females, and if females in better body condition arrived and bred earlier than others, preferred males would gain an advantage by mating with the earlier females. Fisher extended this theory by providing a numerical example (Fisher's 'diamond of monogamy') and by suggesting that female body condition had to be environmentally determined in order to facilitate continuous selection for earlier breeding date.

The 1940s and 1950s were characterized by dramatic progress in evolutionary biology, primarily due to the 'Modern Synthesis' of ecology, evolution, genetics, and palaeontology. Sexual selection was barely mentioned in the books published as a result of the synthesis. For example, Mayr (1942) only mentioned that secondary sexual characters mainly had a function in species recognition, and the same applies to Dobzhansky's (1937) treatment of sexual selection. Julian Huxley had continuously played down the role of sexual selection during the early part of the twentieth century, and it is perhaps not surprising that he does not even mention sexual selection in his 1942 book. Similarly, sexual selection was scarcely mentioned in the volume published in association with the centenary of *On the origin of species by means of natural selection* (Barnett 1958). One major, but largely ignored development was the study by Bateman (1948) of variance in reproductive success of fruitflies. By using genetic markers in *Drosophila melanogaster* he was able to show that the variance in reproductive success of females was significantly smaller than that of males. This was interpreted to suggest that some males were favoured while others were not, and variances in mating success thus apparently provided a measure of the intensity of sexual selection. This conclusion was only recently questioned following a re-analysis which demonstrated that the mating success of males apparently did not differ from a random, binomial distribution (Sutherland 1985).

The rebirth of sexual selection in the 1960s was restricted to scattered publications. First, Peter O'Donald (1962, 1967, 1980a) modelled sexual selection and confirmed that the coevolution of a male trait and a female mating preference was a feasible solution to the problem of female mate choice, as already suggested by Fisher (1915, 1930). Williams (1966) was apparently the first to suggest that females may choose mates entirely for the sake of 'good genes' which improve the viability of their offspring.

The 1970s and 1980s were characterized by large steps forward in the empirical testing of sexual selection theory and by the development of the two main theoretical schools which became ever more polarized. The arbitrary traits school of sexual selection re-derived the runaway process (O'Donald 1962, 1967, 1980a; Lande 1981, 1982; Kirkpatrick 1982), and other indirect mechanisms of mate choice were dismissed as infeasible (Davis and O'Donald 1976; Arnold 1983; Kirkpatrick 1985, 1986a), or suggested to play a minor role in the evolution of secondary sexual characters (for example, Tomlinson 1988). The 'good genes' school of indirect fitness benefits of female mate choice received support from more empirically oriented biologists (Trivers 1972; Emlen 1973; Zahavi 1975, 1977, 1987; Andersson 1982a, 1982b, 1986a; Kodric-Brown and Brown 1984). One of the most hotly debated models of good genes effects is that of the handicap principle which suggests that extravagant secondary sexual displays are reliable signals of male quality because only males of the highest quality can survive the handicapping effect of an ornament (Zahavi 1975, 1977, 1987). A number of opponents claimed that the handicap principle could not work (Davis and O'Donald 1976; Maynard Smith 1976; Arnold 1983; Kirkpatrick 1986a), while others were less sceptical (Bell 1978; Eshel 1978; Motro 1982;

Dominey 1983; Nur and Hasson 1984; Pomiankowski 1987*a*, 1987*b*). The main theoretical issues concerned how genetic variation in viability could be maintained and how the process could get started. The first issue was partly resolved by the proposal that coevolution between hosts and their parasites (or other biotic players) might result in evolution of secondary sexual characters that reliably revealed genetic variance in viability caused by resistance to parasites (Hamilton and Zuk 1982). Later models of the handicap principle demonstrated that the process could in fact get started and result in an evolutionarily stable exaggeration of male traits and female mate preferences for signals reflecting direct fitness benefits (Heywood 1989; Grafen 1990*a*, 1990*b*; Price *et al.* 1993) and for signals reflecting good genes (Pomiankowski 1987*a*, 1987*b*; Michod and Hasson 1990; Iwasa *et al.* 1991).

The controversy over the handicap principle may also have been a clash between personalities. The ideas of the handicap principle were finally made acceptable to a large part of the scientific community in the early 1990s by dressing them in a proper theoretical outfit. One of the major contributors to modern evolutionary biology, John Maynard Smith, who had been sceptical about the handicap mechanism during the 1970s and most of the 1980s, reviewed sexual selection in an address to the Third International Conference on Behavioural Ecology in Uppsala, Sweden, in August 1990. After having elaborated on the theoretical arguments about the handicap principle he formally apologized to the inventor of the handicap principle, Amotz Zahavi, for not having understood earlier the simple mechanism of reliable signalling. Zahavi stood up after the lecture, acknowledged the apology, and said that he still did not think that Maynard Smith had understood the handicap principle!

The history of sexual selection has been characterized by a number of controversies, as previously described. The relative roles of current ideas in society and those of individual scientists for the development of a science are hotly debated by historians of science. The development of sexual selection may provide an example of how a few individuals have shaped an entire field of research.

1.3 Why is sexual selection interesting?

Sexual selection deals with the evolution of secondary sexual characters which give certain individuals of one sex an advantage over others in relation to reproduction. Recent meetings in the behavioural or evolutionary sciences have been flooded with papers and posters dealing with various aspects of sexual selection, and currently every issue of scientific journals in ecology, evolution, and behaviour contains one or more papers about sexual selection. Why are so many evolutionary biologists interested in sexual selection? This question is not easily answered. One possibility is that sexual selection was virtually dormant as a subject of scientific research for almost a century after the publication of Darwin's book in 1871. Most other aspects of evolutionary and behavioural biology made vast advances during that period while the major advances in the study of sexual selection can be numbered very easily.

A second answer to the question of why sexual selection is of such great current interest is that it is closely related to some of the major remaining problems of evolutionary biology. It is easy to give some examples. First, females may benefit directly or indirectly in terms of fitness from their mate choice. Indirect benefits include the choice of mates with so-called good genes which enhance the viability of the offspring of choosy females. Much scepticism has been directed towards the possibility of indirect fitness benefits in terms of good genes, and good genes, one of the main explanations for the evolution of costly female mate preferences, was out of fashion for more than ten years until its recent revival. The main problem with the idea of female choice for good genes is how heritability of fitness is maintained, although a number of solutions to this problem have since been advanced. The relative importance of various direct and indirect fitness benefits of mate choice has never been assessed, and one of my objectives in this book is to evaluate this problem extensively.

Second, the evolution of sex is the major unresolved problem of evolutionary biology. Sexual selection relates to the problem of the evolution of sex in a number of different ways. For example, it has been suggested that parasites play a major role both in the evolution of sex, because recombination gives hosts an advantage in their fight against parasites (Jaenike 1978; Bremermann 1980; Hamilton 1980, 1982), and in the evolution of female mate preferences, if the displays of mates chosen by females directly reveal the ability of males to resist parasite attacks (Hamilton and Zuk 1982). Another example concerns sex differences in the rates of recombination and the benefits of mate choice (Trivers 1988). Female choice has usually been assumed to favour traits in males that are useful for their sons' mating success. However, sexual selection and female choice may often be adaptive, with a bias towards the genetic interests of daughters rather than sons (Seger and Trivers 1986). The heterogametic sex usually shows less recombination across its autosomes than the homogametic sex. This may be related to adaptive sexual selection. Males may have been selected to link their genes more tightly in order to preserve their more intensely selected beneficial gene combinations.

Third, sexual selection is communication between males and females. Communication involves signalling which is the transfer of information (true or false) to conspecifics. How are bluff and cheating controlled, and how does signalling remain reliable? This area of research has evolved in close association with the study of sexual selection (see for example, Zahavi 1975, 1977, 1987; Grafen 1990*b*).

Fourth, sexual selection considers coevolution between a male secondary sexual character and a female mate preference. The interests of the two sexes virtually never coincide because of sex differences in reproductive invest-ments and the resulting different sex roles during reproduction: the study of sexual selection is therefore also the study of how sexual conflicts of interest are resolved (Trivers 1972; Parker 1983*b*; Hammerstein and Parker 1987).

A final answer to the question of why sexual selection is of such great current interest is that it deals with characters of extreme beauty. It is no coincidence that scientists engaged in field studies of sexual selection concentrate on animals that use the same sensory modalities as humans; that is, vision and hearing. Anybody with a minimum of curiosity who has seen a peacock displaying would like to know the importance of its extravagant train in its life.

1.4   The role of sexual selection in animal life

One of the aims of this book is to demonstrate how sexual selection permeates almost every aspect of the lives of animals. Sexual selection is usually equated with mating success which arises as a consequence of the two processes of female mate choice and male–male competition. Darwin explicitly wrote that secondary sexual characters arose as a consequence of the reproductive advantages which some individuals have over others of the same sex (Darwin 1871). Most of what Darwin wrote and what has been written subsequently by others about sexual selection concerns mating success in the most limited sense of the word. For example, sexual selection arises as a result of non-random variance in the number of females acquired by different calling male toads, or the non-random variance in the number of females attracted by different pheasant *Phasianus colchicus* cocks. How-ever, there are a multitude of advantages that can lead to sexual selection. These include mating success, success in sperm competition, variance in mate quality, differential reproductive effort, and differential abortion and infan-ticide (see Table 1.1 for a full list). Thus, the process of sexual selection does not cease at the moment when all females have mated, but continues throughout the reproductive season.

The consequences of sexual selection affect the life of animals almost as much during the non-reproductive as in the reproductive season (see Table 1.2). Males of many animals also carry their secondary sexual characters during the non-reproductive season, when they have to pay the full costs of the sex traits without receiving any of the advantages. Extravagant sexual

Table 1.1. A list of processes that may lead to sexual selection in organisms

Component of sexual selection	Processes in the two sexes	
	Females	Males
Mating success	mate preferences	male–male displacement
Timing of success	mate preferences	male–male displacement
Reproductive mortality	mate preferences	male–male displacement
Sperm competition	extra-pair copulations, differential sperm use	extra-pair copulations, male–male displacement, paternity guards (e.g. mate guarding)
Mate quality	mate preferences	male–male displacement
Selective fertilization	differential abortion, suppression	Bruce effect
Selective embryo development	infanticide	infanticide
Reproductive effort	differential effort	differential effort
Parental care	differential parental care, perception of cuckoldry	differential parental care, perception of cuckoldry

displays during a brief, but intensive mating season often leave males emaciated and stressed. Even though reproduction usually takes place before or during the main growing season, worn-out males may still suffer from their sexual display because predators and parasites also time their reproduction to the season when resources are most readily available. Males that have lived under very stressful conditions may form easy meals for predators. Most animals are host to one or more parasites continuously or intermittently throughout their lives. The ability of potential hosts to defend themselves against debilitating parasites depends on an efficient immune system. Poor body condition and a stressful life are two of the main factors reducing the efficiency of the immune defence. High levels of circulating androgens during the formation of secondary sexual characters and sexual display are a two-edged sword. Hormones like testosterone enhance the expression of secondary sex traits, but simultaneously have an immunosuppressive activity. Thus, excessively displaying males may acquire many reproductive advantages, but at the same time they may make themselves vulnerable to severe parasitemias and diseases (Grossman 1985; Folstad and Karter 1992).

Secondary sexual characters are produced anew annually in animals such as many birds and deer, and the expression of these traits will form the basis of sexual selection in the following season. Secondary sexual characters are often large traits of great intricacy of design, and a sufficient amount of

Table 1.2. Elements of animal life influenced by sexual selection.

Factor	Cost
Females	
Mate preferences	time, mortality, disease
Correlated response to male secondary sexual character	production, maintenance, transport, mortality
Body condition	pre-nuptial acquisition, post-nuptial replenishment
Stress	disease, mortality
Hormone levels	production, disease
Emergence, arrival	timing, mortality
Males	
Secondary sexual character	production, maintenance, transport, mortality
Supporting structures	production, maintenance, transport, mortality
Body condition	pre-nuptial acquisition, post-nuptial replenishment
Stress	disease, mortality
Hormone levels	production, disease
Emergence, arrival	timing, mortality
Display ground (territory)	establishment, maintenance, mortality
Ontogeny	timing, development, mortality

resources must be acquired during the non-breeding season to allow development of perfect sex traits (Table 1.2). The timing of growth of the secondary sexual character has to be fitted into the other activities of the annual cycle for the sex trait to be ready for use at the start of breeding, but not before. Sexual display is usually very costly and close to the maximum limit of energy use. Resources that have been used for display have to be replenished after the breeding season, and new extra resources have to be accumulated well in advance of the forthcoming reproductive season at a time of the year when resources may not be readily available. Males have to emerge or arrive at the breeding grounds before the females in order to be able to attract mates from the very beginning of the reproductive season. Females usually time their emergence or arrival to the optimal time for reproduction, and males may therefore have to emerge or arrive at a suboptimal time, before resources become readily available. Many animal species have breeding territories where females are acquired and reproductive activities take place. Males have to establish a territory before any females can be encountered, and this may result in severe competition with other males.

The ontogeny of secondary sexual characters starts relatively early in life. The new generation of males experiences some of the consequences of sexual selection long before the sexual displays have developed. Elevated levels of circulating androgens during development may affect the immune system, and energy resources have to be built up well in advance of the period of ornament growth. This will mean different goals for the overall growth process of the male body right from the beginning of life. It is not only the secondary sexual character that is costly. Many different aspects of the entire *Bauplan* of male organisms have to be modified in order to make it possible for males to carry the secondary sexual characters (Møller 1993*o*). For example, male barn swallows *Hirundo rustica* do not develop just their sexually size-dimorphic tail feathers. A whole suite of muscles has been modified as a concession to the demands of ornament transportation. Similarly, many other birds have grossly enlarged muscles that serve to erect ornaments, and tail and wing feathers may have been modified to allow transportation of extravagant feather ornaments during flight.

With these few examples I wish to emphasize that sexual selection has pervasive consequences for almost all aspects of animal life, and that the study of sexual selection is also a study of how extravagant displays used under specific reproductive activities are integrated into the overall phenotype and the annual cycle. The following chapters will describe in detail how one animal species has managed to accomplish this task.

1.5 Evolution versus maintenance of sexual displays

Evolution can be studied at different levels and with different methods. This is a book about sexual selection in action and the importance of intraspecific variation in the expression of secondary sexual characters. In other words, it is a book about the maintenance of secondary sexual characters, and about micro-evolutionary processes that may act on phenotypic variation in a particular population. Male sex traits and female mate preferences demonstrate considerable phenotypic and genetic variation, and selection on these characters is often intense. Both the male sex trait and the female preference would soon disappear in the absence of a sexual selection advantage. Male secondary sexual characters are rarely if ever at their natural selection optimum, and the same applies to costly female mate preferences. The maintenance of costly male sex traits and female mate preferences thus begs an explanation. The origin of extravagant secondary sexual characters may be studied from a macro-evolutionary perspective by relying on phylogenetic information (for example, Brooks and McLennan 1991; Harvey and Pagel 1991). The phylogenetic study of macro-evolution of characters involves reconstruction of a phylogeny and mapping of character states on the phylogenetic tree. Sequences of evolutionary events can then be studied by comparative methods. These two approaches are complementary rather than alternative ways of resolving a single problem. The phylogenetic approach to sexual selection will be discussed briefly here.

Darwin knew that even closely related species often differ dramatically in their secondary sexual characters. The birds that are the main subject of the study presented in this book follow this pattern. The barn swallow belongs to the avian family Hirundinidae which consists of some species with and some species without elongated tail feathers. The outgroup of the family (a taxon outside the monophyletic group of interest) does not have tail ornaments (the closest relatives include the nuthatches (Sittidae), treecreepers and wrens (Troglodytidae), tits (Paridae), long-tailed tits (Aegithalidae), kinglets (Regulidae), bulbuls (Pycnonotidae), African warblers (Cisticolidae), and Old World warblers (Sylviidae)) (Sibley and Ahlquist 1990). Sexual dimorphism must therefore have evolved one or more times within the family Hirundinidae, and the factors that resulted in the evolution of the sex traits must lie somewhere in the evolutionary past among the common ancestors of present hirundines with and without ornaments. The phylogenetic approach has not been deliberately excluded from this book, but in fact phylogenetic information was until recently virtually non-existent for the family Hirundinidae. A couple of classifications, which is something different from phylogenies, are based on ecological and behavioural differences between taxa (Sharpe and Wyatt 1885–1894; Mayr and Bond 1943; Brooke 1972, 1974). Many of the similarities between taxa which were used to lump species together in the classifications may not be ancestral character states, but rather derived states that are products of convergent evolution.

Information from DNA–DNA hybridization of avian egg-white proteins has been used to construct phylogenies of all the living birds, including the Hirundinidae (Sibley and Ahlquist 1982, 1990). The phylogeny based on similarities in the DNA–DNA hybridization clearly demonstrates that extravagant tail ornaments have evolved in the Hirundinidae at least twice, once in the subfamily Pseudochelidoninae and once in the subfamily Hirundininae (Fig. 1.1a).

A more extensive analysis based on DNA–DNA hybridization in hirundines also revealed multiple evolutionary origins of tail ornaments (Fig. 1.1b; Sheldon and Winkler 1993; Winkler and Sheldon 1993). The analysis of 18 different species revealed one evolutionary event in each of the genera *Pseudochelidon, Psalidoprocne, Cecropis,* and *Hirundo.* Together with D. W. Winkler, I have constructed a phylogeny of the hirundines based on external morphological characters (Winkler and Møller 1994). This phylogeny included all extant hirundine species with the exception of a recently

Three common hirundines, barn swallow, house martin, and sand martin

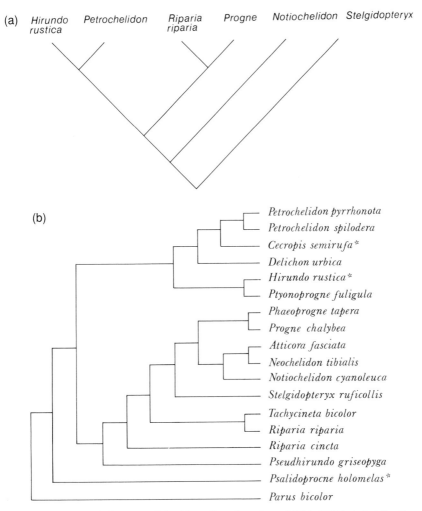

Fig. 1.1 (a) A phylogeny of the hirundines based on DNA–DNA hybridization. Adapted from Sibley and Ahlquist (1982, 1990). (b) A phylogeny of the hirundines based on DNA–DNA hybridization. * denotes independent evolutionary events of tail ornaments. *Parus bicolor* comprises an outgroup. Adapted from Sheldon and Winkler (1993) and Winkler and Sheldon (1993). Nomenclature follows Turner and Rose (1989).

described species which was identified from the remains of a single specimen (Fry and Smith 1985). Again, it is clear that extravagant tail ornaments have evolved several times in the Hirundinidae. Ecological or evolutionary factors which may have resulted in the evolution of the secondary sexual tail character may potentially be determined in the future from comparative phylogenetic studies.

1.6 Outline of the book

This is a book about many facets of sexual selection. I have used a single long-term study of one particular animal species, the barn swallow *Hirundo rustica*, to illustrate the many different aspects of sexual selection. This species may seem an odd choice for such a study because males and females are alike with the exception of the outermost tail feathers which are considerably longer in males than in females. There are relatively few common bird species with extreme sexual size dimorphism, and the barn swallow is the obvious choice for a European scientist. A number of additional reasons for choosing this species as a study object are given in Chapter 3. The theoretical background for the study of sexual selection is basically outlined in Chapter 2 and in introductory paragraphs to the remaining chapters. The different aspects of sexual selection in the barn swallow are treated in Chapters 3–13. First, a brief presentation of the study animal and its natural history is given in Chapter 3. The various mating advantages that accrue to the most extravagantly ornamented males are analysed in Chapter 4. Female mating preferences and the benefits that accrue to females from their mate choice are discussed in Chapter 5. The many genetic and environmental factors that determine the expression of one male secondary sexual character are presented in Chapter 6.

Barn swallows are long-distance migratory animals, and the advantages for males of an early arrival to the breeding grounds and the determinants of arrival date are outlined in Chapter 7. A fraction of the males never acquire a mate, and the characteristics of these birds and their mating options are considered in Chapter 8. Parasites often play an important role in the lives of their hosts and barn swallows are no exception. The role of parasites in sexual selection and the effects of parasites on mate choice in the barn swallow are described in Chapter 9. Males of some animal species play an important role in reproductive activities and the raising of offspring. Male barn swallows provide various kinds of parental care, and their importance for sexual selection is discussed in Chapter 10. Barn swallows are socially monogamous with a pair bond which may last throughout life. However, females regularly copulate with males other than their mates, and the role of sperm competition and various anti-sperm competition strategies in sexual selection is outlined in Chapter 11. Both adult male and female barn swallows have longer tail feathers compared with those of juveniles, and the importance of sexual size dimorphism in tail length is considered in Chapter 12. Barn swallows breed in most parts of the Holarctic region, and there are clear patterns of clinal variation in the size of secondary sexual characters and body size. Sexual selection in a cline and geographic variation in the costs and benefits of extravagant sexual ornaments are analysed in Chapter 13. The role of sexual selection in the life of the barn swallow is synthesized in Chapter 14. The various kinds of fitness benefits that accrue to females from their mate choice are summarized, and sexual selection is considered to be a continuously reinforcing process permeating many different aspects of animal life.

Many of the results and conclusions are based on extensive statistical analyses. I have avoided presenting these tedious details in the main text and present the results of analyses either in the original publications, in the captions to figures, in the tables, or in a separate statistics section at the end of each chapter which relates to numbered superscripts in the main text.

1.7 Summary

Sexual selection is a consequence of variation in mating success being related non-randomly to phenotypic characters which are advantageous during reproduction when certain individuals interact with conspecifics of the same sex.

Non-random variation in mating success gives rise to sexual selection. A whole suite of aspects of mating success, such as acquisition of mates, differential abortion and infanticide, and differential parental effort, may play a role (see Table 1.1 for a summary). The consequences of sexual selection are not restricted to extravagantly ornamented males during the mating period. The consequences of sexual selection in the life of animals are pervasive and affect activities during ontogeny and many parts of the annual cycle of both sexes (see Table 1.2 for a summary). Sexual selection therefore may permeate most aspects of animal life.

Secondary sexual characters and mating preferences can be studied empirically in two different, not mutually exclusive ways: (1) the phylogenetic study of macro-evolution, and (2) the phenotypic study of current micro-evolutionary processes. The latter involves estimation of environmental and genetic components of phenotypic variance, and determination of the optimal solution to evolutionary problems. This is the approach used in this book.

2

Models of sexual selection and monogamy

2.1 Introduction

Until quite recently the study of sexual selection was dominated by theoretical issues. However, empirical research on female choice and male combat has increased dramatically, with field and laboratory studies of individually identifiable organisms being the most frequent approach. Every empirical study has to address the question of whether male–male competition or female choice generates the process of sexual selection. Although these two modes of sexual selection were clearly defined by Darwin, it is not always easy to resolve the relative importance of the two processes, nor to decide whether a given structure or behaviour demonstrates one or the other, or both. Different taxa appear to rely on male–male competition and female choice to different degrees, but the reason for interspecific differences in the relative importance of the two modes of sexual selection remains largely unknown. These issues will be discussed further in section 2.2.

The next two sections describe models of the two modes of sexual selection. In section 2.3 the different models of male–male competition will be analysed in detail. The success of a competing male does not depend on its absolute level of armament, but rather on the size of its armament relative to that of potential opponents. This problem has been analysed in phenotype-dependent game theory models developed by Parker (1983a). Male armament can also be viewed as a handicap (Zahavi 1975, 1977), and therefore the evolution of male weaponry will also be discussed in the light of current handicap models.

The question of female mate preferences has attracted a great deal of intense theoretical interest and generated one of the fiercest debates in evolutionary biology. It was obvious that females had a reason for choosing particular kinds of males if such individuals provided direct benefits. The problem really arose when theoreticians started to consider animal species in which females did not obtain any obvious material benefits such as a high-quality territory from their mate choice. These species included lekking birds such as the ruff *Philomachus pugnax* in which females visit particular sites with displaying males, often demonstrate a clear preference for one or a few males, and subsequently reproduce entirely on their own without any male help. The very intense directional sexual selection arising from the female mate preference would, according to conventional quantitative genetics theory, reduce any genetic variance in quality properties of males within relatively few generations. What could a female gain from being choosy in such a situation? This is the so-called lek paradox.

To account for this paradox, three different kinds of female mate preferences have been suggested while a fourth process concerns mate choice for direct fitness benefits:

(1) the Fisher process;

(2) the sexy son process;

(3) the good genes process; and

(4) the good parent process.

First, according to Darwin (1871) males preferred by females would gain a mating advantage and thus leave more offspring. This suggestion was elaborated by Fisher (1915, 1930), who suggested that the female mating preference and the male sex trait could coevolve to ever more extreme expressions in a self-reinforcing runaway process. The Fisher process of mate choice was subsequently analysed in explicit genetic models with variable assumptions by O'Donald (1962, 1967, 1980a), Lande (1981), Kirkpatrick (1982), and Pomiankowski *et al.* (1991), and was until recently by far the most thoroughly studied theoretical model of a female mate preference. The fitness advantage accruing to preferred males in the Fisher process is the attractiveness of their sons and the preference of their daughters for attractive males.

Second, very attractive males may be poor fathers, since a number of empirical studies have demonstrated that females suffer a fitness cost in terms of offspring production by mating with the most preferred males. This apparent cost could potentially be overcome if the attractiveness of sons compensated for the loss in offspring production because of reduced paternal care (Weatherhead and Robertson 1979). This so-called sexy son hypothesis is in fact a variety of the Fisher model because the indirect fitness benefit is the same; the attractiveness of sons and the preferences of daughters for attractive males. It is different because it introduces a particular kind of cost to the female. Females have to pay for their mate preference with a fecundity cost.

Third, females may prefer particular males if such individuals have a superior genetic constitution (Pomiankowski 1988), enhancing their general viability or resistance to parasites. Good genes models of sexual selection incorporate the various handicap models of female mating preferences.

Fourth, perhaps the most obvious reason why females should choose a particular mate is direct fitness benefits, for example, in terms of quality of paternal care or absence of contagious parasites. Several models of female mate preferences have focused on direct fitness benefits. All these four different classes of models of female mating preferences will be discussed in detail in section 2.4.

Monogamy poses a special problem for the theory of sexual selection because variance in mating success, which is the main fuel for the Fisherian selection process, should be zero in a monogamous population with an equal sex ratio. This puzzle was solved by Darwin (1871) who suggested that females may enter breeding condition at different times depending on their body condition. As a result, the most preferred males would first acquire a mate in better body condition than other females and thus gain a mating advantage. This idea was subsequently developed by Fisher (1930), O'Donald (1972, 1980a, 1980b), and Kirkpatrick *et al.* (1990). Sexual selection by female choice in monogamous mating systems may also evolve by the handicap process. These different models of sexual selection under monogamy are analysed in section 2.5.

2.2 The two modes of sexual selection

Sexual selection occurs as a result of competition between conspecifics of one sex in relation to reproduction. Two modes of sexual selection have been distinguished: choice of partners by individuals of the choosy sex, usually females, and competition between individuals of the chosen sex, usually males. This distinction between the two modes of sexual selection was made as long ago as 1871 by Darwin, but there are great practical difficulties in separating them in reality.

Different secondary sexual characters have traditionally been awarded separate roles in sexual selection. Structures designed to drive away or kill conspecific males have been referred to as weapons and are assumed to have evolved in the context of male–male competition. These include huge antlers, horns, canines, and spurs, but also more subtle kinds of weaponry such as vocalizations, odours, and sheer force. Other structures have been thought to excite or charm individuals of the opposite sex and persuade them to select the individual exhibiting that particular display. These include extravagant feather plumes of birds and bizarre display dances, but also characters such as vocalizations and odours. Even this seemingly straightforward classification of secondary sexual characters is not really foolproof because weapons sometimes play a role in female choice. Thus, the spurs of ring-necked pheasants are used in male–male fights, but also play a role in female mate choice as determined by experimental manipulations of spur length (von Schantz *et al.* 1989). A second example concerns the golden pheasant *Chrysolophus pictus* where males use their elaborate tails as physical support during fights over mating access to females (Harvey and Bradbury 1991). It is difficult to doubt that spurs originally evolved in the context of male combat whereas extravagant tails evolved in the context of female mate choice. These secondary sexual characters may subsequently have acquired a derived role in mate choice and male combat, respectively. Comparative analyses of the role of various secondary sexual characters in the two modes of sexual selection will allow empirical tests of the conditions that determine a switch for secondary sex traits from one mode of sexual selection to another. Alternatively, if reliable signals evolve, they will be informative to conspecifics of both sexes and even to heterospecifics (Zahavi 1987, 1991). The role of many secondary sexual characters cannot easily be distinguished because of their dual function in the two processes of sexual selection. A classical example concerns bird-song, which plays a role not only in territorial defence (which is a kind of male–male competition), but also in attraction of females (White 1770; Searcy and Andersson 1986). A second example concerns the black grouse *Tetrao tetrix* where males lek on communal display arenas. Males fight for access to territories, and males of high quality in terms of survival probability are also most successful in male combat and acquire the majority of matings (Alatalo *et al.* 1991). Male–male competition and female choice are therefore best considered to be descriptions at a mechanistic level. The behaviour or the display structure plays a role in fighting or in sexual display. However, what really matters in evolutionary terms is how genetic variation in males and females covaries with fitness. In pure male–male competition, which is probably an entirely theoretical construct, there should be no genetic variation in females with the fitness of their mating partner. If such covariance exists, then there will be Fisherian or good genes evolution of female mate preferences.

The relative roles of male–male competition and female choice in sexual selection have been assessed in only a few cases, but there is little doubt that while male combat is of overwhelming importance in some species, mate choice predominates in others (Bradbury and Davies 1987). It is not always easy to be sure that females do not demonstrate mate choice in a situation where males fight intensively for access to females, because female behaviour is often more subtle compared with the conspicuous male activities.

The key to the understanding of the relative importance of male–male competition and female choice in different species probably lies in the different reproductive interests of females. Females usually make a larger investment in reproduction because of their production of eggs and embryos, and thus they will be more in control of activities related to reproduction than males. When females play the larger role in reproduction, there will be very strong direct selection on females to enhance their reproductive success in terms of direct fitness benefits, such as male parental care, or indirect fitness benefits, such as genes enhancing offspring attractiveness or viability. Females may circumvent the constraints on their choice of mate in a number of different ways, of which copulation with multiple partners is one of the most powerful (Møller 1992a). Female copulations with multiple males occur in many species, and clear relationships have been documented between mating systems and frequency of multiple copulations (Birkhead and Møller 1992; Møller and Birkhead 1993a). Females will often be in control of fertilizations because of their ability to copulate with more than one male. The control of fertilization can be achieved indirectly by allowing copulations with specific males at specific times of the reproductive cycle, but more direct ways of fertilization control within the female reproductive tract are also possible (Birkhead et al. 1993; Birkhead and Møller 1993a). If males assort themselves with respect to quality through a process of male–male competition, females may accord with the outcome of this assortment because only males of the highest quality will be able to get access to females. There is then no reason for females subsequently to make an active mate choice if males have resolved in advance all problems associated with assortment of male qualities and paid the costs of this activity. The ranking of males according to their quality by females will be congruent with the outcome of male–male competition in this situation (Borgia 1979).

Male and female interests are not always in accordance because of conflicts of interest over reproductive roles in evolutionary games between the sexes (Hammerstein and Parker 1987). Males may sometimes be able to win fights over their competitors and subsequently acquire access to a herd of females, even though the females may be unwilling to copulate only with the resident male. The properties of male quality in terms of male secondary sexual characters, accompanied by success in male–male competition, may not be ones for which females have evolved mate preferences. Apparently, this is the case even in some highly polygynous or lekking species where females usually are assumed to be able freely to execute their mate choice. Females within a harem of the polygynous ring-necked pheasant sometimes copulate with males other than the harem male, resulting in mixed paternity of the brood (Grahn 1992). Similarly, females of the lekking waterbuck *Kobus ellipsiprymnus* sometimes copulate with males other than those on whose territory they are resident (Wirtz 1982). These examples suggest that females do not always accept the assortment resulting from male–male competition and they may subsequently adjust their mate choice in accordance with their own mate preferences.

2.3 Models of male–male competition

Male combat is almost ubiquitous in nature, and Darwin (1871) suggested that this would favour the evolution of different specialized structures used as weapons, but also large body size and other traits, such as specific vocalizations and odours, designed to intimidate potential opponents. It is clear from Darwin's writings that he assumed that individuals possessing such structures were favoured in fights over individuals lacking them, and that individuals with larger or more elaborate weaponry would be at a selective advantage.

Some structures assumed to have evolved as a consequence of male–male competition must be costly to produce or maintain in terms of time or energy use, and thus males may have to pay for their armament by reduced viability. Males fight with other males in the neighbourhood, and the mating success of a particular male therefore depends on its amount of armament relative to that of others. A male needs only to produce weaponry larger than that of his neighbours, not a given amount thereof, as discussed in the following paragraphs. This situation has been modelled several times (Gadgil 1972; Haigh and Rose 1980; Parker 1983a; Charlesworth 1984; Maynard Smith and Brown 1986).

When the success of an individual male depends on its own level of armament and that of potential competitors, an ideal situation exists for applying game theory models. Parker (1983a) modelled the game between competitor males differing in levels of armament. His conclusions were that the evolutionarily stable phenotypic distribution would be polymorphic and the mean male phenotype different from the optimum as dictated by natural selection. Competition between males readily leads to a costly exaggeration of secondary sex traits. Stable phenotypic distributions arise (1) if the cost of armament increases in an exponential or otherwise increasing fashion with each increment in size, and (2) if there is environmental variance in the expression of armament (see also Gadgil 1972; Haigh and Rose 1980; Parker 1983a; Charlesworth 1984). The first of these two assumptions may not be all that self-evident, as we shall see later, but the second is bound to be fulfilled. The outcome of the game is different when these two assumptions are not fulfilled. It can result in cycles, with ever larger males invading until large males are so rare because of the costs of armament that small males are able to reinvade and initiate the process again. Alternatively, the arms race between males may continue until large males are so rare that females have difficulties finding mates.

A similar conclusion was reached by Maynard Smith and Brown (1986) whose model is based on the life history of red deer. They assumed that males grow to a specific size determined completely or partially by genes and then start to reproduce. Males compete with each other for access to females and this may last for years. Fitness was assumed to be the product of survival prospects and reproductive success. Breeding success is determined by the relative size of a male. For example, a male's success may be a linear function of the fraction of the population being smaller than its adult body size. Their simulation model readily leads to runaway selection in the sense that mean male size will increase continuously until all alleles for body size are fixed, even though mean fitness of males is reduced considerably because of the viability costs of increased size. The outcome of male–male competition is different when there is environmentally determined variance in male size, and when male survival rate falls off rapidly with increasing body size, as may usually be the case in nature. For example, male survival prospects may drop with increasing body size as a result of small individuals being more efficient intraspecific competitors for food. The outcome of the simulation model by Maynard Smith and Brown is a stable size distribution of males. Viability selection against large males thus appears to halt the runaway process. Although Maynard Smith and Brown (1986) assumed that selection operated in the same way in the two sexes, their results are qualitatively similar if the genes are expressed only in males (Maynard Smith 1987).

Handicap models have only rarely been developed explicitly for intrasexual selection, perhaps because the evolution of reliable signalling is not dependent on whether the receivers are conspecifics of one sex or another, or even heterospecifics (Zahavi 1987, 1991). The amount of advertising in male–male competition has been analysed by Grafen (1990*b*), and the same rules apply as in other reliable signalling contexts. Grafen used competition between red deer stags as an example. A harem stag wears antlers, which represent a certain level of armament advertising its individual strength. Challenging stags without a harem also possess a certain level of armament reflecting their strength. In this context the strength or quality of a stag is reflected in its ability to fight, or the size of its energy reserves, which influences its capability of winning a series of challenges by one or more competitors. The level of advertising by a challenging stag is the size of its antlers, and the assessment of the quality of the competitor is based upon the expression of its secondary sexual character. A challenging stag will be better off if it has a large energy reserve, and if the harem stag assesses the energy reserves of the challenger as high. The size of antlers is assumed to reduce the viability of stags, and antlers are supposed to be more expensive to produce or maintain for stags with low energy reserves. The fitness gain from a better assessment by the harem stag of a challenger's energy reserves is as great for a challenger with high as for one with low reserves. A harem stag should assess the energy reserves of a challenger correctly, because it will lose by both under- and over-estimation. There should therefore be a graded response in which challengers with more energy reserves should grow larger or better developed antlers, and harem stags should interpret this signal and treat a challenger accordingly. Thus, the conditions for reliable signalling of quality by secondary sex traits used in male–male competition are that the male trait is costly and that it is more costly to low than to high quality individuals (Zahavi 1975, 1977; Grafen 1990*a*, 1990*b*). The pheno-type-dependent evolutionarily stable strategy, therefore, is that males will wear armament reflecting their quality in the sense that males with the largest degree of armament are of higher quality than males with a low degree of armament. It is likely that males with small secondary sex traits used for male combat are in worse body condition before as well as after development of their sex traits compared with animals with large traits. The expression of armament often demonstrates larger degrees of condition-dependence when the male traits are small compared with when they are large (Møller 1992*e*).

In conclusion, male displays used in male–male competition may evolve according to a number of different processes. The evolutionarily stable outcome depends on the given model. Therefore, male phenotypic distribu-tions may, be:

(1) a stable size distribution of males reflecting or not reflecting male quality;

(2) cycles with ever larger males invading until large males become so rare owing to the costs of armament that small males are able to reinvade and initiate the process again; or

(3) the evolutionary arms race between males may continue until large males are so rare due to the costs of armament that females have difficulties finding mates.

2.4 Models of mate preferences

Models of female mate preferences differ according to the kinds of fitness benefits acquired by females, but also according to the modelling approach taken. For example, handicap models have been developed for female mating preferences with both direct and indirect fitness benefits. Here I have grouped the different models in relation to the fitness benefits accruing to females, because this distinction is of crucial interest to the student of sexual selection. I have chosen to describe in detail only the most recent and most advanced models of each type, while most early models are mentioned just briefly. A summary of the assumptions and the conclusions of all the various models is provided in Table 2.1.

Table 2.1. The assumptions, selection pressures and major findings of models of female mating preferences.

Fitness benefit	Ploidy	Genetics	Female preference Model	Female preference Initial level	Selection on male trait	Selection on male condition	Selection on female preference
Good parent process							
Kirkpatrick (1985)	h,d	p,q	p,a,r	high	v,m,f	—	f
Heywood (1989)	h	p	a	high	v,m	v,f	f
Hoelzer (1989)	h	p	fixed	high	v,m	—	—
Grafen (1990a, 1990b)	h	p	o	high	v,m	v,f	f
Price *et al.* (1993)	d	q	p	high	v,m,f	v,f	f
Good genes process							
(1) *Pure epistasis handicap*							
Maynard Smith (1976, 1978)	h	p	a	low	v,m	—	f
Davis and O'Donald (1976)	d	p	a	high	v,m	—	f
Bell (1978)	h	p	a	low	v,m	—	f
Pomiankowski (1987a, 1987b)	h	p	r	low	v,m	—	f
Iwasa *et al.* (1991)	h	q	p	high	v,m	—	f
(2) *Conditional handicap*							
Andersson (1982a, 1986b)	h	p	a	low	v,m	v	f
Andersson (1986a)	h	p	a	low	v,m	v	f
Maynard Smith (1985)	h	p	a	low	v,m	v	f
Kirkpatrick (1986a)	h	q	a,r	high	v,m	v	—
Pomiankowski (1987a, 1987b)	h	p	a	low	v,m	v	f
Tomlinson (1988)	d	p	a	low	v,m	v	f
Michod and Hasson (1990)	h	p	o	high	v,m,f	v	—
Iwasa *et al.* (1991)	h	q	p	high	v,m	v	f
(3) *Revealing handicap*							
Kirkpatrick (1986b)	h			low			
Maynard Smith (1985)	h	p	a	low	v,m	v	f
Pomiankowski (1987a, 1987b)	h	p	a	low	v,m	v	f
Tomlinson (1988)	d	p	a	low	v,m	v	f
Iwasa *et al.* (1991)	h	q	p	high	v,m	v	f

Mating system	Mutation	Initiation of preference	Increase in male trait	Increase in female preference	Equilibria
p	no	—	yes	yes	stable or unstable, runaway
p	no	no	yes	yes	stable
p	no	—	yes	—	stable
p	no	—	yes	yes	costly trait and preference
p	no	—	yes	yes	costly trait and preference
p	no	no	no	no	none
p	no	no	no	no	none
m,p	no	no	sometimes	sometimes	sometimes
p	no	no	no	no	none
p	biased on viability	—	no	no	none
m	no	yes	yes	yes	sometimes
m	no	yes	yes	yes	costly trait and preference
p	no	no	yes	yes	costly trait and preference
p	no	no	no	no	none
p	no	yes	yes	yes	costly trait and preference
m,p	no	no	yes	yes	none
—	no	yes	yes	yes	costly trait
p	biased on viability	yes	yes	yes	costly trait and preference
	no	no	yes	yes	costly trait and preference
p	no	no	yes	yes	costly trait and preference
p	no	yes	yes	yes	costly trait and preference
m,p	no	no	yes	yes	sometimes
p	biased on viability	yes	yes	yes	costly trait and preference

Table 2.1. *continued*

Fitness benefit	Ploidy	Genetics	Female preference		Selection on		
			Model	initial level	male trait	male condition	female preference
(4) *Non-genetic handicap*							
Kirkpatrick (1986a)	h	q	a,r	high	v,m	v	—
The Fisher process							
O'Donald (1962, 1967, 1980a)	d	p	a	low	v,m	—	—
Lande (1981)	d	q	p,a,r	high	v,m	—	—
Kirkpatrick (1982)	h	p	a,r	high	v,m	—	—
Heisler (1984a)	d	q	p,a	low	v,m	—	f
Heisler (1985)	d	q	p,a,r	low	v,m	—	f
Seger (1985)	h	p	n	high	v,m	—	—
Bulmer (1989a)	h	p	a	high	v,m	—	f
Gomulkiewicz and Hastings (1990)	h,d	p	a	high	v,m	—	—
Pomiankowski et al. (1991)	h	q	p	high	v,m	—	f
The sexy son process							
Kirkpatrick (1985)	h	p,q	p,a,r	high	v,m, f	—	f
Curtsinger and Heisler (1988)	d	q	r	low, high	v,m,f	—	f
Pomiankowski et al. (1991)	h	q	p	high	v,m	—	f

Mating system	Mutation	Initiation of preference	Increase in male trait	Increase in female preference	Equilibria
p	no		no	no	none
p	no	sometimes	yes	yes	sometimes
p	no	—	yes	yes	stable and unstable, runaway
p	no	—	yes	yes	stable and unstable
p	no	yes	yes	yes	—
p	no	yes	yes	yes	stable and unstable, runaway
p	no	—	yes	yes	stable and unstable, runaway
p	on trait or preference		yes	yes	point
p	no	—	yes	yes	stable and unstable, runaway
p	biased on trait	—	yes	yes	stable point
p	no	—	no	no	none
p	no	—	yes	yes	stable and unstable, runaway
p	biased on trait	—	yes	yes	point

Ploidy: haploid (h) or diploid (d). *Genetics*: population genetic (p) or quantitative genetic (q) model. *Female preference model*: Female mate preferences were either (1) a psychophysical preference according to which perceived intensity of the male trait is proportional to the power of the actual preference (p), (2) an absolute preference for males of a specific phenotype independent of male phenotypes in the populations (a), (3) a relative preference where the phenotype of a male is scaled to the frequency distribution of phenotypes in the population (r) (Lande 1981), (4) the best of *n* males (n), or (5) another preference model (o). *Female preference level*: Initial level either low or high. *Selection on the male trait*: The male trait is subject to viability (v), mating (m) or fecundity (f) selection. *Selection on male condition*: The male condition is subject to viability (v) or fecundity (f) selection. *Selection on the female preference*: The female mate preference is subject to viability or fecundity selection (f). *Mating system*: The mating system is either monogamous (m) or polygynous (p). *Mutation*: There is either no assumption about directional mutation on the male trait, the female mate preference or the viability trait, or mutation is biased. *Initiation of the preference*: The model leads to an initial increase (yes) or does not lead to an initial increase (no) in the mate preference. *Increase in the male trait*: Whether the model leads to an exaggeration of the male trait after an initial increase in its frequency. *Increase in the female preference*: Whether the model leads to an exaggeration of the female mate preference after an initial increase in its frequency. *Equilibria*: The existence of stable or unstable lines or points of evolutionary equilibria, and whether they may result in a runaway process.

2.4.1 *The good genes process*

It has repeatedly been suggested that female mate preferences evolve as a way of acquiring mates of high genetic quality (Williams 1966; Trivers 1972; Emlen 1973; review in Pomiankowski 1988). If the expression of secondary sexual characters reliably reflects the genetic quality of males, females will indirectly benefit by mating with the most extravagantly ornamented individuals. The possible mechanisms behind this so-called good genes process were later proposed as various versions of the handicap principle first introduced by Zahavi (1975). His basic idea was elegant and simple. Secondary sex traits are usually large, extravagant characters, which impose costs on their possessors in terms of predation risks or use of time and energy. A high cost of male sex traits is a prerequisite for signalling to be reliable, because only males of the highest quality will be able to survive the handicapping effect of the most extravagant expressions of secondary sexual characters. Consequently, by preferring to mate with elaborately orna-mented males, females will automatically acquire a mate of high phenotypic, and probably also of high genotypic quality. The secondary sex trait was seen by Zahavi as a device of testing male quality that had evolved in response to the female mate preference. Handicaps would lower male viability and also female fecundity because handicapped males would not be so able as less adorned males to assist in the rearing of offspring. The theory did not require any assumption about linkage disequilibrium between the male trait and male viability because once a male had passed the test of surviving with a handicap, it had already proven its phenotypic and genotypic quality (Zahavi 1975).

The original handicap idea generated intense criticism apparently because it was difficult to accept that extravagant secondary sexual characters had evolved as costly test devices of male quality (Davis and O'Donald 1976; Maynard Smith 1976; Bell 1978). These early models assumed that the reliability of the male signal was determined by epistatic fitness interactions between the male trait and a general viability trait. These models assumed that any male would be able to survive with a small ornament, but that the intolerably high costs of a large ornament to low-quality males resulted in a greater proportion of males with a large ornament being of high viability, because low-quality males with the handicap would disappear due to natural selection. The expression of a male sex trait therefore does not directly reflect the quality of a male, although on average females preferring males with larger ornaments still will be mated to males of superior viability.

Zahavi (1977) subsequently responded to this criticism and proposed a second type of handicap, the so-called conditional handicaps. The secondary sexual character was now assumed to evolve relative to the quality of an individual, and a given level of display was assumed to impose different costs on low- and high-quality individuals. If males of high phenotypic quality pay less for a handicap of a certain size, this will enforce reliability in the signalling system. Zahavi also suggested that 'any genetic system in which low-quality individuals are continuously produced from the high-quality ones may benefit from the use of handicaps'. As we shall see later, this suggestion resembles mutation bias on male viability, one of the mechanisms that results in evolution and maintenance of costly female mating preferences for males with handicapping sex traits (Iwasa *et al.* 1991).

A third kind of handicap is the revealing handicap. Hamilton and Zuk (1982) suggested that the state of male secondary sexual characters may allow females to discriminate against heavily parasitized males. This kind of handicap has evolved as a means of revealing a particular male quality, and not as a means of signalling general condition. The suggestion is therefore that secondary sexual characters have been designed to be particularly sensitive, for example, to parasite infection and hence directly reveal the ability of males to resist parasite attacks. Variation in revealing handicaps would thus allow females to assess male quality directly from the state of the secondary sexual character.

A fourth kind of handicap has already been alluded to in the early discussions of the handicap principle. Even though the initial models of the handicap principle assumed that the male trait had a genetic basis, this was never stated explicitly by Zahavi (1975, 1977). It was suggested that handicaps had quite low or no heritabilities (Maynard Smith 1976; Zahavi 1977; Bell 1978; Andersson 1982*a*; Dominey 1983; Kodric-Brown and Brown 1984; Nur and Hasson 1984). This could form the basis for a handicap mechanism that maintained a female preference, because the condition for the initial increase in the female mate preference is that the heritability of the male trait is low relative to that of the viability trait.

Another twist to the handicap principle is the concept of so-called amplifiers (Hasson 1989, 1991). There are problems associated with the evolution of male displays unless the female mate preference is associated with the male trait from the very beginning (Kirkpatrick 1982; Heisler 1984*a*). However, if females base their choice directly on observed differences in male viability, a male trait may evolve because it amplifies such perceived viability differences. Thus, the expression of the male trait as a result of the amplifying effect depends on male viability if mating success is increased for more viable, but decreased for less viable males. An amplifier is therefore a property of a signal that makes quality differences expressed in signals easier for receivers to perceive. The evolution of an amplifier is likely to produce a positive relationship between the expression of a male trait and male viability and therefore lead to the evolution of mate preferences based on the expression of the male trait. Whereas amplifiers result in an exaggeration of a male trait and thus increase perceived differences between individuals, reliable quality indicators (such as handicaps) emphasize a male trait's expression. In other words, amplifiers of male traits serve as cues of quality whereas indicators emphasize the expression of a male trait. Hasson (1990) has subsequently also incorporated amplifiers into models of the Fisher process. Some animal decorations that can be interpreted as amplifiers are described by Hasson (1991).

Models of the four kinds of handicaps, pure epistasis, conditional, revealing, and non-genetic handicaps, will be further discussed in the sections to follow.

Pure epistasis handicaps

Pure epistasis handicaps assume that the reliability of the male signal is determined by epistatic fitness interactions between the male trait and a general viability trait. The first models of the handicap principle were of the pure epistasis type, and none of these resulted in stable exaggeration of the female mate preference and the male trait (Maynard Smith 1976, 1978; Davis and O'Donald 1976; Bell 1978). There were two main reasons why there were no effects of the pure epistasis handicap in these early models. Epistasis is a weak force creating linkage disequilibrium between a male trait and viability; linkage arises mainly through natural selection on males, but not due to female choice. The problem is that if natural selection is weak, then genetic variance in fitness is maintained, but only a very weak linkage disequilibrium occurs. However, if natural selection is strong, genetic variance in fitness rapidly disappears even though linkage disequilibium is strong. There were therefore very weak or no effects of the handicap principle.

Early models did not consider costly female mate preferences. Pomiankowski (1987a, 1987b) analysed a haploid major gene model of the evolution of costly female mate preferences. The novel approach is that the female mate preference was considered to be costly and reduce female viability, and that additive genetic variance in viability was supposed to be maintained by host–parasite interactions. The effects of the male trait and the viability trait on natural selection are not independent, but act in an additive fashion, so that males with a low viability genotype suffer higher mortality from having a male trait compared with males with the high viability genotype. The dynamics and equilibria of this genetic system were analysed in simulations. Only one stable equilibrium was found, namely, when the preference and the ornament had frequencies of zero. Females were unable to distinguish between males when the male trait is close to fixation, since the realized male trait does not directly reflect general viability, which is only reflected indirectly through the male trait (Iwasa *et al.* 1991). The costly female mating preference therefore drops to a level below that needed for maintaining the male trait. The conclusion was that the pure epistasis handicap could not cause evolution of a stable and costly female mate preference.

In the pure epistasis handicap models the general viability trait is assumed only to affect viability and not the expression of the male trait. The reason for the covariance between the male trait and the viability trait is that viability selection has eliminated low quality males with large ornaments. Iwasa *et al.* (1991) modelled this in a three-locus polygenic model incorporating a costly female mate preference. As in the previous models there was no stable exaggeration of a costly female preference. The reason is that although the genetic correlation between preference and viability is positive, the preference and the viability traits are correlated only via the male trait. Therefore, at equilibrium, the product of the genetic correlations between trait and preference and between trait and viability equals the genetic correlation between preference and viability, and then the conditions do not

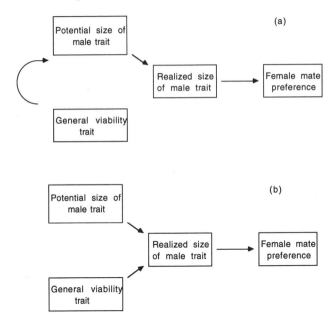

Fig. 2.1 Paths diagrams showing relationships between a male trait, a female mate preference, and a general viability trait for a pure epistasis handicap (a), and a conditional or revealing handicap (b). Females choose mates through the realized size of the male trait. In the pure epistasis model the realized size of the male trait depends entirely on the potential size of the male trait and not on viability. In the other two handicap models the realized size depends both on the potential size of the male trait and the viability trait. Positive relationships arise between the female preference and both the potential size of the male trait and the viability trait in the conditional and revealing handicap models. Adapted from Iwasa *et al.* (1991).

exist for the evolution of a costly female mate preference (Fig. 2.1a). The realized size of the male ornament is not directly affected by the general viability trait, but only by the potential size of the male trait. Therefore, the pure epistasis handicap does not work.

In conclusion, for all models of the pure epistasis handicap it has not been possible to identify conditions required for an initial increase in and a stable exaggeration of the female mate preference and the male trait (Table 2.1). Early models of the handicap were all of the pure epistasis type, and the general conclusion from these models was that handicaps did not have any effect on the exaggeration of male traits (Davis and O'Donald 1976; Maynard Smith 1976; Bell 1978), or that they could have an effect under extreme conditions (Eshel 1978) that are unlikely to apply in nature. Later models demonstrated that this negative view of pure epistasis handicaps was upheld when a continuous source of variation in fitness (Pomiankowski 1987a, 1987b) and when a cost of the female mate preference (Iwasa et al. 1991) were introduced.

Conditional handicaps

A conditional handicap is one in which the expression of a secondary sexual character depends on the quality of an individual, and a similar level of expression impose higher costs on low- compared with high-quality individuals (Zahavi 1977). For example, secondary sexual characters are often expressed more extravagantly in individuals in good body condition, and such individuals may also pay lower costs of the handicap than low-quality individuals. If males of high phenotypic quality were to pay relatively less for the same-sized handicap, the signalling system would be reliable. Williams (1966), Emlen (1973), and West-Eberhard (1979) independently suggested that such 'proportional' handicaps could develop if they reflected male size or other characters contributing to or indicative of male viability. Reliability of signalling would then automatically be built into the male structure, and females (as well as males) could use the expression of the male trait as a reliable estimator of male quality. Kodric-Brown and Brown (1984) claimed that male traits developing as handicaps might not necessarily be costly in terms of viability if they provided males with a natural selection advantage, for example, during intraspecific competition for food or other limiting resources. However, positive correlations between a male trait and two fitness components such as male mating success and male viability would not necessarily support their notion because condition-dependent expression of a male trait would automatically result in the apparent absence of life-history trade-offs. The only reason for this apparent absence is obviously that condition may mask existing trade-offs (Tuomi et al. 1983; van Noordwijk and de Jong 1986), which can only be detected by perturbation experiments of the male trait or male condition (Møller 1989*a*; Schluter *et al.* 1991).

Andersson (1982*a*, 1986*a*) modelled the condition-dependent handicap in a three-locus haploid genetic model based on the assumptions that the mating system is monogamous and a Fisherian mating advantage therefore is impossible, and that the viability of males increases with phenotypic quality and decreases with the degree of ornamentation. Under monogamy the handicap mechanism becomes more powerful as a selective force affecting the evolution of the male trait. However, Andersson was unable to find conditions that could increase the gene for ornamentation above its initial frequency. This was caused by the viability allele rapidly becoming fixed because of intense selection. Genetic variance in viability thus could not be maintained. With polygamy the situation was quite different. The assumption of a higher viability of the most ornamented males resulted in an increased rate of spread and a higher equilibrium value of the male trait. The handicap mechanism increased the rate of spread of the male trait above that produced by the Fisherian mating advantage of the most ornamented males. Maynard Smith (1985) later questioned any additional effect of the viability difference on the increasing frequency of the handicap. An absolute preference would lead to a very high mating success for handicapped males if they were rare relative to females with the preference. Andersson (1986*b*) replied to this critique by demonstrating that the handicap trait would increase in frequency if the handicap effect were added to the Fisherian mechanism. When the viability difference was absent, a pure Fisherian mechanism was at work, and the frequency of the handicap trait increased only slightly. In the presence of a viability trait there was an extra effect on the mating success, which resulted in a higher equilibrium level of the handicap trait. When both viability and condition-dependence were modelled by assigning males in poor condition an extra mortality cost, the handicap trait rapidly rose to very high levels and went to fixation because of genetic hitch-hiking due to linkage between the male trait, the female preference, and the viability trait. The handicap mechanism thus enhanced an increase in the male trait and the female preference because the male trait tended to reduce the viability of low quality individuals.

Using a haploid three-locus genetic model, Andersson (1986*a*) modelled the evolution of a condition-dependent secondary sexual character as a result of a female mating preference for the most extravagantly ornamented males. The condition-dependence of the secondary sexual character was thus the only difference from the previous models, which had failed to demonstrate an initial increase in the male sex trait and the female mating preference. Both the allele for the ornament and that for the female preference increased in frequency under certain conditions. The assumption that the female mate preference carried no direct costs is unrealistic (Parker 1983*b*; Pomiankowski

1987*b*), and was therefore relaxed by decreasing the viability of females with the preference allele because of the extra time and energy needed to locate a preferred mate. This viability cost of the mate preference reduced the frequency of the preference allele, and the frequency did not exceed the starting value if the cost of the mating preference was as high as the viability cost of the ornament. A smaller cost for the female mate preference could cause the preferred male trait to reach a higher frequency because an initial reduction in the frequency of the preference allele resulted in a more rapid increase in the ornament allele. Continuous selection on genetically beneficial alleles results in fixation. The results of the model therefore rely on the assumption that beneficial alleles with which the preference allele and the ornament allele can hitch-hike arise at other loci at a sufficient frequency due to mutation, migration, or fluctuating selection. Both the ornament and the female mating preference would increase if novel viability alleles arose regularly. The conclusion of Andersson's model was that indicator mechanisms might play a role in the evolution of a female mate preference and a male sex trait.

Kirkpatrick (1986*a*) developed a three-locus quantitative genetic model of the handicap mechanism. The evolutionary dynamics of the model gave rise to equilibria for the mate preference, the handicap and the viability trait on a line, and this might result in the viability and the handicap trait being close to or far from their optima under natural selection reached in the absence of a handicap. The line of equilibria could be either stable or unstable depending on the genetic covariance between the handicap and the preference. When there was directional rather than stabilizing selection on the viability trait, no evolutionary equilibria could be identified. The male trait and the female preference could equilibrate at points deviating widely from their natural selection optima. When the mating system was monogamous rather than polygamous, the handicap and the viability trait would equilibrate at their natural selection optima, which is their location in the absence of the handicap mechanism. An absence of sex-limitation in the expression of the male trait would increase the strength of natural selection on the handicap which would be driven to its natural selection optimum. Kirkpatrick (1986*a*) therefore concluded that the handicap mechanism did not work.

Pomiankowski (1987*a*, 1987*b*) modelled the handicap process for a costly mate preference in a three-locus genetic model. The condition-dependent handicap model assumed that the development of the male trait would act as a reliable indicator of viability to females, and ornaments thus were only present if males had high viability. Consequently, the ornament would become less costly because only males which actually developed the ornament would pay the costs. The results of simulations revealed that females would be able to pick males with viability genes even when the male trait was close to fixation. There existed a stable equilibrium at which a costly female preference was maintained and the male trait was exaggerated above its natural selection optimum.

Tomlinson (1988) simulated the condition-dependent handicap in a three-locus diploid genetic model. When a single viability allele was introduced, there was a small transient increase in the male trait and the female preference while the viability gene went to fixation. When a succession of viability genes were introduced, a transient increase occurred in the frequency of the handicap and the female mate preference, with a subsequent decrease in first the handicap trait and then the preference trait. The handicap and the female mate preference thus would not go to fixation or be maintained in the population. When the mating system was polygynous rather than monogamous, the handicap mechanism increased the probability of the male trait becoming fixed in the population if there should be a high initial frequency of the preference gene. If the handicap gene went to fixation, this was accompanied by fixation of the preference gene. The handicap mechanism was thus concluded to facilitate the Fisherian process.

The correlation between the male trait and viability depends on the direct effects of the trait on viability, which may be negative if the trait is costly to produce or maintain, or on an indirect positive effect of condition on viability. This scenario was modelled by Zeh and Zeh (1988), who investigated the relationship between the expression of the male trait and male survival when the variation in the male trait has a strong environmental basis. The expression of the male trait is assumed to affect male survival negatively, whereas body condition has a positive effect on the expression of the male trait. The effect of phenotypic quality on male viability depends on assumptions about the female mate preferences. Male viability was found under several conditions to increase with the expression of the male trait. A similar conclusion was reached in a graphic model developed by Nur and Hasson (1984). This conclusion differs from that of Fisherian models of female mate choice, which do not assume a positive relationship between male condition and male viability.

Michod and Hasson (1990) analysed the evolution of reliable fitness indicators and used male secondary sexual characters as an example. Their model considers the initial evolution and maintenance of reliability in the expression of a male trait, but assumes that the female preference is already present in the population. The haploid model has two loci coding for viability and a modifier, which controls the expression of a male trait. The male trait evolves under conflicting effects of two components of fitness, that is the male trait has a negative effect on viability, but a positive effect on mating success. The conditions for the initial increase in the frequency of the modifier trait are that modifiers act independently on the male trait in the two viability classes and allow the male trait to develop different optima which maximize individual fitness. The conclusion, that the male trait does not need to have differential costs in the different viability classes for reliable signalling of fitness to evolve, differs from that of other models of the handicap principle (Zahavi 1975; Andersson 1982*a*; Kodric-Brown and Brown 1984; Nur and Hasson 1984; Grafen 1990*a*, 1990*b*). As both greater absolute costs and differences in relative costs of the handicap can facilitate the modification of the male trait in the viability classes, the cost of the male trait could be one important factor contributing to maximization of fitness at different trait values for the different viability classes. Because the optimum for the trait in the high-viability class is greater than that for the low class, the traits could be used as reliable indicators of genetic quality.

Finally, Iwasa *et al.* (1991) modelled condition-dependent handicaps in a quantitative genetic model. The assumptions were that the expression of the male trait depends both on the potential size of the male ornament and the general viability trait (Fig. 2.1b), and females use the realized size of the male trait as influenced by condition when they choose mates. The condition for evolution of a costly mate preference is that an increase in the realized size of the male trait is associated with an increase in general viability. Calculations of additive genetic correlations demonstrated that this condition was always fulfilled unless there was a perfect correlation between the potential size of the male trait and the viability trait. Thus, a costly female mate preference could only evolve for a condition-dependent handicap when the preferred male trait directly reflected general viability.

In conclusion, out of ten models of the condition-dependent handicap, several suggest that a female mating preference can evolve for a male trait that reliably reflects viability (Table 2.1). The inadequacy of some of the early models may be explicable in terms of their restrictive assumptions about the maintenance of genetic variance in viability.

Revealing handicaps

Hamilton and Zuk (1982) suggested that the state of male secondary sexual characters may allow females to discriminate against heavily parasitized males because male sex traits have evolved directly to reveal this quality property. The state of a revealing handicap allows females directly to assess male quality, since the male trait will be constructed irrespective of general male viability. Revealing handicaps have thus evolved as a means of signalling male quality, and not as a means of signalling male condition. Revealing handicaps would therefore allow females to assess male quality directly from the current state of the secondary sexual character.

Kirkpatrick (1986*b*) used a three-locus haploid genetic model to simulate the revealing handicap process. He assumed that each resistance allele was favoured for a number of generations and then replaced by another favoured resistance allele and so on. The result was that there exists a threshold frequency of the preference gene, below which the handicap gene is eliminated, and above which it goes to fixation. The revealing handicap mechanism appeared unable to account for the initial rise in the frequency of the mate preference.

A second model of the revealing handicap process is a three-locus genetic model of a costly female mate preference developed by Pomiankowski (1987*a*, *b*). The model assumed that the development of the male trait acts for females as an accurate and direct indicator of whether a male has high or low viability. Simulations revealed that females would be able to pick males with viability genes even when the male trait was close to fixation. There existed a stable equilibrium at which a costly female preference was maintained and the male trait exaggerated above its natural selection optimum. This equilibrium was stable for a wide range of viability differences at the general viability locus even when the maximum cost of the female mating preference was relatively low.

Tomlinson (1988) simulated the revealing handicap in a three-locus diploid genetic model. Males with the sex trait were assumed to be at a selective disadvantage without the resistance allele. When a succession of resistance genes were introduced, the result was a transient increase in the frequency of the handicap and the female mate preference, with a subsequent decrease in first the frequency of the handicap trait and then the preference trait. When the mating system was polygynous rather than monogamous, the handicap mechanism increased the probability of the male trait becoming fixed in the population if there was a high initial frequency of the preference gene. If the handicap gene went to fixation, this was accompanied by fixation of the preference gene. Thus, the handicap mechanism was concluded to facilitate the Fisherian process.

Iwasa *et al.* (1991) analysed the revealing handicap in a polygenic model of costly female mate preferences. The assumptions for this model were that the realized size of the male trait would increase with both the potential size of the male trait and male viability (Fig. 2.1b), and females used the realized size of the male trait as a cue in their mate choice. Since the male trait directly reflects general viability, the conditions for the evolution of a costly female mate preference are fulfilled. A revealing handicap and a condition-dependent handicap differ in their production costs. The costs of a revealing handicap depend on the potential size of the male trait while the costs of a condition-dependent handicap depend on both the potential size and male viability. This will affect the mean expression of the two kinds of handicaps at the evolutionary equilibrium.

In conclusion, there have been several models of the revealing handicap mechanism, the more recent of which all suggest that this process can give rise to a stable exaggeration of a female mate preference and a handicapping male trait (Table 2.1).

Non-genetic handicaps

Several authors have suggested that handicaps actually may have quite low or no heritabilities (Maynard Smith 1976; Zahavi 1977; Bell 1978; Andersson 1982a; Dominey 1983; Kodric-Brown and Brown 1984; Nur and Hasson 1984). For example, Maynard Smith (1976) suggested in his first model that there would be no reason to doubt that the handicap mechanism could work if the handicap itself was not inherited. The major problem with models of the pure epistasis handicap is that the heritability of the handicap is required to be much higher than that of viability under natural conditions. Dominey (1983) solved this problem by suggesting that the Fisher process might result in depletion of additive genetic variance in the male trait, and that the handicap principle could subsequently take over the sexual selection process when the heritability of the male trait was low or absent. Females that preferred ornamented males would still obtain the viability advantage provided by the handicap mechanism without paying the cost of producing brighter than average sons.

There is ample evidence to suggest that secondary sexual characters often have statistically significant heritabilities which may not differ in magnitude from those of other morphological traits (Winge and Ditlevsen 1947; Haskins *et al*. 1960; King 1975; Cade 1981; Kirpichnikov 1981; Bubenik 1982; Harmel 1983; Templeton *et al*. 1983; Hedrick 1988; Møller 1989b, 1991a; Grant 1991; Hill 1991; Lemel 1993; Norris 1993). It is thus unlikely that non-genetic handicaps are common.

Concluding remarks on handicap models

In the preceding pages I have reviewed the various handicap models (Table 2.1). There is no theoretical evidence to suggest that the pure epistasis handicap process can result in stable exaggeration of female mate preferences and male traits. However, both the conditional handicap, the revealing handicap, and the non-genetic handicap process produce the initial increase and the maintenance of female mate preferences for male handicapping traits. The reason is that the male trait is costly, that the cost of the male trait is higher for low- than for high-quality individuals when the trait is condition-dependent, and hence the expression of the male trait directly reflects male quality. In the revealing handicap it is assumed that the reflection of male quality is guaranteed in the expression of the male trait, and a female preference for males with the handicap therefore predictably confers an indirect fitness benefit.

2.4.2 *Fisher and the sexy son*

The Fisher process

Darwin (1871) suggested that male secondary sexual characters could evolve as a result of a female mate preference for the most elaborately ornamented males which were best able to charm females. This idea was developed by Fisher (1915, 1930), who suggested that the female mating preference itself would evolve as a consequence of differential mating success produced by the preference. The female mate preference could be assumed to have a genetic basis as could the male trait. Fisher (1915, 1930) also suggested that this initial mate preference would favour males possessing an advantageous trait reflecting some component of male quality. Females preferring certain male phenotypes would produce sons which tended to carry genes for both the male sex trait and the female preference. They would soon start indirectly to select on the mating preference itself when choosing a preferred mate because of the genetic correlation between the two traits. In other words, the preference genes will hitch-hike with the genes for the preferred male trait. This could result in a self-reinforcing runaway process. Eventually, the runaway would come to a halt because of the effects of oppositely directed natural selection, and the secondary sex trait would gradually disappear because females had switched their preference to other male traits. Fisher never explicitly analysed his ideas of the evolution of a female mate preference in a genetic model, but several people have later confirmed that mate preferences may indeed result in exaggeration of male secondary sexual characters far beyond their natural selection optima.

At least three questions have to be answered before the Fisherian process can be stated to work. These are:

(1) whether the process can get started when the preference genes are rare;

(2) how much of a selective effect will the preference genes acquire as an indirect result of selection on the male trait; and

(3) will the hitch-hiking effect of the mate preference caused by selection on the male trait be strong enough to result in a runaway process?

The first of these questions may not be as crucial as the other two if the female preference evolved before the male trait, as suggested by some studies (Heisler *et al.* 1987; Kirkpatrick 1987; Basolo 1990; Ryan 1990; Ryan and Rand 1990; Ryan *et al.* 1990). One or more of these questions have been addressed by nine different models.

O'Donald (1962, 1967, 1980*a*) analysed a two-locus genetic model of the Fisher process. His simulation study showed that the mating preference gene is selected in association with the gene for the male trait, and the preference gene subsequently increases in frequency to an equilibrium which is only a few times its initial frequency. This raises the problem of how a preference arising by mutation is able to raise in frequency to a level where it will start exerting its own selective effects. Unless the male trait confers some additional advantage, for example, in terms of increased viability of offspring, the female mating preference will not increase much in frequency. The Fisher process will therefore be extremely slow and unable to account for the widespread evolution of secondary sexual characters. However, even an independent selective advantage of the male trait would result in only a small increase in the mating preference.

One solution to the problem of the low frequency of the mate preference is further selection on the male trait. If the male secondary sexual character was subject to a succession of new mutations, of which each produce more extreme phenotypes, the female mate preference would be able to reach very high frequencies and carry the male trait along.

Genetic models with few loci are restrictive because the evolution of a male trait comes to a halt when one allele is fixed at the trait locus. O'Donald partially circumvented this problem by entering new mutations on the male trait. Secondary sexual characters are often quantitative traits with a presumed polygenic inheritance, and a polygenic model also allows for the maintenance of genetic variance by mutation and recombination (Lande 1975). Lande (1981) modelled the evolution of a secondary sexual character and a female mate preference in such a polygenic model. The results of this model are that the male trait and the female preference may evolve to exaggerated levels along lines of equilibria where the mean female mate preference is matched by a mean size of the male trait for which the viability loss and reproductive gain are balanced. The male trait will then be greatly displaced away from its optimum under natural selection. Evolution of the two characters depends on the stability of the equilibria. The exaggeration of the male trait may eventually come to a halt when the effects of the female mate preference is balanced by opposing natural selection. Variance in the female mate preference may, however, exert a disruptive effect on the male trait, and this effect may sometimes be so large that it overcomes the stabilizing effect of natural selection.

Natural populations are often relatively small, and random genetic drift may then play a role during evolution. Drift in a female mate preference may be important when it arises, and it may produce random selective forces on the male trait, and thus indirectly on the female preference because of the genetic correlation. Random genetic drift could trigger a runaway process of sexual selection when the line of equilibria is unstable, but it may be unimportant for a stable line of equilibria because populations starting from the same point may drift to either side of the line but evolve in the same direction. This could set the stage for rapid divergence of populations even in the absence of geographic barriers, and sympatric speciation as a result of pre-mating isolation mechanisms may be the final outcome (see Chapter 13). The results of the polygenic model by Lande (1981) differ somewhat from those of the two-locus model by O'Donald (1962, 1967, 1980a). The rapid runaway evolution of the male trait and the female preference during the unstable phase in the polygenic model was not demonstrated by the simple two-locus model, which only resulted in exaggeration when the male trait was subject to continuous input of genetic variance due to mutation.

Lande's model starts out with the presence of the female mate preference and the male sex trait. The major problem of how the runaway process starts from an initially low frequency of the preference gene is thus not solved. Perhaps it is more critical that the condition for the line of equilibria, which was not envisaged by Fisher (1915, 1930), depends on the relationship between the covariance of the male trait and the female preference and the variance in the male trait, which has to exceed the slope of the line of equilibria. However, the two sides of this inequality are not independent (O'Donald 1990). The conditions for the runaway process have not been determined because the genetic correlation between the male trait and the female preference has not been obtained analytically in terms of the mating preference parameters (Karlin and Raper 1990). The runaway process will only ensue if the slope of the line of equilibria is greater than one which means that sexual selection on males must produce a greater response on the preference carried by the males than on the male trait (O'Donald 1990). This condition will only be fulfilled if there is a close genetic correlation between the male trait and the female mate preference, which could occur as a result of females mating with males close to their preferred phenotype.

The problem of the origin of female mating preferences was addressed specifically in quantitative genetic models by Heisler (1984a). Fisher (1915, 1930) suggested that females initially would favour male traits that were favoured by natural selection. Female choice of such 'good genes' male traits was analysed in a model that assumed expression of the trait in both sexes, and that there was stabilizing natural selection on the trait. The conditions for the initial increase in the female mate preference were that a high heritability and an ability by females to discriminate against males deviating from the optimum male phenotype would improve the increase in offspring viability caused by the female mating preference. For non-equilibrium populations, directional female mating preferences for mates exceeding the optimum phenotype would be favoured and could result in a rapid evolutionary response. The evolutionary history of male traits would therefore determine the extent to which they were close to the optimum phenotype, and the recent selection history could thus determine the likelihood of sexual selection affecting a particular trait (Heisler 1984a). A second model of the initial increase in a female mate preference was based on the assumption that the male trait subject to the mate preference may not itself be related to viability, but rather become the target of a female preference because of genetic correlations with other fitness-related traits (Heisler 1985). Selectively arbitrary traits may thus become the target of sexual selection via a female mate preference if they are relatively immune to random environmental variation and genetically correlated with another trait that affects viability.

A different modelling approach from that used by Lande was developed by Kirkpatrick (1982), who analysed a haploid model with single genes for the female mate preference and the male trait. The reason for choosing a haploid model is that such models are much more tractable and can be analysed mathematically. The haploid model showed that the frequency of the female mating preference allele changes only in response to changes in the male trait allele. A second result was a line of equilibria along which the female preference and the male trait remained in stability; a result similar to that of Lande's model.

All the previous models have relied on the unlikely assumption that the female mating preference is uncostly. This is not the case if the mating preference results in delayed breeding, extra energy expenditure, increased risks of predation or disease transmission (Parker 1983*b*; Pomiankowski 1987*b*). The Fisher process under costly female mate preferences has been modelled by Pomiankowski *et al.* (1991) in a two-locus polygenic model. When a female mate preference is costly, the magnitude of the cost can be assumed to be an increasing function of the intensity of the female preference, and randomly mating females minimize their costs. When the change in the male trait and the female mate preference has come to a halt at equilibrium the isoclines intersect at the point where female fitness is maximized. The line of equilibria obtained by Lande (1981) only occurs as a result of the absence of any costs due to the female mating preference. A Fisherian process cannot result in stable exaggeration of costly female preferences.

The assumption that mutation on the male trait is unbiased is probably not valid. Mutation pressure on exaggerated secondary sexual characters is likely to be biased because random alterations of a complex structure will generally result in less effective displays. When secondary sex traits are close to their limit of production, any mutation is likely to result in reduced performance. The great intricacy of design of sex traits may require well-controlled developmental pathways for their production, and these can probably be disrupted in numerous ways. Secondary sexual characters are subject to strong directional female mate preferences and this history of directional selection predisposes such characters to biased mutation because any genetic change caused by mutation would very probably result in an inferior display (Mukai 1964). It is important to note that all previous models of the Fisher process have tacitly assumed that mutation is unbiased.

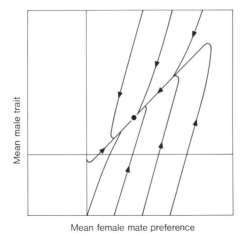

Fig. 2.2 Equilibrium mean phenotypes for the male sex trait in relation to the mean female preference (dot). The male trait is under stabilizing natural selection and directional sexual selection by females. The female mating preference, which is under viability selection due to the costly mate preference, evolves as a correlated response to selection on males along lines with arrowheads. Adapted from Pomiankowski *et al.* (1991).

When mutation bias on the male trait is entered as an assumption in the polygenic model, the isoclines for the mean female mate preference and the mean male trait intersect at an equilibrium differing from zero (Fig. 2.2; Pomiankowski *et al.* 1991). The equilibrium values of both the male trait and the female preference lie above the natural selection optima. Female fitness under a costly mating preference is not maximized at equilibrium because of the costs of the mate preference. The equilibrium female preference increases with mutation bias on the male trait, but decreases with the costs of the preference.

The polygenic model analysed by Pomiankowski *et al.* (1991) suffers from some of the same weaknesses as the models by Lande (1981) and Kirkpatrick (1982), although it is much more realistic because of the assumptions about the costly female mate preference and biased mutation on the male trait. The male trait and the female preference are assumed to be present from the very beginning, even though it is the initial increase in the female preference which is the crux in the Fisherian process. Given that there is biased mutation on male secondary sexual characters, the preference genes will acquire a substantial effect due to indirect selection because of the genetic correlation between the two traits. The model of the evolution of costly female mating preferences under the Fisherian process resulted in a stable exaggeration of the female preference and the male trait at equilibrium, but did not result in the runaway process of ever increasing exaggeration of the two traits.

In conclusion, the Fisherian process is theoretically well developed in a number of different models (Table 2.1). All models except from those of O'Donald (1962, 1967, 1980*a*) and Heisler (1984*a*, 1985) assumed that the female preference was already established from the very beginning. This assumption may not be unrealistic if female preferences are present before the evolution of male traits, as suggested by some studies (Heisler *et al.* 1987; Kirkpatrick 1987; Basolo 1990; Ryan 1990; Ryan and Rand 1990; Ryan *et al.* 1990). The polygenic model by Lande (1981) and the haploid model by Kirkpatrick (1982) resulted in a runaway process, but were based on unrealistic assumptions such as uncostly female preferences. O'Donald (1962, 1967, 1980*a*) was only able to demonstrate an exaggeration of the female preference from an initially low frequency by continual mutational input to the male trait, which would otherwise go to fixation. This mutational input actually resembles biased mutation on the male trait, which was established as a necessary condition for stable exaggeration of the female preference and the male trait when mate choice was costly (Pomiankowski *et al.* 1991). Thus, there is still little theoretical analysis of the initiation of the Fisher process as well as the hitch-hiking effect of the female mating preference, which should result in the runaway process.

The sexy son hypothesis

The Fisher process assumes that males only provide sperm to females and therefore in theory may be able to acquire an infinite number of mates. This situation may almost apply to organisms that display on leks or in mating swarms. Empirical studies of polygynous birds, particularly the red-winged blackbird *Agelaius phoeniceus*, demonstrated that females sometimes mated with the most preferred males, but at a fitness cost. They reared fewer offspring in their current brood than females mated to less preferred males because attractive males provided less parental care. This observation prompted Weatherhead and Robertson (1979) to propose the sexy son hypothesis, which suggests that females that mate with preferred males may still acquire a net fitness advantage relative to those mating with less preferred males. In other words, females mated to the most preferred males were able to compensate for their fitness loss during reproduction by rearing particularly attractive 'sexy' male offspring, which acquired their attractiveness from their father. This so-called sexy son hypothesis has received considerable attention from empiricists and theoreticians (Heisler 1981; Searcy and Yasukawa 1981; Weatherhead and Robertson 1981; Wittenberger 1981; Kirkpatrick 1985, 1988; Curtsinger and Heisler 1988, 1989; Pomiankowski *et al.* 1991).

Kirkpatrick (1985) was the first to analyse the sexy son hypothesis in explicit genetic models, which assumed that males have a secondary sexual character that affects the probability of their survival, their mating success, and the fecundity of each mating. The sexy son hypothesis would be confirmed if there is an evolutionary equilibrium at which females mate with males that differ from the phenotype with highest average fecundity per mate. The single stable equilibrium for the male trait occurs at the phenotype with the maximum fecundity per mating. Female fecundity is therefore maximized at equilibrium (Fig. 2.3). The male secondary sexual character may, however, evolve to extreme expressions which seriously impair male survival. The alternative outcome is an unstable evolutionary equilibrium from where the male trait and the female preference can evolve to ever more extreme expressions by the runaway process of self-reinforcing feedback. Female fecundity is continuously impaired as a result of exaggeration of the male trait, but the increasing attractiveness of sons can never keep up with the decreasing female fecundity. The conclusion was therefore that the sexy son hypothesis could not work because female fitness had to be maximized at the evolutionary equilibrium. Kirkpatrick's (1985) model relies on a number of assumptions which may have produced his very pessimistic 'demise of the sexy son', as discussed later.

Curtsinger and Heisler (1988) used a two-locus diploid genetic simulation model to analyse non-additive genetic effects in a sexy son situation. Polymorphism of the male trait was maintained at a stable equilibrium when there is random mating and heterosis in viability and/or fertility, or when there is a balance between opposing viability and fertility selection among

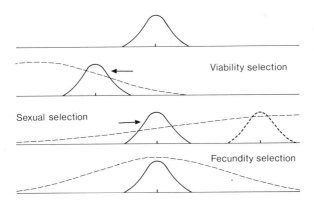

Fig. 2.3 The evolutionary equilibrium for the male trait assuming unlimited male reproductive potential in Kirkpatrick's model of the sexy son process. Frequency distributions of the male trait (full drawn lines) change their position because of the action of viability, sexual and fecundity selection (dashed lines). Adapted from Kirkpatrick (1985).

males. When there is genetic variation in the female preference locus, the mate preference evolves in response to differences in female fertility caused by matings with more or less fertile males. The mate preference also evolves in response to differences in the total fitnesses of males, which cause change in the frequency of the preference through linkage disequilibium between the male trait and the female preference. The results of the model by Curtsinger and Heisler (1988) were that:

(1) Polymorphism at the male trait locus can be maintained by several different kinds of balances between viability, fertility, and sexual selection;

(2) attractive males having low fertility can increase from low frequency to intermediate frequency or fixation;

(3) the heritable tendency to mate preferentially with attractive males can increase from a low initial frequency; and

(4) the maximization of female fertility and viability at the evolutionary equilibrium can be violated.

In other words, under some conditions a stable equilibrium develops between the fecundity costs of the female mating preference and the indirect fitness advantages of mating with the most attractive males.

Kirkpatrick (1988) stated that this result of the diploid model is consistent with those of his previous haploid models in three respects. First, there is a stable evolutionary equilibrium where female fecundity is maximized. This equilibrium does not allow the female mate preference to be costly. The second possibility is an unstable equilibrium where the runaway process increases the female mate preference and the male trait indefinitely. There would thus not be any possibilities for an evolutionary equilibrium where females pay a cost for their mate preference. Third, the mean male trait and the mean female mate preference do not respond to selection in the absence of additive genetic variance, and the population may therefore be constrained from evolving toward increased female fecundity even in the presence of fecundity selection. This conclusion was questioned by Curtsinger and Heisler (1989) in a reply to Kirkpatrick's critique. Diploid genetic models were claimed to give rise to complex results among which one outcome is a stable evolutionary equilibrium with a costly female mate preference where female fitness is not maximized. Both Curtsinger and Heisler's (1988) study and Kirkpatrick's (1988) reanalysis are based on simulations alone, and conditions for the sexy son may therefore exist for some parameter values but not for others. Kirkpatrick may not be right in his critique that stability of evolutionary equilibria between a female mate preference and a male trait requires maximization of female fitness. One way in which male traits may become exaggerated and displaced from the natural selection optimum at the same time as female fitness is not maximized is described in the following model.

The assumptions of the sexy son hypothesis resemble those of the Fisher process when female choice is costly (Pomiankowski *et al.* 1991). The only difference is that the cost of the female mating preference now is a fecundity cost rather than a viability cost. Fecundity costs are complicated to model because of their dependence on the distribution of male phenotypes and how harem size affects female fecundity. A stable equilibrium between a male trait and a female mate preference can be found if there is mutation bias on the male trait because losses in fecundity as a result of the female preference are offset by the mating success of attractive sons. It is an empirical question whether mutation bias on the male sex trait ever will be strong enough to establish such stable equilibria.

In conclusion, a number of different models of the sexy son hypothesis have been developed. Some of these models suggest that conditions exist for a stable exaggeration of the male trait and the female mate preference even when female fitness is not maximized (Table 2.1).

2.4.3 *The good parent process*

The simplest possible situation for the evolution of a female mate preference and a male trait occurs when the female benefits directly from its mate preference. It is obvious that females should choose males with the best resources or high parenting ability. The problem for the female is how to choose a mate when these direct fitness benefits are not obvious. Should a female choose by assessing male ornaments? And if that is the case, can this process explain the evolution of extravagant secondary sexual characters? In this so-called good parent process females may acquire material benefits or male parental care, or avoid contagious parasites by assessing the level of advertising in the male secondary sex trait. Several models have been developed to deal with this situation.

Heywood (1989) modelled the evolution of a female mate preference and a male handicapping trait when there is non-heritable variation in paternal investment. The model was based on the assumption that there are two environments, such as absence or presence of parasites, and individuals in the good environment are more viable and better parents than individuals in the poor environment. Under certain conditions the male trait and the female preference evolve to either of two equilibria: (1) a complete loss or fixation of the male trait, or (2) polymorphic equilibria of the male trait and the female preference along a line. If the male trait is a condition-dependent handicap, it generates a very strong association between the male trait and paternal investment. This enhances the exaggeration of the male trait and the female preference even with a small variance in paternal investment. Furthermore, males remain phenotypically variable after the male trait has gone to fixation, and the handicap mechanism thus continues to favour the female mate preference until it is also fixed in the population. Heywood (1989) also analysed the situation when heritable rather than environmental variation in viability is the source of variation in mate quality. The results were, however, qualitatively similar to those described previously.

Hoelzer (1989) developed a simulation model of the good parent process where the male trait was supposed to advertise a non-heritable component of parental quality. The male trait was able to increase in frequency because of a temporal association between parental quality and presence of the male trait following sexual selection in each generation. The evolution of the male trait is dependent on a linear trade-off between the benefit that male care confers on offspring viability and the viability cost of the male trait. Extreme exaggeration of the male trait would thus impair further evolution because of its negative effects on the quality of male parental care. Phenotypic plasticity in the male trait would potentially allow low-quality males not to reveal their quality. However, phenotypic plasticity further enhances the evolution of a good parent trait because when the cost of the expression of the male trait to low-quality males is small, the mean fitness of males with the sex trait relative to males without the sex trait is greater at small values for the male parental care trait.

Grafen (1990a, 1990b) modelled the evolution of a male sex trait which reliably reflects male non-genetic quality. He analysed this problem by determining the phenotype-dependent evolutionarily stable strategies as first formulated by Parker (1982), and by analysing explicit genetic models. The haploid population genetic model of sexual selection was based on the assumptions that females enjoy higher fecundity if they mate with a male of higher quality, each male achieves an advertising level relative to its quality, and each male survives the pre-breeding stage relative to its quality and its level of advertising. The equilibrium was found by determining the set of advertisement and preference rules where no alternative strategy could invade. There are two pairs of strategies: (1) the non-signalling equilibrium at which males advertise at the lowest possible level and at which females mate independent of male phenotype, and (2) male advertising is a continuously increasing function of male quality, and the female preference for higher quality males is costly. Males as well as females may pay a cost at this second equilibrium because the male trait reduces male viability, and females may experience reduced fitness if male survival is reduced to such an extent that females are less able to mate at the best time of the season. This model has no Fisher process incorporated because the two strategies are uninvasible, there is no local structure, and all offspring are equivalent.

Price *et al.* (1993) analysed the joint evolution of a condition-dependent male trait and a female preference when the female benefited directly from its preference. The number of offspring produced by a pair is assumed to decrease with the expression of the male trait and increase with male condition. The results of the model were that natural and sexual selection favour the evolution of condition-dependence of the male trait, and that female preferences for condition-dependent traits evolve even if the male trait has negative effects on male viability and female fecundity. The female mating preference is under stabilizing selection at the evolutionary equilibrium when the positive effects of male condition balance the negative effects of the male trait on fecundity. The mean fecundity of females is not maximized at the equilibrium and may continuously decline as the male trait and the female preference evolve. This differs from the results of models where the male trait is not condition-dependent since female fecundity is then maximized at the equilibium (Kirkpatrick 1985). If there are no direct negative effects of the male trait, female mate preferences will increase continuously. At the evolutionary equilibrium the male trait will be unrelated to fecundity because the positive effects of male condition are balanced by the negative effects of the male trait (Fig. 2.4). When the assumption of a non-heritable condition is relaxed, female mate preferences may evolve as a result of the direct benefits due to fecundity selection and the indirect fitness benefits due to higher offspring condition. Thus, direct and indirect selection would act in a reinforcing manner.

In conclusion, the good parent process may lead to increases in and stable exaggeration of male traits and female preferences (Table 2.1). Male fitness is not maximized at the evolutionary equilibrium. Female fitness may or may not be maximized at the equilibrium depending on whether the male trait demonstrates condition-dependence.

2.5   Sexual selection and monogamy

Monogamy may pose special problems to the theory of sexual selection because theoretically all males should be able to acquire a mate under an unbiased tertiary sex ratio. Non-random variance in male mating success, which is the main fuel for the Fisherian sexual selection process, is thus not available. Other mechanisms may cause variation in male mating success and several theoretical models have been developed for sexual selection under monogamy. These are discussed in the following sections.

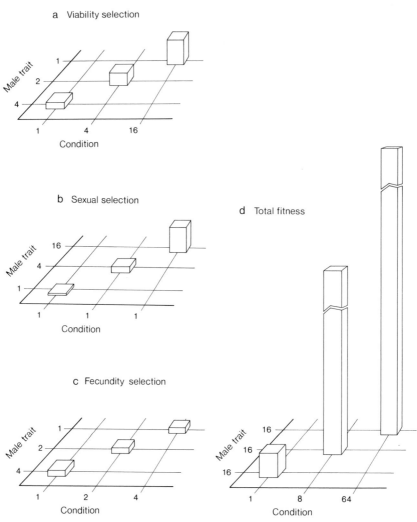

Fig. 2.4 Patterns of selection at equilibrium on males with an extravagant sex trait. Numbers at the margins represent the fitness effects of condition and the male trait. Numbers within the figure represent the net fitness at each life stage. Fitnesses were assumed to be multiplicative. The male trait and condition are positively correlated (a), and mating success is correlated with the expression of the male trait (b). The two conditions for an equilibrium are that (1) the male trait and fecundity are uncorrelated (c), and (2) there is no net directional selection on the male trait when condition is held constant (d). Adapted from Price *et al.* (1993).

2.5.1 *The Darwin–Fisher mechanism of sexual selection under monogamy*

Darwin (1871) was well aware of the fact that monogamous organisms might become an objection to his theory of sexual selection. The majority of all bird species are socially monogamous, but some still exhibit extravagant ornamentation which cannot easily be accounted for by processes other than sexual selection. The sex ratio is usually equal in sexually monomorphic species, and in a monogamous mating system there should in theory be a mate for every female. Darwin suggested that sexual selection by female choice could still operate because males usually arrive before females at the breeding grounds. The mechanism for sexual selection by female choice in monogamous species was based on two assumptions: (1) females differed in fecundity because of variance in body condition, and (2) females in prime body condition were able to breed early and thus had first choice of a mate. This idea could lead to exaggeration of a male display by a female mate preference because the most preferred males would leave more offspring, and because of the resultant genetic correlation between the male sex trait and the female mating preference. The earliest arriving females had a larger variation of male phenotypes to choose among, and they would also rear more offspring if the conditions for successful rearing of offspring deteriorated as the season progressed. The most preferred males therefore achieved a mating advantage by mating with the earliest females.

Fisher (1930) suggested that Darwin's theory of sexual selection in monogamous mating systems could work only if the correlation between breeding date and breeding success was environmental. A genetic correlation would result in continued selection for earlier breeding, and this was not feasible given Fisher's fundamental theorem of natural selection which states that the mean fitness of a population under natural selection increases at a rate equal to the additive genetic variance in fitness (Fisher 1930). Fisher also provided a numerical example of this process of mate choice in monogamous mating systems (Fig. 2.5). This example is based on three parameters: a male sex trait, a female body condition, and the resulting reproductive success. Males, which arrive at the breeding sites before females, vary in their expression of a secondary sexual character, and females vary in their body condition, which is environmentally determined. Females in prime body condition will arrive at the breeding sites before other females and have first priority in mate choice. The most preferred males are therefore mated to females in the best body condition. Female fecundity is directly related to body condition, and preferred males therefore acquire a reproductive advantage by mating with females in the best condition.

The process of sexual selection in monogamous mating systems by a female mating preference as envisaged by Fisher was subsequently modelled by O'Donald in single-locus genetic models (O'Donald 1972, 1980*a*, 1980*b*).

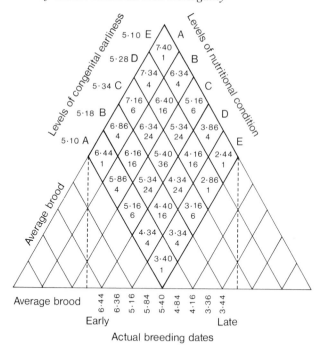

Fig. 2.5 Fisher's 'diamond of monogamy' which illustrates the relationships between female body condition, timing of breeding, and reproductive success. Adapted from Fisher (1930).

The models determined whether a male character could become exaggerated above its survival optimum as a result of a fixed female mate preference and an environmental component of female body condition. The analyses clearly suggested that the male character could evolve away from its survival optimum because of the female mating preference. The models of mate choice by the Darwin–Fisher mechanism were subsequently fitted to field data obtained from the dichromatic Arctic skua *Stercorarius parasiticus*. Early-breeding females preferentially mated with males of the dark colour morph and these females also had higher reproductive success (O'Donald 1980*c*, 1983). However, it was never tested whether plumage colour was the direct target for the female mate preference.

The simultaneous evolution of a male trait, a female mate preference, and breeding date was analysed for a monogamous mating system in a quantitative genetic model by Kirkpatrick *et al.* (1990). The model is based on the assumptions that the breeding date of a female is determined by the sum of an additive genetic component, a non-heritable component attributable to the female's nutritional state, and the reproductive success of a female is determined both by its condition and the breeding date.

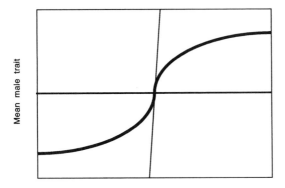

Mean female mate preference

Fig. 2.6 Equilibrium mean phenotypes for the male sex trait in relation to the mean female preference (heavy curve). The survival optimum is indicated by the horizontal, intermediate line, while the curve of equilibria under polygyny with no variance in nutrition is indicated by a light line. Note that monogamy results in a limited degree of exaggeration of the male trait. Adapted from Kirkpatrick *et al.* (1990).

The evolutionary equilibria of the polygenic model of sexual selection in a monogamous mating system are a line of mean male traits corresponding to the possible values of the female mating preference (Fig. 2.6). The major difference between the line of equilibria in polygynous and monogamous mating systems is that the degree of exaggeration of the male trait is much more limited under monogamy. The reason for the limitation of exaggeration of the male trait under monogamy is twofold. First, the exaggeration process of the male trait becomes limited by the variance in female fecundity. This component of variance in male mating success may be smaller than that caused by variance in the number of mates acquired by polygynous males. Second, direct selection acting on the female preference may also result in a limitation of the exaggeration process. Direct selection on the female preference will result if males with extreme sex traits for example attract predators to the offspring or the female. Less conspicuous males may then acquire a mating advantage and select for less extreme sex traits. The line of equilibria is a theoretical abstraction arising from the unlikely assumption that female choice is an uncostly process (Pomiankowski *et al.* 1991). However, biased mutation on the male trait may render stable exaggeration of the male trait away from the survival optimum and the female preference for ornamented males evolutionarily stable (Pomiankowski *et al.* 1991).

Even in the absence of nutritional variation among females and the resulting variance in breeding date, there will still be sexual selection on males in the model by Kirkpatrick *et al.* (1990; see also O'Donald 1972). Females will breed at the optimal breeding date if there is no nutritional variation among females. Some males will, however, mate nearer to the optimal breeding date, which gives them enhanced reproductive success.

Except for the condition and the breeding date mechanisms of sexual selection, at least nine others may give rise to sexual selection in monogamous mating systems:

(1) differential male mortality during the breeding season;

(2) a skewed tertiary sex ratio;

(3) occasional polygyny;

(4) differential mate quality;

(5) differential extra-pair copulation success;

(6) selective fertilization;

(7) selective embryo development;

(8) differential parental care; and

(9) differential reproductive effort (Table 1.1).

Each of these components of sexual selection may apply to the situation of social monogamy and result in much wider limits to the exaggeration of the male trait than under the pure Darwin–Fisher mechanism. The extent to which these components of sexual selection may result in greatly exaggerated male sex traits depends on their relative magnitude. These mechanisms are treated extensively in the following chapters of this book.

2.5.2 Handicap models under monogamy

There are no explicit handicap models of sexual selection in monogamous mating systems although Bell (1978), Maynard Smith (1978), Andersson (1982a, 1986a), and Tomlinson (1988) have used a monogamous mating system in order to study a viability-based mechanism of mate choice separated from a Fisherian mating advantage. Most models of the handicap principle have been developed to investigate the effects of the handicap mechanism in the absence of the Fisher process (see section 2.4.2, The Fisher process). The general conclusions from these models are thus also valid for monogamous mating systems.

In Andersson's (1986a) model it was assumed that the mate preference was costly and decreased the viability of females with the preference allele. This viability cost of the mate preference reduced the frequency of the preference allele, and its frequency did not exceed the starting value if the cost of the mating preference was as high as the viability cost of the ornament. A smaller cost for the female mate preference could make the preferred male trait reach a higher frequency because an initial reduction in the frequency of the preference allele resulted in a more rapid increase in the ornament allele.

Tomlinson (1988) analysed by means of simulation a diploid, three-locus model of the condition-dependent handicap and the revealing handicap in a monogamous mating system. In the condition-dependent handicap model, the handicap trait and the female mate preference initially increased in frequency while the viability trait went to fixation. The handicap trait and the female preference were subsequently lost from the population. It is possible that these traits would be maintained given genetic differences in viability.

It has been claimed that condition-dependent secondary sexual characters could not evolve in males under monogamy because there is no force for sexual dimorphism in the absence of an initial asymmetry in the expression of the characters in males and females (Heisler *et al.* 1987). For example, mate choice could occur in both sexes as a result of a parasite-dependent expression of a morphological character in both males and females. An initial sexual asymmetry could, however, arise in a number of different ways. Males and females always differ in their reproductive roles, even in monogamous species with shared parental duties. Males may, for example, play a more important role in territory defence or in defence of a fertile mate. Different roles in reproduction might render individuals of the chosen sex more susceptible to parasitism because of immunosuppressive effects of circulating hormones (Grossman 1985; Folstad and Karter 1992). High levels of circulating androgens necessary for male sexual displays may therefore simultaneously increase the susceptibility of males to parasite infections. Sex differences in reproductive roles could also directly affect the costs of sexual dimorphism, even during the initial stages of the process of sexual selection.

2.6 Summary

The two modes of sexual selection, male–male competition and female choice, are well defined, but their relative importance is often difficult to evaluate in practice. The reason for this is that particular traits rarely have a fixed role in either mate choice or male–male competition. Secondary sexual characters that are reliable indicators of general quality may be so for conspecifics of both sexes, and even for heterospecifics. It is suggested that the evolution of male traits is reinforced by the two modes of sexual selection if females agree with the assortment of male phenotypes during male combat. Selection pressures by the two modes of sexual selection may be counteractive if females value properties of male quality that are unimportant in male combat.

Game theory models of male–male competition suggest that male traits may evolve stable polymorphisms with the mean male phenotype displaced away from the natural selection optimum. Alternatively, the male phenotype may demonstrate cyclical changes in size, or arms races between males until large males become so rare that females have difficulty finding mates. Secondary sexual characters used in male–male competition may also evolve as a result of the handicap mechanism.

A large number of different models of female mate preferences have been reviewed in this chapter. Their assumptions, the selection pressures on the male trait, the male condition, and the female mate preference, together with the main conclusions of the models are summarized in Table 2.1. It is immediately evident from this summary that several mechanisms could potentially result in stable exaggeration of a female mate preference and a male trait. The male trait is usually displaced from the natural selection optimum, and that is also the case for the female preference in some of the models. Stable exaggeration can occur because of good genes (conditional and revealing handicaps), the Fisher process, or the good parent process. It is an empirical question to what extent the different models of female mate preferences have contributed to the diversity of extravagant secondary sexual characters in nature.

The outcome of the many different models depends to a large extent on the assumptions, which are often quite different from the natural conditions under which secondary sexual characters have evolved. Theoreticians have often favoured haploid models because they are analytically tractable. However, haploid models may not be particularly realistic because many extravagant secondary sexual characters are found in diploid organisms. Many haploid models also exaggerate the genetic correlation between the male trait and the female preference, and a female mate preference therefore appears to be selected more readily than in the much more realistic, but analytically intractable, diploid models (O'Donald 1990). Discrete genetic models are advantageous because genetic covariances between characters such as a male trait and a female mate preference can be estimated as they evolve, but genetic variance cannot be maintained because of fixation of alleles. Quantitative genetic models superficially appear to create greater realism because the genetic variance in a male trait and a female preference is not depleted as a result of directional selection, but the variance and the covariance structure of the models has not yet been analysed mathematically. Therefore, the genetic variances and covariances usually have been assumed to be constant. This assumption of constancy may not be particularly sound from a theoretical or empirical point of view (O'Donald 1983; Pomiankowski 1988).

Many genetic models make restrictive assumptions about female mate preferences. It is, for example, usually assumed that a female mating preference is uncostly, although recent models by Andersson (1986*a*), Iwasa *et al.* (1991), and Pomiankowski *et al.* (1991) are exceptions. Females probably always incur fitness costs from their mate preference in terms of time and energy use, increased risks of predation and parasitism, or lost opportunities for acquiring a preferred mate (Parker 1983*b*; Pomiankowski 1987*b*). A second problem is that the specific mate preferences used in genetic models may not reliably reflect the kinds of female preferences adopted by real females. Female mate preferences have either been assumed to be:

(1) a psychophysical preference according to which perceived intensity of the male trait is proportional to the power of the actual preference;

(2) an absolute preference for males of a specific phenotype independent of male phenotypes in the population;

(3) a relative preference where the phenotype of a male is scaled to the frequency distribution of phenotypes in the population (Lande 1981); or

(4) a best of *n* males preference.

Real animals appear to use rules like the best of *n* males (Janetos 1980; Trail ·and Adams 1989; Dale *et al.* 1990; Petrie *et al.* 1990; Bensch and Hasselquist 1992). The outcome of particular models often relies strongly on assumptions about the mate preference as stated by Seger (1985). For example, the assumption of an absolute preference for particular males will lead to a very high mating success for males of this phenotype, if such males are rare relative to females with the preference.

Many models also encounter problems because of their unrealistic assumptions about the maintenance of genetic variance in a viability trait or a male sex trait (Bulmer 1989*a*; Gomulkiewicz and Hastings 1990; Barton and Turelli 1991). Genetic variance rapidly becomes depleted in many models as a result of intense directional selection, although this may rarely if ever be the case in nature. It is likely that genetic variance can be maintained, for example, by biased mutation on the viability trait or the male trait (Iwasa *et al.* 1991; Pomiankowski *et al.* 1991). Theoretical considerations suggest that biased mutation may be the rule rather than the exception for secondary sexual characters with a history of intense directional selection.

Sexual selection may result in the stable exaggeration of male traits and female preferences even under social monogamy. Darwin and Fisher suggested that this could be the case if preferred males acquired mates in better condition earlier than less preferred males. This would result in exaggeration of a male trait because of variance in female fecundity. Handicap models may also result in the stable exaggeration of a female mate preference and a male trait under monogamy. The Fisher process of self-reinforcing male traits and female preferences will usually be less important under monogamy. However, a number of additional mechanisms of sexual selection, such as differential extra-pair copulation success and differential reproductive effort, may increase the variance in success of socially monogamous males, as analysed in detail in the following chapters of this book.

3

The study organism

3.1 Introduction

When scientists choose an organism in order to study a particular scientific problem, a number of different factors are often involved. Considerations like abundance and easy access to a large number of individuals are usually important for the potential success of a project. Genetic factors are most readily studied if individuals can be followed across generations. Most aspects of modern biology have an observational and an experimental approach, and study organisms amenable to experimentation usually greatly facilitate scientific progress. Observations can usually suggest whether potential causal relationships exist, but their existence can only be verified through properly conducted experiments.

When I started the detailed part of my study of barn swallows (*Hirundo rustica*) in 1983, the main objective was to study the behavioural and life-history consequences of sperm competition — a previously neglected topic in most animal studies. Initially, the study of sperm competition did not appear to be closely related to sexual selection, but later analyses demonstrated that they were two sides of the same coin. Sperm competition could be viewed as just sexual selection at a later stage than pair formation. Simultaneous study of different subjects may therefore sometimes result in positive feedback. The main reason for choosing the barn swallow as a study organism was that it is abundant, easily caught, nests are easily located, and there is considerable variation in social dispersion from solitary pairs to relatively large colonies with sometimes over one hundred breeding pairs; an ideal setting for a study of the effects of the social environment on behaviour. Barn swallows are almost unique among common European bird species by being sexually highly size dimorphic in one character, tail length. This character is a linear trait which can be easily manipulated. The causes of variation in the male secondary sexual character and the fitness costs and benefits of tail length for both sexes might therefore be assessed. This offered a unique research opportunity, and accordingly I started to collect data on morphology and potential fitness costs and benefits of long tails as a side project during my study of sperm competition. An important breakthrough came when I successfully learned to manipulate tail length following a telephone conversation with Malte Andersson, a pioneer in the methodology of sexual ornament manipulation studies. This opened up a novel way in which the costs and benefits of secondary sexual characters could be assessed directly.

Many scientists also have irrational reasons for studying particular organisms. Sheer coincidence may give opportunities which otherwise would not have arisen. I was unemployed for half a year following my PhD (on the costs and benefits of coloniality in barn swallows), and I was able to continue my studies of sperm competition and sexual selection only because I got a position as assistant professor of animal ecology at Uppsala University, Sweden. Although the job was restricted to four years, it gave me almost unlimited opportunities to do whatever I wanted — a wonderful chance for a young scientist with vague ideas about a large-scale study.

Study organisms are living beings and most of them have some appeal to the investigator. A fascination with beautiful living organisms is, I think, crucial in making a particular study feasible. Engagement with a scientific problem and with the study organism are necessary for success.

In the following sections I introduce the main actor in this book, the barn swallow. I describe its morphology in section 3.2, and its natural history in section 3.3. The behaviour, in particular the sexual behaviour of the barn swallow, is described in section 3.4, while I address the question of foraging and other activities as constraints on sexual selection in section 3.5. Finally, I briefly describe the behavioural, genetical and biometrical research methods which I have employed in my large scale field study of the barn swallow in section 3.6.

3.2 Morphology of the barn swallow

The barn swallow is an aerial insectivorous passerine, weighing *c*. 20g. It belongs to a genus with 34 species (Turner and Rose 1989), many of which are sibling species replacing the barn swallow in other parts of the world. Like many other aerial insectivorous birds, it is easily recognized by its small beak, wide gape, streamlined body shape, very long pointed wings, and a long forked tail. Characteristic plumage features include a shining metallic blue-black dorsal side, a chestnut forehead, chin and throat patch, and elongated outermost tail feathers which form a distinct V-shape. Near their base, the tail feathers have conspicuous white markings, which become fully visible when the tail is spread like a fan during courtship displays. The function of the white spots on the tail feathers is probably to allow visual communication in dark nest sites as hollow trees or caves or, in more recent times, human habitations. The size and the position of the white spots on the tail feathers are strongly and positively related to tail length, because long-tailed male barn swallows have larger and more posteriorly situated spots. The size and position of these spots therefore provide information about tail length (A. P. Møller unpublished). The colour of the ventral side varies from buff-white to red-buff, and usually females are paler than their mates. There is considerable variation between subspecies. For example, the North American subspecies *erythrogaster* is chestnut or cinnamon-rufous whereas the nominate *rustica* is whitish. However, within subspecies there is also considerable variation. Almost 4% of nominate *rustica* in my Danish study area have a chestnut ventral side. The adult male is most easily distinguished from the adult female by its much longer outermost tail feathers. Juveniles during their first autumn differ from adult barn swallows by having shorter wings and tails and an orange-brown fore-face. The sexes of juveniles cannot be separated on external morphology. Sexing first becomes possible following their first complete annual moult in the African winter quarters.

Male, female, and juvenile barn swallows

The size of a number of morphological traits varies according to age and sex. As already mentioned, males have considerably longer tails than females, but sex differences are otherwise very small (Table 3.1). Slight sexual size dimorphism, with males being larger than females, occurs in several morphological traits. However, sexual size dimorphism is reversed for four traits; females have slightly longer tarsi, shorter central tail feathers, larger throat patch areas, and greater body mass than males (Table 3.1). Juveniles have considerably shorter wings, and particularly tails, than adult barn swallows, but otherwise differ very little from adults (Table 3.1).

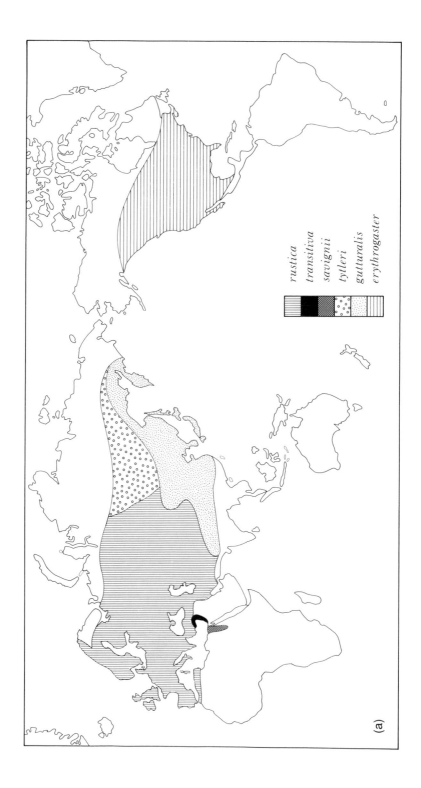

rustica
transitiva
savignii
tyleri
gutturalis
erythrogaster

(a)

Fig. 3.1 Breeding (a) and winter (b) distribution of the different subspecies of the barn swallow. After Turner and Rose (1990).

Table 3.1. Morphology of sex and age classes of the barn swallow from the Kraghede study area. Values are means (SE).

Morphological trait	Males	Females	Juveniles
Wing length (mm)	126.83 (0.11) ***	124.77 (0.10) ***	121.83 (0.12)
Wing span (mm)	328.87 (0.02) ***	324.18 (0.02) ***	321.97 (0.28)
Tail length (mm)	107.66 (0.32) ***	90.01 (0.21) ***	65.65 (0.22)
Short tail length (mm)	44.28 (0.07) ***	44.68 (0.07) *	44.11 (0.12)
Throat patch area (mm^2)	206.75 (1.00) *	213.52 (1.05)	210.68 (0.04)
Beak length (mm)	7.66 (0.02) *	7.72 (0.02)	7.70 (0.02)
Beak width (mm)	11.99 (0.02) ***	11.88 (0.02) *	11.91 (0.02)
Beak depth (mm)	2.94 (0.01)	2.92 (0.01)	2.93 (0.01)
Tarsus length (mm)	11.42 (0.02) *	11.49 (0.02)	11.48 (0.03)
Keel length (mm)	21.70 (0.02) ***	20.98 (0.02)	21.26 (0.04)
Body mass (g)	19.20 (0.04) ***	20.38 (0.06)	19.81 (0.78)
n	781	814	516

Differences between sexes (asterisks between columns for males and females) and age (asterisks between columns for females and juveniles) classes, respectively, according to *t*-tests, * $P < 0.05$, ** $P < 0.01$, *** $P < 0.001$

3.3 Natural history of the barn swallow

The barn swallow is a migratory insectivorous passerine in most parts of its range, which includes most of the Holarctic; namely, North America, Europe and Central Asia with the exception of the Arctic. The range reaches North Africa including Egypt and parts of the Middle East (Fig. 3.1). The migratory populations of barn swallows spend three to ten months in the breeding areas depending on latitude. Breeding areas in the far north are visited for only three months whereas breeding areas in Southern Europe are visited for ten months. The barn swallows of the subspecies *transitiva* in the Middle East and *savignii* in Egypt are residents. The number of broods reared in different parts of the breeding range varies from one to three, the largest number being raised in the southern part of the range. Barn swallows leave the breeding areas shortly after the end of the reproductive season and migrate rapidly southwards towards the winter quarters, a trip that may take more than three months. The start of the single annual moult of the feathers depends on latitude (Kasparek 1976). Thus, it starts before the autumn migration in the majority of individuals from southern populations, but after the end of the migration in most individuals from northern populations (Fig. 3.2). The moult continues throughout the stay in the winter quarters, and a few of the birds returning to the breeding grounds have not yet completed their moult. The spring migration appears to be more rapid

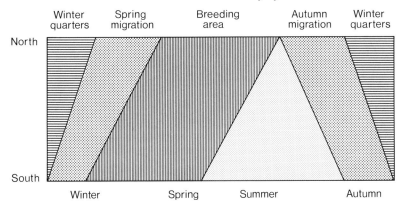

Fig. 3.2 Timing of moult and reproduction during the annual cycle of barn swallows breeding at different latitudes in Europe. After Kasparek (1976).

than the autumn migration, lasting from two to three months depending on the distance between the winter and the breeding ranges.

3.3.1 *Reproduction*

Barn swallows resemble many monogamous passerines in respect of their reproductive cycle and the sex roles during reproduction. The following account is based on my own observations combined with extensive information from the literature (von Vietinghoff-Riesch 1955; Glutz von Blotzheim and Bauer 1985; Cramp 1985). Males on average arrive at the breeding grounds before their mates, the difference in my study area at Kraghede being 4.80 days (SE = 0.31, $n = 198$). They establish a small breeding territory of a few square metres with potential nest sites and perches. The territory is established by singing and defending the boundaries against the intrusions of other males. Breeding territories were originally located in caves or rock exposures in mountain areas or in large hollow trees, but today most barn swallows breed in close association with human habitation. Large hollow trees nesting sites are nowadays rare, and only caves or rock exposures are in frequent use. Today, the majority of barn swallows breed inside barns, sheds, stables, outhouses, or porches, less often in culverts and under bridges. The close association between barn swallows and humans has lasted more than two thousand years in Denmark, and it must have resulted in a tremendous release from nest-site limitation and a manifold increase in population size. Barn swallows nesting in farms form colonies of a similar size to those found in natural sites (Erskine 1979; Speich *et al.* 1986). It has been suggested that modern barn swallow breeding aggregations are larger than natural aggregations because of the association with humans, which could have important consequences for our views of sexual selection (Koenig *et al.* 1992). However, colony sizes on farms do not differ from those recorded in natural sites which suggests that the social environment has not changed by the association with humans. Many bird studies are currently performed on species breeding in nest-boxes, and scientists may unintentionally affect their study organisms by their design and arrangement of boxes (Møller 1992*b*). This situation differs from that of the barn swallow where the association between birds and humans and the distribution of individuals between farms are outside the control of the field biologist (Møller 1992*b*).

When females arrive, males attempt to attract mates by performing aerial displays exposing their out spread tail feathers while singing vigorously. If the male succeeds in attracting a female to its territory, it will start spreading out its tail against a vertical surface which may serve as a nest site, while vigorously singing a special mate-attraction song. The female barn swallow may leave the territory several times while the male continues attempting to attract her. Finally, the female may perch at the nest site together with the male, and then the pair bond for the first breeding attempt has been established. If the female decides to stay with the male, however, this does not necessarily imply that it will stay and breed. Some females desert their mate after 1–16 days and subsequently establish a pair bond with another male. The abandoned male barn swallow then attempts to attract a new female following such cases of mate switching. Unmated males are quite indiscriminate in their attempts to attract a mate and persistently follow neighbouring paired females and may perform pair formation displays in front of such females many times each day over several weeks.

Polygyny is very rare in the barn swallow and accounts for much less than one per mille of all social relationships, the rest being monogamous. Fewer than ten cases have been described in the literature despite the fact that barn swallows have been intensely studied in a large number of sites (review of polygyny in Møller (1986)). I have never recorded polygyny in my Danish study population during ten years of intense study of colour-ringed birds. Polygyny occurs as a result of a male disappearing from its territory early during the reproductive cycle or one male displacing another from a nearby territory. The neighbouring male may subsequently occupy both territories. However, males often contribute to paternal duties only at their first nest, and as a result the secondary female rears considerably fewer offspring than the primary female.

Following establishment of the pair bond, the barn swallow pair spends a considerable amount of time in the territory, and roosts at night on perches within the territory. The pair starts to copulate following pair formation and copulations usually continue until late in the egg-laying sequence. The male barn swallow also follows its mate closely, guarding it from the approaches of other males during this part of the breeding cycle. Some male barn swallows succeed in copulating with other females, particularly neigh-bouring ones, and these extra-pair copulations sometimes result in fertiliza-tions. A number of paternity guards, such as mate guarding, frequent copulations, and the use of deceptive alarm calls to disrupt extra-pair copulation attempts, are used by male barn swallows with a fertile mate.

Shortly after pair formation, when the spring weather is fine, both partners start nest construction by collecting material from nearby ponds or puddles. A mixture of straw and small balls of mud is added to the nest site: some 750–1400 mud pellets are used to make the cup-shaped nest. Barn swallows may use other building materials, such as cow dung, if mud is unavailable, or no mud at all if the nest site is a hole in a wall. Mud pellets for nests are sometimes transported over large distances: up to 1500m in drought periods. Old nests from previous years may be refurbished if they are of a sufficiently high quality. Barn swallow nests may persist for many years, and their average life expectancy in a Belgian study was seven years (Vansteenwegen 1982). All nests are subsequently lined with straw and provided with a fine lining of hair or feathers as available. Nest building lasts from three days to more than three weeks depending on the availability of nesting material, and during spells of dry weather nests may take very long to build. They vary considerably in size, and pairs that subsequently lay a large clutch build larger nests (Møller 1982). Barn swallows lay two or three broods per year. A new nest is often built for the second and third clutch if the previous nest is heavily infested with ectoparasites (Barclay 1988; Møller 1990*c*). Both males and females build the nest, although males do less than females; instead, when not bringing any nest material, males invariably follow the mate singing.

The pre-laying period, which is the interval between pairing and laying of the first egg, lasts 16 days (SE = 0.5, *n* = 210). The clutch size ranges from two to seven eggs, one egg usually being laid each day. The modal size of first clutches is five eggs, on average 4.94 eggs (SE = 0.03, *n* = 801 clutches), and the average date of laying of the first egg in my Kraghede study area is 5 June (SE = 0.4, *n* = 801). During periods of cold and rainy weather females may interrupt laying for one day. Clutch size varies with latitude and longitude, increasing from south to north and from west to east (Møller 1984*a*).

Female barn swallows sometimes engage in brood parasitism and lay eggs in the nests of other pairs. This occurs more often in large colonies and during years with high population densities (Møller 1987*a*, 1987*b*, 1993*a*). Females will remove eggs laid in their nest before they have started laying themselves, but not afterwards. Both males and females guard their nest against potential conspecific brood parasites. Nest guarding is particularly intense during the laying period and early incubation when parasitic females are most likely to lay an egg in a neighbouring nest. Guarding is also intense in large colonies and during years with high population densities, that is during conditions when brood parasitism is likely. Nest guarding is an efficient deterrent to potential intraspecific brood parasitism as determined from the higher incidence of brood parasitism in experimental nests without attending pairs. Nest guarding is also less intense at nests suffering from intraspecific brood parasitism (Møller 1987b, 1989e, 1993a).

The female starts to incubate after the laying of the penultimate or ultimate egg, that is after the third or the fourth egg in my study. There is intraspecific variation in the participation of sexes in incubation duties. Males of the North American subspecies *erythrogaster* incubate (Ball 1983; Smith and Montgomerie 1992), but in all other subspecies males never incubate. The incubation period, which is the period from the laying of the last egg to the hatching of the last nestling, on average lasts 15 days (SE = 0.05, $n = 290$), and the female spends 80.2 % (SE = 6.6, $n = 12$) of the time on the eggs during this period. The female regularly leaves the nest during incubation in order to feed. On average 90.4% (SE = 0.9, $n = 801$ clutches) of the eggs hatch, the main reason for hatching failure being egg infertility.

Some male barn swallows are unable to attract a mate because of a male-biased sex ratio. These males attempt to acquire a mate by performing infanticide (Crook and Shields 1985; Møller 1988a). Unmated male barn swallows start to visit nests in neighbouring territories around the time of hatching. If the nest is not intensely guarded by the owners, the unmated male may be able to injure and remove the nestlings. The female nest owner will subsequently divorce its mate if still present. Otherwise, infanticide may take place as a consequence of the male mate having disappeared. The female is thus unable by itself both to guard the nest and provision the nestlings with food. It then usually forms a new pair bond with the infanticidal male, and more than half of these newly formed pairs are able to rear a brood of nestlings during the same season. Nest guarding is an efficient way of avoiding infanticide as is clear from the higher incidence of infanticide in colonial nests with temporarily removed males during the early part of the nestling period. Barn swallow nests suffering from infanticide are also guarded at a lower intensity than other nests (Møller 1988a). Unmated males that are unable to obtain a mate by committing infanticide usually leave the colony shortly after the last of the first clutches has hatched, when the latest opportunity for infanticide has passed.

Nestlings hatch during a period lasting from one to three days. The average brood size at hatching is 4.49 nestlings (SE = 0.05, $n = 801$). Hatching asynchrony usually leads to a clear size hierarchy among nestlings, and frequently, the smallest one or two nestlings die during periods of food shortage caused by unfavourable weather. The nestlings are brooded by the female for the first week, but with a decreasing frequency. Barn swallow nestlings are fed by both parents, which contribute approximately equally to the raising of young. The nestling period, which is the period from hatching of the last egg to fledging of the last nestling, lasts on average 21 days (SE = 0.13, n = 178). On average 92.4% (SE = 0.8, n = 801 broods) of all hatched nestlings fledge, the main reasons for fledging failure being infanticide, ectoparasites, and starvation during unfavourable weather. Brood size at fledging is on average 4.14 (SE = 0.06, n = 801). Fledglings are fed with a gradually decreasing proportion of their daily requirements for the first one to two weeks following fledging. In the first few days following fledging they roost in or near the nest. The first days following independence are spent near the hatching colony, but fledglings soon disperse and start to join nearby communal roosts.

Nest with nestlings

Barn swallows may lay a second and even a third clutch. The frequency of second and third clutches decreases with increasing latitude (Møller 1984a). In my Kraghede study area 60.3% (n = 812) of the breeding pairs lay a second clutch, on average on 20 July (SE = 0.5, n = 488). Barn swallows usually remain paired with the same mate during subsequent clutches in the same breeding season and also during subsequent breeding seasons. The size of the second clutch is usually smaller than that of the first clutch: in my Kraghede study area second clutches average 4.41 eggs (SE = 0.03, n = 488), brood size at hatching is on average 4.14 nestlings (SE = 0.05, n = 488) and at fledging, is 3.69 fledglings (SE = 0.07, n = 488). In Central and Southern Europe where a third clutch is sometimes laid it is one egg smaller than the second clutch on average.

3.3.2 *Communal roosting*

Barn swallows are gregarious outside the breeding season and spend the night at communal roosts, which sometimes contain more than one million birds. Roosts occur in the breeding areas in Denmark during spring in April–June and again during autumn from late July until early October. They also occur regularly during migration, and enormous roosts have been

Swallows over a reedbed

recorded in the winter quarters. Most barn swallows roost in reeds, sedges, or willow shrub in wetlands, but other habitats are also used.

Barn swallows begin to assemble at the roost site up to two hours before sunset. The large dense flocks perform spectacular manoeuvres by rapidly swirling and passing low over the roost site. At last light individual swallows start to detach themselves from the passing flocks and drop into the roost, soon followed in large numbers. The remaining birds form new flocks, which continue the spectacular flight displays, before finally settling. Barn swallows give continual twittering calls while performing these flight manoeuvres and while perched at the roost. In the morning, after sunrise, they leave the roost in wave upon wave, and spend a considerable amount of time in comfort behaviour (sun-bathing, bathing, preening) in post-roost gatherings.

The composition of communally roosting flocks varies during the season (Aarestrup and Møller 1980). Unmated males and a few juveniles attend the roosts in late July, but the fraction of juveniles from the first clutch increases rapidly during August. Adults having reared only a single brood also start to attend the roosts at this time of the year. Juveniles from the second brood appear from late August, and maximum numbers are reached in September. Juveniles dominate the roost in late autumn.

The function of communal roosts has attracted considerable scientific attention. One possibility is that the roost functions as a selfish herd in which individuals seek protection against predators. As long as the predation risk per individual is lower at the roost than during dispersed roosting, the roost will be maintained (Hamilton 1971). I have visited communal roosts on many evenings and have found predation attempts to be very rare events. It is therefore unlikely that the main function of communal roosts is anti-predation behaviour. A second possibility is that communal roosts function as information centres (Ward and Zahavi 1973). This hypothesis assumes that successful and less successful foragers can be distinguished by their time of arrival or from their behaviour during pre-roost and post-roost gatherings. Less successful foragers are assumed to be able to assess the foraging success of other birds and thereby find more profitable foraging sites the next day by following the successful foragers. It should pay all individuals to attend the communal roost as an insurance against low foraging success. There is no evidence for or against the information centre hypothesis as far as the barn swallow is concerned.

3.3.3 *Migration*

Most populations of the barn swallow perform diurnal transequatorial long-distance migration. As they migrate by day, barn swallows are able to intersperse periods of migration with foraging. The winter quarter of migratory barn swallow populations includes parts of Central and South

America, most parts of Sub-Saharan Africa, Iran, India and South-east Asia including Indonesia, Phillipines, and Papua-New Guinea (Fig. 3.1). Adult barn swallows that produce a single annual clutch usually leave the breeding grounds in my Danish study site between late August and mid-September whereas adults that produce two clutches leave during September. Juveniles usually depart from mid-August to early October.

The Danish population of barn swallows travels directly southwards across the Alps or south-west over the Iberian Peninsula. The migration then proceeds across the Sahara. The southerly migration direction was documented very clearly in October 1974 when hundreds of thousands of barn swallows died in the Alps during an unusually early cold spell. A total of 29 Danish barn swallows were then recovered from the Alps and surrounding areas, which is considerably more than the total number of birds recovered abroad in all other years (Møller 1978). The breeding population in my study area fell to less than half in the subsequent year. Similar cases of mass mortality in the Alps occurred in 1740, 1770, 1829, 1855, 1881, 1931, and 1936 (Glutz von Blotzheim and Bauer 1985). Incidents of mass mortality will obviously influence the migration pattern of the remaining population provided there is a genetic basis for the choice of migration direction. It is likely that birds using a south-westerly migratory direction predominated in the years following the 1974 event, since they were not hit by the meteorological catastrophy.

Barn swallows from different parts of the Western Palearctic breeding range winter in different parts of Africa. Birds from the British Isles mainly winter in South Africa, birds from Germany in Zaire, birds from Eastern Europe in East Africa, and birds from Scandinavia in Zambia, Namibia and South Africa (Zink 1970; Asbirk 1971).

Barn swallows from Europe start arriving in the transequatorial African winter quarters from the end of September with large numbers from November and a peak is reached by early December. South Africa is not reached by large numbers until late October, but juveniles do not predominate until January. Barn swallows are particularly often found in open savanna, wet grasslands, swamps, steppe, and agricultural areas in the winter quarters. The majority of the annual mortality takes place during migration and in winter. Precipitation in the winter quarters affects the survival of barn swallows indirectly by affecting the size and timing of mass insect abundance. Very dry winters are accompanied by extensive mortality; larger proportions of the barn swallow population survive after particularly wet winters (Møller 1989c). Spring migration begins in South Africa in late February and most birds have left by late April. More northerly populations leave considerably earlier, departing from Zaire as early as from mid-January passing through Zambia in mid-February. Peak migration in North Africa and the Mediterranean occurs from mid-March to late April.

The migration of individual barn swallows can be remarkably rapid with a maximum of more than 300km travelled per day. The distance from South Africa to Scandinavia can in theory be covered in approximately 30 days.

3.3.4 *Moult*

A bird's feathers deteriorate continuously because of abrasion, and as a result the energy costs of flight increase as the plumage ages. Barn swallows have a single complete annual moult. It starts in autumn either before or immediately after migration and continues for 4.5 to 6 months; barn swallows sometimes still have growing feathers when they arrive at the breeding areas. Different feather tracts are replaced at different times. For example, the body feathers start to moult first together with the feathers of the wings. The tail feathers start to be replaced when the fifth primary is in moult, and they are the last to be fully grown, sometimes not until arrival at the breeding areas. Barn swallows rely on their flight ability for catching prey, and the rate of feather replacement is therefore relatively slow in order not to impair foraging efficiency. The wing and tail feathers are replaced in a characteristic sequence as in most other passerine birds. The primaries are moulted symmetrically from the middle of the wing and outwards, the secondaries symmetrically from the middle of the wing and inwards towards the body, and the tail feathers from the centre outwards. There are often signs of incomplete tail moult at the start of the breeding season, and this is more often true for males with short outermost tail feathers than long-tailed males (Fig. 3.3).

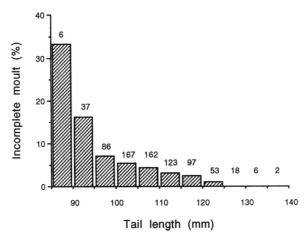

Fig. 3.3 The relative frequency of incompletely moulted tail feathers in male barn swallows in relation to their tail length (mm). Numbers are number of individuals. The difference between groups is statistically significant ($G^2 = 8.88$, $df = 4$, $P < 0.05$).

Juvenile barn swallows differ from adults in the colour of their fore-face and the length of their outermost tail feathers. They start moulting their orange fore-face before migration relatively early during autumn, and many juvenile barn swallows have growing feathers in the fore-face just before the autumn departure. The other feather tracts are replaced in the winter quarters slightly later than in adult barn swallows, for example, in South Africa 2–3 weeks later, in Namibia 3 weeks later, and in Zaire 5–6 weeks later (Mendelsohn 1973; Møller *et al.* 1993; Herroelen 1960; de Bont 1962).

3.3.5 *Aerial insectivory*

The barn swallow, like all other hirundines, is an aerially insectivorous bird. Its foraging behaviour therefore is particularly sensitive to adverse weather conditions, such as persistent rain or cold spells. Foraging barn swallows either fly around solitarily or in loose groups of up to several hundred individuals. Individual prey items are captured in aerial pursuit, but occasionally picked up from the ground or the vegetation when temporarily abundant, or when conditions are unsuitable for aerial pursuit. Most foraging takes place in open areas like fields, savannas, swamps, and plains. A considerable proportion of the time is spent foraging on the leeward side of trees during strong wind and rain because insects are particularly abundant in such sites. Foraging also often occurs above open water with high insect concentrations or near cattle or other large mammals which attract and disturb resting insects. Individual prey items are pursued in active flapping flight, and barn swallows make rapid turns in response to the movements of the evading prey. The proportion of time spent in flapping flight during active foraging is 88%. Birds feed up to 600m from the nest site, but the average distance is only 170m (Bryant and Turner 1982). In my Kraghede study area barn swallows foraged within a distance of 500m from the nest site with a median foraging distance of 100m (Møller 1987*a*).

Foraging swallows

Barn swallows take considerably larger prey items (mean dry mass 6.6mg) than do the two other abundant European hirundines, the sand martin *Riparia riparia* (1.3mg) and the house martin *Delichon urbica* (1.9mg) (Bryant and Turner 1982). Large prey items take longer to catch and require faster flapping flight than do small items, but still are more profitable because the net energy gain is higher (Turner 1982). Therefore, large, fast-flying prey items such as large Diptera are preferred over small, slow-flying items such as Hemiptera. Barn swallows prefer prey of a specific size rather than prey of a certain taxonomic category. Parent barn swallows in my Kraghede study area bring 1–155 items, on average 15 items ($n = 362$), per bolus to their nestlings. The average mass of food boluses brought to nestlings is related positively to air temperature and negatively to rainfall (Bryant and Turner 1982). Bolus mass also increases significantly both during the breeding season and with nestling age.

The majority of prey items by number in the Kraghede study area is flies (Diptera) (52.7%, $n = 3160$ prey items) and aphids (Hemiptera) (39.8%), with smaller numbers of beetles (Coleoptera), bees (Hymenoptera), butterflies (Lepidoptera), and other insects. Most of the Diptera are hoverflies (Syrphidae) and true flies (Muscidae). Diptera predominate as food for the first brood whereas Hemiptera are more abundant for the second brood, reflecting seasonal patterns of abundance (A. P. Møller unpublished). Similar results have been obtained in the United Kingdom, Germany, Czechoslovakia, USSR, and Spain. The food in the African winter quarters mainly consists of ants (Hymenoptera), termites (Isoptera), and flies (Diptera) (Glutz von Blotzheim and Bauer 1985).

3.4 Sexual behaviour of barn swallows

In the barn swallow, male attraction of a mate is closely associated with occupation of a nest site. Communal song chorus displays consist of unpaired and paired barn swallows (up to 30–40 individuals) which fly high, gliding slowly or with smooth wing-beats, sing or give contact-calls. These displays usually occur in early morning, particularly in spring, and they may last for more than one hour. Males regularly spread their tails out in a fan-like fashion in front of a female, the white spots in the tail clearly visible, and then swoop down, calling, towards a potential breeding site. The female may subsequently follow the male and enter the breeding site, or the male may return and make more displays in front of the female before it eventually succeeds in attracting the female. The male will continually display its tail in a fan-like fashion during the approach to the nest site. The male finally lands at a potential nest site within its territory, spreads out the tail against the surface thereby making the white spots of the tail clearly visible, and gives enticement calls until the female has landed. The male will then sing vigorously, and display the nest site by making pecking movements towards an old nest or a new potential nest site while displaying its yellow gape.

Sexual display

The male may make repeated attempts to attract the female and perform similar nest-site displays at a number of sites. Some males are visited by a number of different females during a breeding season, since unattractive males may be deserted by several females. Females sometimes spend more than two weeks in the company of a given male before switching to another male in the neighbourhood. The female will stay with the male when the pair bond is established, and the pair will roost within the territory at night.

Copulations occur only after formation of the pair bond. When the female is perched near the nest site, the male solicits copulation by singing, often vigorously. The male then hovers with its tail fanned and its legs dangling over the female. The accepting female stays put and bends forwards while the male mounts. During copulation the male gives a characteristic copulation call. Unwilling females either expose their carpal joints, peck at the male or fly away. The rejected male flies a little way off and returns to perch near the female again. Preening of feathers is common following copulation, particularly by the female. In terms of behaviour, extra-pair copulations appear to be more intense than intra-pair copulations.

Male–male competition consists mainly of aggressive interactions. Both males and females attack or display at intruders of the opposite sex, but males are particularly aggressive towards other males, before mate acquisition and especially during the fertile period of their mate. Any intruding male is approached in flight or by walking. Intruders are challenged by vocalizations and threat displays, when males sit next to each other, singing

Copulation

and with carpal joints exposed and the plumage sleeked. Intruders are often chased in flight silently or while alarm calling. Competing males often fight directly in flight, with feet interlocked and the birds sometimes tumbling downwards. Sometimes they hit the ground with their feet interlocked, and such crashes may be fatal because they may fall prey to cats, or because fierce blows may result in deep wounds. Fighting between two neighbouring males may continue for days at a rate of many fights per hour, until one male finally gives way. Males usually defeat any intruder on their own territory.

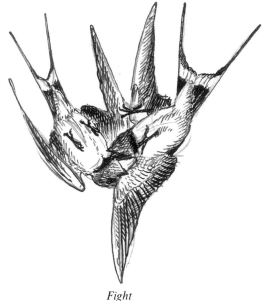

Fight

3.5 Constraints on evolution of secondary sexual characters

Extravagant secondary sexual characters such as the large feather ornaments of pheasants and birds of paradise are relatively rare among birds but have evolved a number of times. Ornaments may not evolve in particular taxa for a number of reasons. For example, there may not be genetic variation for a male trait or a female preference which could cause exaggeration of the male trait. However, it is likely that high costs of production and maintenance of extravagant morphology may often prevent exaggeration of morphological characters. For example, extravagant feather ornaments are unlikely to evolve when birds rely on highly unpredictable food sources or food items that are difficult to obtain. Similarly, feather ornaments are much less likely to evolve when males play an important role in parental care rather than when males play no role at all. Feeding habits, food availability, and male roles in parental care may therefore be constraints on the evolution of extravagant secondary sexual characters.

Extravagant ornaments are usually assumed to be highly costly to produce or maintain, but obviously such costs are not independent of the behaviour of individuals of the ornamented sex. If males have to rely on an unpredictable food source and spend a considerable amount of time and energy on feeding offspring, they are likely to experience higher costs of the ornament than are males that rely on a superabundant food source and do not provide any parental care at all.

The general relationship between extravagant ornamentation and the constraining factors mentioned above are not without exceptions. Males of a few bird species with relatively unpredictable food and extensive roles in parental care nonetheless have evolved an extravagant feather ornament. One of these species is the barn swallow. Tail length of male barn swallows, which is on average one fifth longer than that of females, will considerably increase flight costs because of the increased drag of a long tail (Evans and Thomas 1992). High flight costs due to a secondary sexual character are particularly important in bird species such as the barn swallow which spends a considerable amount of time in flight. When ornaments occur despite the fact that severe ecological constraints usually prevent the evolution of extravagant sex traits under these circumstances, ornaments are likely to be extremely costly. I demonstrate in Chapters 5 and 6 that there is considerable evidence supporting the hypothesis that barn swallow tail ornaments are reliable quality indicators and that ornaments are very costly to produce and maintain.

3.6 Study methods

Any scientific study requires the modification of a number of standard methods and the development of new ones.

I have studied barn swallows in an area of 14km^2 around Kraghede (57°12′N, 10°00′E), Denmark. The barn swallow population decreased during the 1970s, and I therefore had to double the study area in 1982 and increase it by 50% in 1987. The Kraghede area is an open agricultural site with scattered farms, mixed plantations and ponds. The main crops in the study area are barley, potatoes and rape, although there have been marked changes in land use since 1970 (Møller 1983). Fields have generally increased in size, and ditches, hedges, and field boundaries have been removed. Barn swallows are almost completely restricted to farms with domestic animals, particularly dairy farms, the number of which has decreased by more than 75% since 1970 (Møller 1983).

The present study of sexual selection in the barn swallow is based on work carried out during the period 1983–1992. I spent almost the entire summers from early May to early September in the Kraghede area during 1983–1989 and 1992, six weeks in 1990, and ten weeks in 1991. All potential breeding sites (buildings and bridges) were visited regularly during the breeding season in order to determine whether barn swallows were present.

Whenever barn swallows were observed I attempted to capture them as early as possible with mist nets during the day or sweep nets during the night at roosting sites. Each individual was provided with an individually numbered aluminium ring and two colour rings, sometimes also a combination of dyes on the belly feathers. The sex of an individual was determined from the presence (female) or absence (male) of an incubation patch and the size of the cloacal protuberance, which is considerably larger in males than in females. I also took a number of standardized morphological measurements. Beak length, width and depth, tarsus length and keel length were measured to the nearest 0.01mm with a digital ruling caliper. Wing length, tail length (the length of the outermost tail feathers), and shortest tail length (the length of the two central tail feathers) were measured to the nearest millimetre with a ruler. I measured the left and the right tarsus, wing length, tail length and shortest tail length during 1990–1992 in order to study fluctuating asymmetry in these characters. The size of the chestnut throat patch was determined as the product of the maximum length and width when the beak was held at an angle of 90° to the body axis. Wing span was measured as the distance from wing-tip to wing-tip when the wings were stretched out. Body mass was determined to the nearest 0.1g on a Pesola spring balance.

The arrival date of individual swallows in the intensely studied sites was defined as the first day that an individual was recorded. The pairing date was the first day that a male was permanently accompanied by a female. The pre-mating period was defined as the period from arrival until mate acquisition. The duration of the pre-laying period was the period from mate acquisition until start of egg laying.

When a pair of barn swallows had been located, I monitored their reproductive activities throughout the breeding season. The nest site was visited daily in the intensely studied sites and at least weekly in the less intensely studied sites. I noted the time for start of egg laying directly or estimated it from the age of nestlings, assuming an incubation period of 15 days and one egg being laid daily. Clutch size is the number of eggs present after completion of the clutch. The duration of the incubation period is the period from the laying of the last egg until the hatching of the last nestling. Brood size at hatching is the number of nestlings that hatched, and brood size at fledging the number of nestlings present at the last nest check before the nestlings fledged. The duration of the nestling period is the period from the hatching of the last nestling until the fledging of the last fledgling. The interclutch interval is the duration of the period from the start of laying of the first clutch until the start of laying of the second clutch. The reproductive events in the second clutch were followed similarly. Tarsus length and body mass of nestlings were recorded when nestlings were 15 days old, and they were then also provided with an individually numbered aluminium ring. Nestling tarsus length, which is a skeletal character kept throughout life, has reached adult size at that age (Stoner 1935; A. P. Møller unpublished data).

The survival of barn swallows was determined from their return rate to the study area. All unringed birds recorded from the second study year were assumed to be yearlings. This assumption is supported by the recapture of 98 locally ringed yearlings, which all turned up in their first year of life. An individual was considered to be dead if it did not return to the study area in one year. This assumption was supported by the facts that (1) only 3 out of 401 individuals returning to the Kraghede study site ever moved to a new breeding site, and (2) the fact that only 6 of 401 returning individuals ever returned in year $(i +1)$ without being captured in year (i). Longevity of barn swallows was therefore determined in terms of number of years that they had spent in the Kraghede study area.

The parasites of barn swallows were quantified in a number of different ways. The prevalence and intensity of infection by the tropical fowl mite *Ornithonyssus bursa* were determined by:

(1) counting the number of mites on the head feathers of adult barn swallows;

(2) counting the number of mites on nestlings aged 7 days; and

(3) estimating the number of mites in nests on the day following fledging of offspring by placing a hand on the rim of the nest for 10 seconds and then determining whether there were 0, 10, 100, 1000 or 10 000 mites on the hand.

The last method has been verified by comparison with estimates obtained from extraction of mites (Møller 1990*c*). The prevalence and intensity of infection by the swallow lousefly *Stenepteryx hirundinis* were determined by counting the number of louseflies when handling adults and nestlings during capture and measurement. The prevalence and intensity of infection by the feather louse *Hirundoecus malleus* were estimated by counting the number of characteristic round holes left in the tail and wing feathers. This method is reliable because there was a strong positive relationship between this estimate and the number of adult Mallophaga recorded when handling the barn swallows. The prevalence and intensity of infection by the feather louse *Myrsidea rustica* were determined directly by counting the number present between the barbules in the wing feathers. Finally, I took a blood smear whenever a barn swallow was captured. The smear was fixed in absolute alcohol after being dried and later inspected for the presence of Haematozoa by G. F. Bennett. A total of five species were recorded and their intensities of infection estimated as numbers per microscopic field.

Many of the results presented in this book are based on extensive behavioural observations. The intensity of some behavioural categories was assessed by making scan observations every second minute for one hour each morning during a specified period (Altmann 1974). Mate guarding was assessed by calculating the percentage of these scan observations when the male was within a distance of 5m from its mate. Male barn swallows were always able to interfere with male intruders when they were less than 5m away from their mate. An overall estimate of the intensity of mate guarding was obtained by calculating the mean value for the daily estimates for the presumed fertile period, which lasts from five days before start of egg laying until the day when the penultimate egg is laid (the day when the ultimate egg is fertilized). Barn swallows probably have prolonged sperm storage like other bird species, and the shortest sperm storage duration known in any bird is five days (Birkhead and Møller 1992). I therefore assumed that female barn swallows are fertile at least during the period from five days before start of laying. Nest guarding intensity against infanticidal males was determined as the percentage of scans when at least one nest owner was within a distance of 2m from their nest, since nest owners were always able to interfere with intruders at that distance (Møller 1988*a*). An estimate of overall nest-guarding intensity against infanticidal males was determined as the mean nest-guarding intensity for the daily estimates for the first five days of the nestling period when most cases of infanticide occurred (Møller

1988*a*). Song activity of males was determined as the percentage of scans every second minute during daily one-hour observation periods when a male was singing.

Several behaviours were recorded continuously as they were observed (Altmann 1974). The presence of females in territories early during the breeding season was determined daily, and the number of females visiting each male was determined as the number of different females present during morning observations. An alternative measure of mate guarding was the percentage of times that a leaving female was followed by its mate within 30 seconds and the percentage of times that a leaving male was followed by its mate. Copulation behaviour was recorded whenever seen. Copulations or copulation attempts are easy to notice because of the loud copulation calls. The identity of barn swallows participating in copulations was determined whenever possible. The feeding rate of male and female barn swallows was assessed for daily one-hour observation periods throughout the nestling period by recording all visits with food to the nest.

Causal relationships between two variables such as tail length and a female preference can be tested only in experimental studies. I performed a number of field experiments in order to verify causal relationships deduced from observations. Most of the experimental procedures have been described in detail in various publications. In the tail-manipulation experiments I usually used two control groups: one in order to check for the effects of treatment *per se* and another to check for the effects of general disturbance caused by capture and manipulation of birds. I have attempted to record as many detailed observations as possible from barn swallows involved in experiments both from the season when the experiment was performed and from the subsequent breeding season in order to look for long-term effects of the experimental procedure. The effectiveness of mate guarding and nest guarding was assessed in short-term detention experiments in which males were removed for a short period in order to determine whether a certain behaviour (for example, extra-pair copulations) occurred at a higher frequency as a consequence of the absence of guarding. Finally, I manipulated the nest loads of the haematophagous tropical fowl mite early during the egg-laying period by spraying nests with a pesticide, keeping nests as controls, and inoculating nests with *c.* 50 mites. The prevalence of mites on barn swallows was quite low, as is usually the case with parasites, and by random assignment of nests to treatments it was possible to determine the effects of parasitic mites on (1) reproductive performance, and (2) the development of tail ornaments in the subsequent breeding season. Nestlings were exchanged between nests in a partial cross-fostering experiment which allowed determination of the relationship between infestation rates of parents and their offspring when reared under different environmental conditions.

Heritabilities estimate the resemblance between parents and offspring (Falconer 1981). This resemblance is usually attributed to additive genetic variance and common environmental factors and maternal effects. I estimated the heritability for tarsus length, mite abundance, tail length and longevity by means of regression of average offspring values on mid-parents (tarsus), average offspring values on single parents (mite abundance), and offspring values on parent values of the same sex (tail length, longevity). I was only partly able to remove maternal effects by performing partial cross-fostering experiments for the tarsus length and mite abundance estimates. Heritability estimates are also likely to be biased if paternity is not determined with certainty. This will generally result in estimates being lower than the real values.

The reliability of morphological measurements and behavioural observations are usually not assessed statistically in field studies, but I estimated reliabilities by calculating repeatabilities and Kendall coefficients of concordance (Falconer 1981; Siegel and Castellan 1988). The reliability of morphological measurements of the same individual in the same season was estimated from the repeatability. The consistency within individuals in morphological measures in different years was also estimated by calculating the repeatability. Finally, I standardized some traits for growth between years and recalculated the repeatability in order to remove the effects of growth. I tested for the consistency in mate guarding, anti-infanticide nest guarding, song rate, and nestlings' feeding rate during daylight hours by measuring these behaviours for the same pairs in the morning, around noon, and in the evening. All behaviours demonstrated statistically highly significant consistencies (Møller 1988*b* and unpublished). The behaviours were sampled daily throughout the period when they were performed, and overall estimates for each pair were simply the average activity during this period.

I have spent ten years collecting the information on which the present book is based. This has included thousands of hours of observations, capture of birds with mist nets, and performance of field experiments. Several people have asked to what extent I have disturbed the barn swallows. This is a difficult question to answer, although I have two pieces of evidence which suggest disturbance to have been negligible. First, the number of barn swallows that have deserted breeding sites is limited to a couple of unmated males which had their tails manipulated in the tail-manipulation experiments in 1987 and 1991. I have handled barn swallows thousands of times and experimentally manipulated the birds or their nests in a number of ways without any other cases of desertion. Second, disturbance effects should result in desertion, reduced reproductive output and sooner or later a decreasing barn swallow population. Hardly any adult barn swallows have moved from one breeding site to another during or between breeding seasons, which speaks against strong effects of disturbance. It is true that

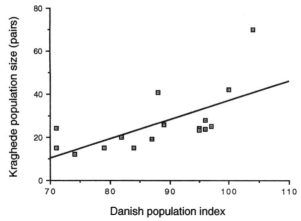

Fig. 3.4 Size of the barn swallow population in a part of the Kraghede study area (pairs) in relation to that of the entire Danish barn swallow population from 1976 to 1991. The line is the model I regression line. The Danish population size indices are based on country-wide point counts with 1976 set to baseline 100. Index values for Denmark kindly provided by the Bird Census Group, Danish Ornithological Society.

the study population of barn swallows has decreased in recent years (Møller 1989c), but that is also the case for the population in Denmark in general and in several other European countries. The relationship between the size of the population in the study area and in the whole of Denmark is clearly positive[1] (Fig. 3.4). The slope of the regression line does not differ significantly from one.[2] This suggests that the size of the barn swallow population in the Kraghede study area changes in parallel with the Danish population. Large-scale factors thus probably determine the changes in barn swallow populations over very large areas.

3.7 Summary

Barn swallows are particularly suitable organisms for studies of sexual selection because of their sexual size dimorphism, their abundance, the ease of finding nests, and their amenability to experimental manipulations. Barn swallows show small degrees of sexual size dimorphism in most body measurements, with the exception of the length of the outermost tail feathers, which are considerably longer in males than in females. These tail feathers play an important direct role in sexual selection as demonstrated by observational and experimental evidence. Males directly display their tail ornaments during mate acquisition activities. Male barn swallows play an important role in the overall reproductive activities, and furthermore, they rely on the relatively unpredictable food resource of aerial insects. Sexual ornaments are usually unlikely to evolve under such circumstances, and any extravagant secondary sexual character therefore is probably particularly costly. The tail ornament is thus likely to show clear signs of condition-dependence because only males of the highest quality may be able to produce large ornaments of perfect design under adverse conditions.

STATISTICS

1. Linear regression: $F = 10.35$, $df = 1,14$, $P = 0.006$.
2. $b = 0.90$ (SE = 0.28), $t = 0.35$, $df = 14$, NS.

4

Male mating advantages

4.1 Introduction

The primary driving force in sexual selection is non-random variation in
mating success. For example, males with the most extreme secondary sexual
characters may enjoy a mating advantage that results in an association
between the expression of a sex trait and a fitness component. However, it
is usually forgotten that sexual selection may take place at a number of
different stages of the reproductive cycle, and that mating success is but one
selection component, perhaps not even the most important one. The several
components of sexual selection are briefly reviewed in section 4.2.

In section 4.3 of this chapter I summarize the evidence for the action of
different sexual selection components in the barn swallow. More specifically,
I describe the relationship between the tail length of males and:

(1) male mating success;

(2) male mortality before mate acquisition;

(3) timing of mate acquisition;

(4) quality of the mate acquired;

(5) extra-pair copulation success;

(6) parental investment;

(7) differential parental investment by the sexes; and

(8) infanticide.

The obvious next question concerns the relative importance of these
various selection episodes, when so many selection components potentially
can be involved in determining the total sexual selection arising from a mate
preference. The relative importance of the selection components are evaluated by expressing them in terms of standardized selection differentials in

section 4.4. This exercise reveals that, while male mating success is an important selection component, there are others of equal importance, and surprisingly the frequency of extra-pair copulations proves to be the most important sexual selection component.

4.2 Components of sexual selection

Sexual selection has usually been equated with the selection processes arising from non-random variance in mating success. However, a number of different components affect sexual selection, and these may best be described as a sequence of selection episodes which are sequentially related to each other. This view of sexual selection can be represented as a path diagram (Fig. 4.1). There are two major routes to fitness improvement by means of sexual selection: the differential mating success route and the differential mate fecundity route. Mating success is determined by the proximate mechanisms of mate choice such as search time and handling time, which give rise to the time required for processing a mate (Fig. 4.1). The fecundity route to sexual selection arises as a result of different components that affect the fecundity of each mate, which may be influenced by courtship feeding, parental investment, differential parental investment, the quality of the mate, differential abortion, and infanticide (Fig. 4.1). Differential total fitness arising from sexual selection is determined as the product of fecundity per mate and mating success.

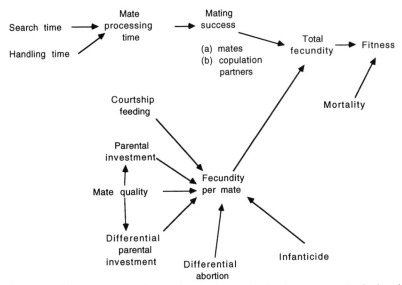

Fig. 4.1 The sequence of sexual selection episodes in an hypothetical animal.

This view of sexual selection has several major advantages. First, the sexual selection process can be perceived as a natural chain of events which each lead onto subsequent events. Second, the process of sexual selection can readily be analysed in terms of the relative importance of different selection components. Third, the different components of sexual selection and their relationships can be analysed by the use of path analysis.

This view of sexual selection as a continuous process consisting of a chain of reproductive events is effectively the definition of a mating system. Mating systems have traditionally been defined by the way in which mates are acquired, mating success, characteristics of the pair bond or the relationship between mates, and patterns of parental care (Davies 1991, 1992). A number of different levels of explanation are entwined in this definition. For example, the proximate mechanisms of mate acquisition are confused with the ultimate explanations of sex roles in parental care. Mating systems have been difficult to investigate and understand because of this confusion. However, the sequential sexual selection concept places the study of mating systems in a sexual selection framework and allows comparison of the mating systems among different populations or taxa.

The different components of sexual selection and thus the mating system of the barn swallow are analysed in the next section.

4.3   Evaluation of sexual selection components

There are a large number of potential sexual selection components, and they will be briefly discussed for the barn swallow in the following sections.

4.3.1 *Mating success*

Barn swallows are monogamous birds, and an unbiased sex ratio therefore should allow all individuals to acquire a mate. This is, however, not the case. The percentage of males remaining unmated for the entire breeding season during the years from 1977 to 1992 has varied between 3.6 and 25.0%, with a mean of 12.5% (SE = 1.6, n = 16 years). The annual survival rate of adult barn swallows is quite low (28.4% in males and 25.5% in females (Section 8.2)), and unmated males therefore do not have a high probability of returning in a subsequent year and then acquiring a mate.

Male barn swallows that remained unmated for a whole season had shorter tails than mated males (Fig. 4.2). Some of this effect can potentially be accounted for by annual differences between tail length of mated and unmated males. Indeed, there was a strong year effect on the length of male tails, but also a strong effect of mating status (Møller 1992c). However, the difference in tail length between mated and unmated male barn swallows

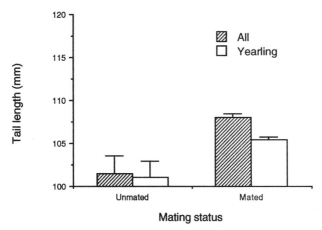

Fig. 4.2 Tail length of male barn swallows differing in mating status. Values are means (SE). The difference in tail length in relation to mating status was statistically significant for all males ($t = 7.09$, $df = 773$, $P < 0.001$) and yearling males ($t = 5.99$, $df = 483$, $P < 0.001$).

did not vary between years, since the interaction between mating status and year effect was non-significant.

Older male barn swallows have longer tails than younger males (Møller 1990*a*, 1991*a*), and the difference in tail length between mated and unmated males potentially may be accounted for by an age difference between these two groups of males, if young males were less likely to get a mate. I tested whether this were the case in two different ways. First, I analysed the relationship between mating status and tail length within the group of yearlings. Unmated males still had much shorter tails than mated males (Fig. 4.2). Second, I determined if male mating success was independently related to tail length and age in a logistic regression analysis. The result of this analysis was clear since tail length, but not age, was an important determinant of mating success (Møller 1992*c*). The difference in tail length between mated and unmated males thus cannot be accounted for by a difference in the age composition of these two groups of males.

A correlation between mating status and tail length does not necessarily mean that there is a causal relationship. A third variable may independently affect both mating success and tail length, and generate the apparent relationship. This problem can be handled in three different ways. First, one may see whether mating success is correlated with any other obvious morphological character that could be used as a cue in mate choice. Among all the different morphological characters in the barn swallow I have measured, only one in addition to tail length correlated with mating success: unmated males had significantly shorter wings than mated males.[1] While the mean difference in tail length between unmated and mated males was 6.4%, the difference in wing length was only 1.4%. As tail length and wing length are positively correlated traits (Møller 1991*a*), selection on one of these characters would result in a response to selection in the other provided the correlation is genetic. The independent effect of tail length and wing length on mating success was determined in a logistic regression analysis, and whereas the effect of tail length remained statistically highly significant, the effect of wing length was only marginally significant (Møller 1992*c*). In other words, tail length of male barn swallows was a more important determinant of mating success than wing length.

The second possibility is to perform an experimental manipulation of tail length before males acquire a mate and subsequently determine the effect of tail length on mating success. I performed such an experiment in the Kraghede study area in 1987 by cutting and glueing the outermost tail feathers with the use of superglue. There were 11 males in each of the four experimental groups; namely, males with shortened tails, males with elongated tails, males which had their tails cut and glued back again and males which were only caught, measured, ringed and released. Only two of these males did not acquire a mate, and both were males with shortened tails. Although this difference is not statistically significant, both unmated males had the shortest tails in the sample. In order to demonstrate that an 18% (2 out of 11 males) difference in mating success is statistically significant at the 5% level, I needed 50 replicates. In 1991 when a large-scale tail manipulation experiment was performed, the proportion of unmated male barn swallows was very low, and all males acquired a mate. Andersson (1982*b*) was able to use a tail-manipulation experiment in the polygynous long-tailed widowbird *Euplectes progne* to demonstrate that mating success was positively related to tail length. However, barn swallows in my study area are monogamous and there is only little variation in mating success.

A third way in which to measure the effect of tail length on mating success is to determine the number of females attracted by a given male but then abandoning it in favour of another male. Male barn swallows succeeded in attracting between zero and three different females, many of which later left their territories and mated with other males. Short-tailed male barn swallows, however, were deserted by significantly more females than were long-tailed males (Møller 1990*a*). Female rejections were causally related to the length of male tails as determined from the tail manipulation experiments. Male barn swallows with shortened tails were rejected on average by 0.73 females (SE = 0.20, $n = 11$), whereas males receiving the two control treatments and the elongation treatment were not deserted by any females (Møller 1988*c*). A later experimental manipulation of tail length, which also involved manipulation of tail asymmetry, revealed a highly significant positive relationship between the number of rejections by females and the manipulated size of the tail (Møller 1992*d*). Female barn swallows thus demonstrated a more long-lasting interest in their mates if the males were wearing a long tail ornament.

In conclusion, male barn swallows with long tails were more likely to acquire a mate than were short-tailed males. Mating success was not affected by age or differences in any other morphological trait measured, perhaps with the exception of wing length which was slightly longer in mated than in unmated males.

4.3.2 *Timing of mate acquisition*

The time available for reproduction is always limited, either because the opportunities for reproduction decrease with time or because the opportunities for rearing more than a single brood deteriorate. It is therefore important not only to acquire a mate, but also to acquire a mate as quickly as possible. It takes male barn swallows between 1 and 31 days to acquire a mate, but a male that is able to get a mate only after several weeks of display is unlikely to reproduce as successfully as a male that almost immediately acquires a mate. The reason is that the probability of recruitment decreases for nestlings reared later in the breeding season (Fig. 4.3), and the probability of producing a second clutch decreases rapidly as the breeding season progresses (Møller 1990*a*).

The duration of the pre-mating period ranged from 1 to 31 days, on average 4.8 days (SE = 0.3, n = 200). It was negatively related to tail length of the male in the unmanipulated situation, and also when the effect of arrival date was controlled statistically (Møller 1990*a*, 1992*c*). Males with long tails took less time to acquire a mate even when the effects of male age and differences between years were controlled statistically (Møller 1992*c*).

An experimental test of the presumed causal relationship between duration of the pre-mating period and tail length of male barn swallows was performed by manipulating the tail length of males shortly after their arrival (Møller 1988*c*). Because they were randomly assigned to treatments, this experiment also controls for any confounding effect of male age and other factors. Males with shortened tails took much longer to acquire a mate than control males, which again took longer than males with elongated tails (Fig. 4.4). There was therefore a direct causal relationship between male tail length and the difficulties experienced in attracting a mate. A later experimental manipulation of tail length, which involved an additional treatment for tail asymmetry, confirmed that it took longer for males with shortened tails to acquire a mate than for males with elongated tails (Møller 1992*d*).

The same tail manipulation experiment, albeit with a slightly different design, was done on the North American subspecies *erythrogaster* of the barn swallow by Smith and Montgomerie (1991). They were not sure whether their males were unmated at the time of the tail manipulations, and therefore

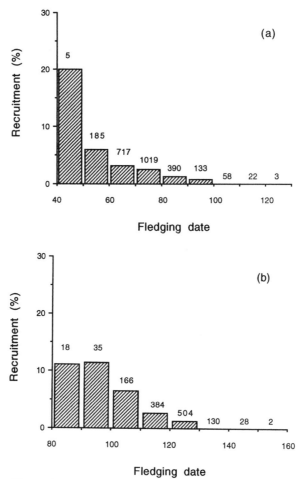

Fig. 4.3 The probability of recruitment of barn swallow nestlings reared at different times of the breeding season for (a) first and (b) second clutches. Numbers are number of nestlings that fledge in a specific 10-day period. Dates are days since 1 May.

measured the intensity of female preference in terms of the duration of the pre-laying period, which was significantly shorter for males with elongated than for males with shortened tails. The results of this experiment are difficult to interpret, however, because the experiment confounds the effects of the female preference for males with long tails, if the males were unmated at the time of manipulation, with the effects of differential parental investment by females in relation to tail length of their mates, if the males were mated at the time of manipulation (see section 4.3.7).

These experiments do not address the question whether differences in duration of the pre-mating period are due to mate choice or male–male

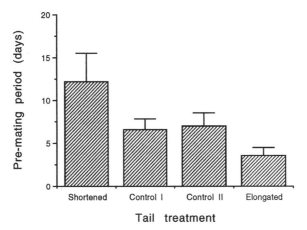

Fig. 4.4 The duration of the pre-mating period in relation to experimental manipulation of the tail length of male barn swallows upon arrival. Males in the Control I group had their tails cut and glued back again, while males in the Control II group were captured but their tails were not manipulated. Values are means (SE). Adapted from Møller (1988*c*).

competition. The fact that females often visited several males before making their choice suggests that mate choice is involved. The frequency of male–male interactions and the outcome of such interactions were unrelated to treatment in three tail manipulation experiments (Fig. 4.5; Møller 1988*c*, 1992*d*, 1993*q*). This suggests that male–male interactions were unaffected by the experiments, and that differential success of male barn swallows was caused by mate choice.

Sexual selection will also be related to temporal component of reproduction other than the duration of the pre-mating period, if there is an optimal time for successful rearing of offspring. Barn swallow nestlings reared at different times of the breeding season are not equally likely to recruit into the population (Fig. 4.3). In general it appears that early reared nestlings of both the first and the second brood are more likely to recruit. If there is an optimal date of reproduction, which results in a maximum number of recruits, some males will be mated closer to this date than others. Such males, which mate closer to the optimal date of reproduction, will leave more descendants than others, and this should result in sexual selection on male traits (Darwin 1871; O'Donald 1980*b*, 1980*c*; Kirkpatrick *et al.* 1990). This effect of optimal breeding date *per se* is not easily evaluated because it would require detailed knowledge about the probability of recruitment during the breeding season in different years. Any effects of parental

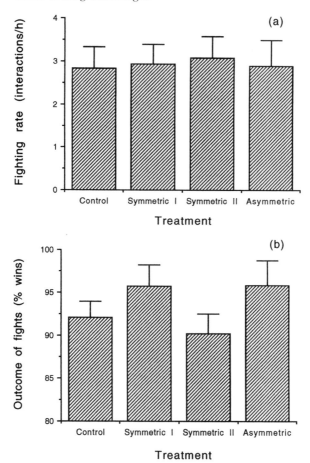

Fig. 4.5 Male–male interactions ((a) fighting rate and (b) relative frequency of wins) in relation to tail treatment. The outermost 10mm of the tips of the outermost tail feathers were painted with black correction fluid in the control group, while the outermost 10mm (symmetric I) or 20mm (symmetric II) of the tips were painted with white correction fluid in the symmetric treatments. In the asymmetric treatment the outermost 20mm of one feather were painted with white correction fluid while the outermost 20mm of the other feather were painted with black correction fluid. Values are means (SE). Adapted from Møller (1993*b*).

investment and differential parental investment (see sections 4.3.6 and 4.3.7) on the probability of recruitment also had to be controlled.

In conclusion, female barn swallows preferred males with long tail ornaments, and as a consequence long-tailed males acquired a mate more rapidly than short-tailed males.

4.3.3 *Male mortality before mate acquisition*

If the duration of the pre-mating period varies among males, those which need a long time to obtain a mate may run a higher risk of dying before acquiring a mate (Fisher 1930). This sexual selection component would particularly affect males that were avoided by females, and evolution of the secondary sexual character therefore may be affected by differential mortality before mate acquisition.

A number of barn swallows die during the breeding season, usually in April and May just after arrival. The main reasons for this in my study area were predation by cats and fatal encounters with automatic ventilation systems. The identity of 'lost' barn swallows was determined from known birds no longer being present in their territories and from findings of carcasses at the breeding sites. A total of 30 birds were 'lost' from breeding territories; 27 were found dead at the breeding sites. Of these 27 barn swallows 17 were males, which is a statistically non-significant over-representation of males. The length of the tails of the dead males did not differ from that of the male population in general.[2]

Cat with swallow

Less preferred male barn swallows with short tails did not suffer a higher mortality than preferred long-tailed males during the breeding season, thus differential male mortality in relation to tail length was not a cause of sexual selection.

4.3.4 *Extra-pair copulations*

Socially monogamous animals are not necessarily faithful to their mates. There are many theoretical reasons why males and females of monogamous species might attempt to copulate with individuals other than their mates (see Chapter 11). Copulations with non-mates may result in extra-pair paternity and as a result increase the variance in realized male mating success. All females are not able to acquire the most preferred mates in a socially monogamous mating system, because the most attractive males will be removed from the pool of available mates very early during the mating season (Møller 1992*a*). The discrepancy between the mate preferred by females and the male they actually acquire is therefore bound to increase during the mating sequence of the population. The latest mating females are bound to acquire the least preferred males even though they actually may have preferred to mate with more attractive males. This constraint on female mate choice caused by the mating system has important consequences for the propensity of females to engage in extra-pair copulations (Møller 1992*a*), which can be seen as a means of female adjustment of mate choice to circumvent this constraint. Extra-pair copulations and the resulting extra-pair paternity could therefore be a powerful selection component affecting the evolution of secondary sexual characters if males with the most extravagant degree of ornamentation are likely to acquire more extra-pair copulation partners.

There was a highly significant difference in tail length between male barn swallows that were successful and unsuccessful in copulating with female non-mates (Fig. 4.6; Møller 1992*c*). This effect of male tail length on success in extra-pair copulations remained even after controlling for age effects and annual differences in the frequency of extra-pair copulations (Møller 1992*c*). Some male barn swallows were very successful and acquired extra-pair copulations with several females, while most males did not succeed at all

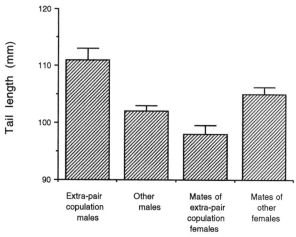

Male status

Fig. 4.6   Tail length (mm) of barn swallow males participating and not participating in extra-pair copulations, and the tail length (mm) of the mates of female barn swallows participating and not participating in extra-pair copulations. Values are means (SE). Adapted from Møller (1992*c*).

(Møller 1992*c*). The distribution of extra-pair copulations among males was thus very skewed, and the distortion in copulation success was remarkably similar to the distribution of copulations among males of many lekking bird species. Females of lek-breeding animals are supposed to be able to choose freely among males with no constraints. As a result, the most preferred males will acquire a disproportionate number of copulations. Females of monogamous birds that engage in extra-pair copulations may experience a similarly unconstrained situation of mate choice, and consequently a few highly preferred males may obtain a majority of all extra-pair copulations.

Female barn swallows that engage in extra-pair copulations may not have been mated to a random sample of all males. For example, females mated to males with a less preferred phenotype may be more likely to engage in extra-pair copulations. Females accepting extra-pair copulations had mates with significantly shorter tails than females resisting copulations with other males (Fig. 4.6; Møller 1992*c*). This difference in tail length between male barn swallows mated to females participating and not participating in extra-pair copulations remained even after controlling for annual differences in the success of extra-pair copulations and age differences among the two groups of males (Møller 1992*c*). In other words, female participation in extra-pair copulations depended on the phenotype of their mates.

The association between tail length of male barn swallows and their participation in extra-pair copulations and that of their mates may not necessarily represent a causal relationship. However, the male tail-manipulation experiments provided an opportunity to determine whether the relationship was indeed causal. When the tail of some males was elongated while that of others was shortened, this clearly affected their engagement in extra-pair copulations (Møller 1988*c*). Males with elongated tails were much more successful in obtaining extra-pair copulations than were males with shortened tails or males in the control groups (Møller 1988*c*; Fig. 11.2). This effect was not caused by differences among experimental groups in the frequency of extra-pair copulation attempts. Since females are able to determine the outcome of extra-pair copulation attempts by accepting or rejecting the male, the difference in success rate of extra-pair copulation attempts must be due to a difference among females' reactions to different males. Female barn swallows accepting extra-pair copulations were not a random sample of females, because females mated to less preferred males, particularly those mated to males with shortened tails, more frequently participated in extra-pair copulations (Møller 1988*c*; Fig. 11.2). The result of the experiment was that the most preferred males were the most successful ones in obtaining extra-pair copulations, particularly with females mated to the least preferred male barn swallows.

Extra-pair copulations may not necessarily lead to extra-pair paternity. It is therefore essential to test whether there is an association between extra-pair paternity and male tail length in the barn swallow. Smith *et al.* (1991) manipulated tail length of males belonging to the North American subspecies *erythrogaster* and subsequently determined the frequency of extra-pair paternity in broods by means of DNA-fingerprinting. The results of their paternity analyses were very clear: (1) male barn swallows with naturally long tails suffered less from cuckoldry than males with short tails, and (2) there was no effect of experimental treatment on the frequency of extra-pair paternity in the broods. Smith *et al.* (1991) interpreted this result as indicating that males of preferred phenotypes were indeed more likely to sire their broods, but that mates of males with experimentally manipulated tail length did not alter their extra-pair copulation behaviour. Although they planned their experiment to mimic mine, there were three important differences.

First, Smith *et al.* (1991) did not use any control groups because they claimed that the design of the experiment was the same as that of my Danish experiment. Since there was no effect of treatment *per se* in my Danish experiment, they did not expect any such effects in theirs. However, their experimental procedure differed markedly from mine because they used feather imping (small insect pins were inserted into the shaft of the feathers). As a result, the outermost tail feathers of experimental birds must have become heavier than those of unmanipulated birds, which might have affected the manoeuvrability and therefore also the foraging ability and display rate of manipulated birds. This effect may have been stronger in males with elongated tails because these birds had a piece of feather added to their outermost tail feathers.

Second, Smith *et al.* (1991) did not watch their birds, and were therefore unable to tell whether males with elongated tails participated in more extra-pair copulations than males with shortened tails.

Third, if the tail length of males was manipulated after their acquisition of mates, females would have been more constrained in their mate choice than if the manipulation had taken place before the females ever saw their mates. An increase in the magnitude of the constraint on female mate choice should result in an increase in the frequency of extra-pair copulations and extra-pair paternity (Møller 1992*a*).

Preliminary results from DNA-fingerprinting of barn swallows from my Kraghede population confirm that males with long tails are more certain of their paternity than are short-tailed males (A. P. Møller and H. Tegelström unpublished data). This difference in susceptibility of males to cuckoldry was not associated with other variables such as male age, other morphological traits, or body mass, and males which guarded their mates more intensely are not more certain of their paternity than the average male barn swallow (A. P. Møller and H. Tegelström unpublished data).

Compared with short-tailed males, long-tailed male barn swallows are able to acquire more extra-pair copulations, and in addition are fathers to a larger proportion of the offspring in their nest.

4.3.5 *Quality of female mates*

All females are not equal, and males that are able to attract high-quality mates should be able to gain higher reproductive success than the average male. Darwin (1871) originally suggested that some males of monogamous species may gain a mating advantage through acquisition of more fecund females. This idea was later modified by Fisher (1930), who suggested that differences in female fecundity could cause sexual selection in monogamous mating systems. The first females to arrive at the breeding sites could potentially choose among all males present. A female mate preference for any particular kind of male would result in such males becoming mated earlier than others. The last male to mate would therefore be the least preferred male. The earliest females to arrive would, on average, be in better body condition than the average female, and this would result in an earlier time of breeding and/or production of more offspring. If females in good body condition breed earlier than others, this could provide males with a mating advantage without causing any evolutionary change in breeding date because of the phenotypic nature of the female condition (Fisher 1930). Males of high phenotypic quality measured in terms of size of a secondary sex trait therefore should acquire mates of high phenotypic quality in terms of reproduction or survival prospects (Parker 1983*b*). This process was subsequently studied in population genetic and quantitative genetic models, and a potential effect of time of breeding by females on the evolution of male ornaments was confirmed (O'Donald 1980*b*, 1980*c*; Kirkpatrick *et al.* 1990).

An alternative to this process of non-random assortment of females among males differing in quality is the differential access hypothesis, which may apply when males and females are already present at the breeding sites from the start of the breeding season. This hypothesis posits that desirable individuals have greater access and therefore can be highly selective when choosing a mate, whereas the least attractive individuals must accept whichever partner is available (Burley 1977, 1981*a*, 1981*b*, 1986). In other words, the reason why high-quality males acquire high-quality mates is that females differ in their access to males because of differences in their relative quality, and not that low-quality females enter breeding condition later than high-quality females.

What is the evidence for preferred male barn swallows acquiring mates in better body condition? Males with long tails did acquire mates with higher body mass: males with very short tails had mates weighing on average 8%

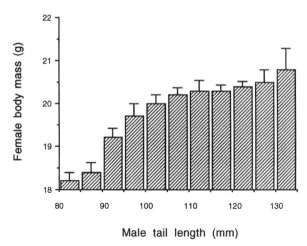

Fig. 4.7 Body mass (g) of female barn swallows in relation to the tail length (mm) of their mates. Values are means (SE). The relationship was statistically significant ($F = 11.51$, $df = 1,612$, $P < 0.001$).

less than males with very long tail feathers (Fig. 4.7; Møller 1992c). This relationship might be confounded by old females arriving earlier and therefore mating with early arriving males with long tails. However, female barn swallows mated with long-tailed males on average were not older than females mated with short-tailed males (Møller 1992c). The relationship between female body mass and tail length of their mates could be caused by heavy females being attracted to males with another attribute, which also happened to be long-tailed. Experimental manipulation of tail length, however, revealed a strong positive relationship between female body mass and male tail length (Møller 1991b). Consequently, the female preference for long-tailed males resulted in preferred males acquiring mates in better body condition.

The higher phenotypic quality of female barn swallows mated to long-tailed males may be reflected by their ability to survive. In fact, female survival prospects were strongly positively related to the tail length of their mate, whereas their own tail length was of less importance (Fig. 4.8; Møller 1991b)! Male tail length thus can be considered 'the extended phenotype' of female barn swallows, although male tail length obviously cannot influence the survival probability of females directly! The intensity of 'selection' on male tail length, which was calculated as the difference between the mean value of the $\log_{10}$-transformed tail length before and after selection, standardized to have unit variance before any calculations, was 0.95 standard deviations, which is an extremely large value in relation to other published intensities of selection (Endler 1986). Female survival prospects were not

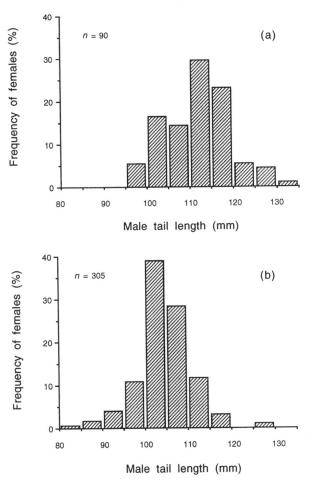

Fig. 4.8 Frequency distribution of tail length (mm) of male barn swallows mated to (a) female survivors and (b) female non-survivors. Adapted from Møller (1991*b*).

related to their own age, body mass or a number of other morphological traits (Møller 1991*b*).

The effects of male tail length on the survival prospects of females could be estimated more directly in the tail-length experiment. The prediction was that females mated to long-tailed males should have higher future survival prospects than females mated to short-tailed males. There was indeed a very strong positive relationship between male tail length and female survival prospects, and both natural tail length and experimentally altered tail length had a highly significant positive effect on female survival prospects (Møller 1991*b*). The univariate directional 'selection' differential was 1.30 standard deviations, which is considerably larger than the 'selection' differential in the large, natural data set. Females therefore paid attention to both natural tail length as well as experimental tail length when making their mate choice. The extremely high 'selection' differentials both under natural conditions and in the experiment suggest that the distribution of high condition females among males closely reflects male attractiveness. The enhanced survival prospects of females mated to long-tailed male barn swallows are remarkable particularly in view of their high rates of reproduction. Females mated to the most preferred males survived much better than other females, despite the fact that they reproduced earlier in the season and at a higher rate than other females.

Male barn swallows also benefited directly from the better survival of their mates. Pairs of monogamous birds with previous breeding experience often produce more offspring compared with similarly aged pairs without common breeding experience (Rowley 1983). Barn swallow pairs consisting of individuals with previous breeding experience bred earlier, laid more eggs and produced more fledglings in their second and later reproductive seasons than did experienced pairs consisting of individuals without previous common breeding experience (Møller 1991*b*). This effect was not confounded by age-dependent reproductive rates (Møller 1991*b*). Long-tailed male barn swallows therefore also experienced a direct reproductive advantage because of a higher probability of remating with a previous mate.

4.3.6 *Parental investment*

Male barn swallows may gain an advantage in terms of sexual selection if they acquire mates that are able to invest more energy in parental investment. Such parental investment may arise because (1) females mated to the most preferred males are of higher phenotypic quality than females mated to the average male, or (2) females mated to the most preferred male barn swallows allocate a relatively larger proportion of their reproductive effort into rearing of offspring. The latter aspect of parental investment, which is often termed differential parental investment (Burley 1986), will be treated in section 4.3.7.

The effects of female body condition on the reproductive success of the preferred male phenotype can be measured in terms of time of breeding, clutch size, and propensity to lay a second clutch. Mates of long-tailed males laid earlier than mates of short-tailed males (Møller 1992*c*). The duration of the pre-laying period, which is the period from mate acquisition until start of egg laying, was significantly shorter for females mated to long-tailed males (Fig. 4.9). Such males therefore started to reproduce earlier than other males (Møller 1992*c*). There was no direct effect of female body mass on the size of first clutches, when the confounding effects of male age and year were controlled statistically, but second clutches were larger among females mated to long-tailed males (Møller 1992*c*). The propensity to lay a second

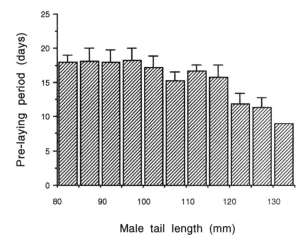

Fig. 4.9 Duration of the pre-laying period of female barn swallows in relation to the tail length (mm) of their mates. Values are means (SE). The relationship is statistically significant ($F = 4.47$, $df = 1,207$, $P = 0.036$).

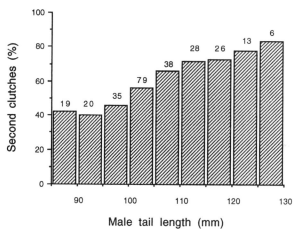

Fig. 4.10 Percentage of female barn swallows raising two broods during the same breeding season in relation to the tail length (mm) of their mates. Numbers are number of females. Adapted from Møller (1992c).

clutch among female barn swallows was higher for females mated to males with long tails (Fig. 4.10; Møller 1992c). Brood size of both first and second clutches was larger for mates of long-tailed male barn swallows, and seasonal reproductive success therefore increased with male tail length (Møller 1992c). The later timing of reproduction during the season because of a higher frequency of second clutches among females mated to long-tailed males was partially compensated for by more rapid reproduction. These females took less time to incubate their second clutch, and their nestlings fledged earlier, despite the fact that they on average reared more offspring (Møller 1992c).

The increased reproductive success of females mated to male barn swallows with long tails may be because of:

(1) superior female body condition;

(2) older age and greater experience of female; or

(3) earlier start of breeding by such females.

When the effects of these three factors were partitioned statistically, the main factor was found to be the breeding date, whereas female body mass and female age were only of marginal importance (Møller 1992c).

There is one additional piece of evidence for particularly high parental investment by female barn swallows mated to long-tailed males. Across the entire nestling period, females contributed a larger percentage of all feeding visits to the nestlings of their first brood than did their mates (Fig. 4.11). A similar relationship was found for second broods (Møller 1992c). The relative feeding rates of females rose from just above 50% for those mated to males with the shortest tails to almost 60% for those mated to male barn swallows with the longest tails. This positive relationship between female parental investment and degree of ornamentation of their mates may potentially be affected by confounding variables such as brood size and age and hence experience of the female. However, the positive relationship between relative feeding rate by the female parent and the tail length of its

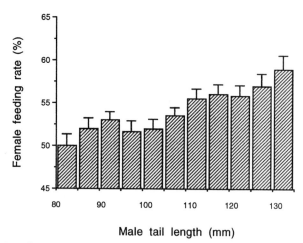

Fig. 4.11 Percentage of nestling feedings by female barn swallows during the nestling period of their first clutch in relation to tail length (mm) of their mates. Values are means (SE). The relationship is statistically significant ($F = 5.90$, $df = 1,165$, $P < 0.02$).

mate remained even after these variables were controlled statistically (Møller 1992*c*).

In conclusion, long-tailed male barn swallows invested less than their mates when raising offspring. Variation in female parental investment could be due to females mated to the most attractive males being of higher quality and therefore able to invest more in reproduction. Alternatively, variation in female parental investment could be due to females mated to the most attractive males investing a higher proportion of their reproductive effort in rearing of offspring. The latter alternative is investigated in the next section.

4.3.7 *Differential parental investment*

Males endowed with large ornaments may gain a reproductive advantage if their mates invest in reproduction relative to the attractiveness of their males, in other words if females mated with highly preferred males increase their reproductive effort as a result of the perceived quality of their mates (Burley 1986). The logic behind this argument is that females invest differentially in reproduction in relation to the attractiveness of their mates in order to obtain or maintain relatively attractive mates (Burley 1986). Alternatively, females may invest relatively more in their offspring if they have acquired highly attractive mates because they will be repaid in terms of the higher genetic quality or the more extravagant degree of ornamentation of their offspring (de Lope and Møller 1993). This differential allocation of female parental investment therefore parallels the sexy son hypothesis, which states that some individuals may enhance their fitness by incurring a short-term cost to obtain the long-term benefits of an attractive mate (Weatherhead and Robertson 1979). I have already demonstrated that preferred male barn swallows acquire mates of higher quality, but this does not demonstrate that males obtain an additional reproductive advantage because of differential investment in reproduction by their females.

As stated in the previous section, there is some evidence that female barn swallows increase their reproductive effort when mated to the most preferred, long-tailed, males. It is not possible to determine statistically from observations whether this increased reproductive effort is a result of an increased effort on the part of females. The only way to do this is by experimental manipulation of the tail length of males that have already acquired a mate. If females invest differentially in reproduction relative to the attractiveness of their mates, females which suddenly after their mate choice obtain a highly attractive mate should invest relatively more in reproduction than do females which suddenly find themselves mated to an unattractive male.

Florentino de Lope performed such an experiment and found that, indeed, female barn swallows invested differentially in reproduction when the attractiveness of their mates was suddenly altered during the pre-laying

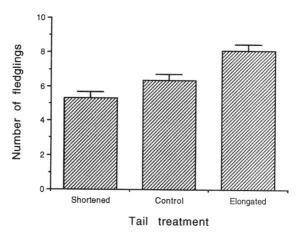

Fig. 4.12 Seasonal reproductive success of barn swallow pairs in relation to experimental treatment of the tail of the male during the pre-laying period of the first clutch. Values are means (SE). Adapted from de Lope and Møller (1993).

period of their first clutch. Females given an attractive mate started to breed earlier, raised more offspring in the first and the second clutch, and raised more broods per season than did females given an unattractive mate (de Lope and Møller 1993). Seasonal reproductive success of males therefore increased with experimental tail length of male barn swallows (Fig. 4.12). This result is surprising because both absolute and relative male feeding rates decreased with increasing experimental tail length (de Lope and Møller 1993; see also Fig. 10.5). Male barn swallows with elongated tails also brought smaller, less profitable but more prey items per bolus than males with shortened tails. The mates of males with elongated tails were able to compensate for the reduced feeding efficiency of their males, and the total feeding rate of the pair and therefore also nestling quality were unaffected by experimental treatment. Females mated to the most attractive male barn swallows despite their higher investment in reproduction during their first clutch were still able to invest significantly more in second and third clutches in comparison with females mated to the least attractive males (de Lope and Møller 1993). A subsequent experimental manipulation of tail length and tail asymmetry also revealed a clear association between differential investment by females and the attractiveness of their mates (Møller 1993*b*). The pattern of differential parental investment by female barn swallows resembles that by female zebra finches *Taeniopygia guttata* mated to males with rings of a preferred colour (Burley 1981*a*, 1986, 1988). Such females reproduced at a higher rate, invested relatively more than their mates in reproduction, and preferred males more often became bigamous. Attractive male zebra finches also lived longer than unattractive individuals.

Female barn swallows clearly increased their reproductive effort relative to their mates, if they suddenly found themselves mated to a male of high perceived attractiveness. This experiment thus clearly supports the hypothesis that females invest differentially in reproduction relative to the degree of ornamentation of their mates.

4.3.8 *Infanticide*

A considerable proportion of unmated male barn swallows acquire a mate by killing the young nestlings of pairs which have not guarded their nest intensely. The infanticidal male usually mates with the female parent of the killed nestlings and thereby acquires a mate (see section 8.6 for a detailed description of infanticide behaviour). Such sexually selected infanticide is a component of variance in male mating success because it redistributes females among males. Does infanticide result in selection on the male tail ornament?

Unmated male barn swallows that were able to commit infanticide had longer tails than unmated males that tried but were never seen to commit infanticide (Fig. 4.13). Infanticidal males did not differ in morphology other than tail length from non-infanticidal males. Successful males had a longer tail than unsuccessful males also when the effects of male age were controlled for by restricting the analysis to unmated yearling males (Møller 1992*c*). Thus, the ability of male barn swallows successfully to commit infanticide was related to their tail length.

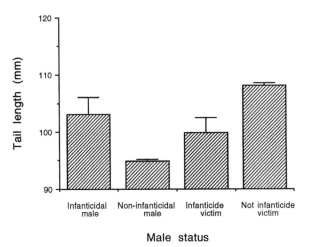

Fig. 4.13 Tail length (mm) of unmated male barn swallows which committed and did not commit infanticide, and tail length (mm) of male infanticide victims and of successfully reproducing males not suffering from infanticide. Values are means (SE). Adapted from Møller (1992*c*).

Infanticidal behaviour may also affect sexual selection among fathers of killed nestlings if males whose young were killed differed from males whose young were unmolested. Males of killed young on average had considerably shorter tails than had males able to avoid infanticide (Fig. 4.13). Bereft fathers did not differ from fathers of unmolested broods in other morphological traits, but were significantly younger (Møller 1992*c*). The difference in tail length between the two groups of males remained when the effects of age were controlled for by restricting the analysis to yearling males (Møller 1992*c*). Infanticide therefore also resulted in sexual selection on the tail ornament of mated male barn swallows because bereft males had shorter tails than had mated males which repelled infanticidal males.

Finally, did replacement of mated male barn swallows with unmated males result in sexual selection on the tail ornament? The tail length of bereft males was shorter than that of successful perpetrators (Fig. 4.13). The difference in tail length was far from statistical significance. In conclusion, there was no statistically significant increase in the tail length of mated male barn swallows as a consequence of sexually selected infanticide.

4.4 Relative importance of mating advantages

Male barn swallows with long tails gain advantages during a number of different sexual selection episodes. What is the net total sexual selection, and what is the relative importance of each of the many different selection components? The total intensity of selection on male tail length can be calculated as the univariate selection differential *s* which is defined as the difference between the mean value of the $\log_{10}$-transformed characters before and after selection, standardized to have unit variance before any calculations (Manly 1985; Endler 1986). The intensity of selection is then expressed in standard deviation units.

The different sexual selection episodes on male tail length and their relative importance are briefly summarized in Table 4.1. Most selection episodes resulted in a statistically significant change in tail length, but the magnitude of the change and the number of males involved differed markedly between episodes. Extra-pair copulation success and extra-pair copulations involving the female mate were by far the most important selection episodes followed by timing of mate acquisition. Selection episodes such as infanticide and male mortality during the reproductive season were relatively unimportant (Table 4.1).

The intensity of several sexual selection episodes on male tail length was estimated both under natural, unmanipulated conditions and under experimental conditions by manipulation of male tail length. The reliability of selection differentials obtained under natural conditions can be directly

Table 4.1. Univariate directional selection differentials (*s*) for sexual selection episodes in male barn swallows. Male tail length (mm) before and after selection is reported as the mean (SE).

Selection episode	Selection differential (*s*)	Tail length before selection	Tail length after selection	*n*
Mating success	0.08	107.71(0.25)	108.26(0.24)	762 709
Timing of mate acquisition	0.23	105.92(0.72)	108.24(0.67)	200 200
Mortality	0.01	108.19(0.20)	108.20(0.21)	762 738
Acquisition of extra-pair copulations	0.96	103.11(0.91)	112.11(1.60)	161 30
Lost certainty of paternity	0.80	104.87(1.03)	98.07(0.94)	161 37
Female quality	0.18	106.14(0.70)	107.94(0.68)	210 210
Female differential parental investment	0.10	106.15(0.85)	107.12(0.79)	131 131
Infanticide success	0.10	108.20(0.21)	109.09(0.19)	738 707
Infanticide victims	0.02	107.94(0.22)	108.10(0.20)	738 726

The duration of the pre-mating period was used as a measure of fitness for the timing of Mate Acquisition selection episode. The duration of the pre-laying period was used as a measure of fitness for the Female Quality selection episode. The proportion of nestling feedings by the female was used as a measure of fitness for the Female Differential Parental Investment selection episode. The Infanticide Success selection episode estimates the change in male tail length as a result of successful infanticide by unmated males. The Infanticide Victims selection episode estimates the change in tail length as a result of some males being evicted from the mating pool due to infanticide. The selection differentials for continuous fitness variables (Timing of mate acquisition, Female quality, Female differential parental investment) were estimated from the standardized regression coefficient of a linear regression of the fitness component on male tail length. The sample sizes are the number of males present before and after selection. All fitness components were not measured for all males, and sample sizes therefore differ for the various selection episodes.

estimated by comparison with selection differentials obtained under experimental conditions. The consistency of these pairs of estimates for a number of sexual selection episodes is reported in Table 4.2. The correlation between the pairs of estimates is positive and statistically highly significant,[3] but the mean values for the two pairs of estimates differ significantly from each other.[4] Selection differentials from observational studies were smaller than those obtained during experiments.[5] This strongly suggests that field estimates of the intensity of selection on tail length of male barn swallows are representative of the relative intensity of selection when potentially confounding variables are controlled in field experiments. The larger selection differentials under experimental conditions suggest that the effects of

Table 4.2. Univariate directional selection differentials (*s*) for sexual selection episodes in the barn swallow under natural conditions and experimentally manipulated conditions.

Selection selection	Observational selection differential (*s*)	Experimental selection differential (*s*)
Mating success	0.07	0.12
Timing of mate acquisition	0.23	0.91
Acquisition of extra-pair copulations	0.96	2.35
Lost certainty of paternity	0.80	1.88
Female quality	0.18	0.55

The experimental selection differentials were obtained during male tail-manipulation experiments, and these differentials were calculated from the relationship between the fitness components and experimental male tail length.

potentially confounding variables on the relationship between male ornament size and various fitness components are reduced when experiments are performed.

4.5 Summary

Contrary to common belief, the mating advantage accruing to the most extravagantly ornamented males is not just their higher mating success. In fact, male barn swallows with particularly long tails enjoyed a whole suite of mating advantages, namely:

(1) higher mating success;

(2) shorter time to acquire a mate;

(3) more extra-pair copulations, while their mates did not engage in so many extra-pair copulations;

(4) higher quality mates;

(5) mates with higher parental investment;

(6) mates that allocated a larger proportion of their reproductive effort to rearing of offspring; and

(7) higher infanticide success, but low probability of suffering from it themselves.

A direct causal relationship between male tail length and several of these sexual selection components was confirmed experimentally for:

(1) time to acquire a mate;

(2) extra-pair copulations;

(3) acquisition of mates of higher quality, and

(4) differential parental investment.

The relative importance of the different directional sexual selection episodes was assessed by calculation of the directional selection differential for each episode of sexual selection. Extra-pair copulations were by far the most important selection episode. A direct comparison of selection differentials for sexual selection in the unmanipulated situation and in experiments, respectively, revealed that they were positively correlated. However, intensities of selection measured under experimental conditions were considerably higher than those measured in observational studies.

STATISTICS

1. Unmated males: 125.1mm (SE = 0.4, $n = 51$), mated males: 127.0mm (SE = 0.11, $n = 725$), $t = 4.28$, $df = 773$, $P < 0.001$.
2. Disappearing males: 108.1mm (SE = 1.8, $n = 17$), other males: 108.2mm (SE = 0.4, $n = 444$).
3. Product-moment correlation, $r = 0.99$, $t = 10.02$, $df = 3$, $P = 0.002$.
4. Paired t-test, $t = 2.98$, $df = 4$, $P = 0.04$.
5. Observational differentials: 0.45 (SE = 0.18), experimental differentials: 1.16 (0.42).

5

Benefits of mate choice

5.1 Introduction

Female barn swallows exert a strong selection pressure on the males because of their mate preferences. Females thus form one of the main selective pressures which cause the evolution of the tail streamers of the males. How can so strong a preference for a single morphological character have evolved? Several solutions to this puzzle have been put forward (see section 5.2). These include random genetic drift, sensory bias in the female sensory system, and an initial mating advantage of good genes. Mate preferences are the basis for the action of one of the modes of sexual selection, namely mate choice, and several models of mate choice make specific predictions concerning the relationship between male secondary sexual characters and the mate preference. The Fisher process assumes coevolution of the female preference and the male character, but similar arguments can also be raised for the handicap process. A female mate preference can be considered a trait which may change as a result of drift and selection. The genetic basis for a mate preference can only be elucidated by testing the preference of both mothers and daughters repeatedly under controlled conditions. Such a test is rarely feasible under field conditions. If there is a genetic basis for the mate preference, females should also demonstrate consistency in their mate choice among mate choice situations, as described in section 5.3. The repeatability of mate choice, which represents the variation in choice within females relative to the total phenotypic variation within and among females, is an estimate of the consistency in female preferences, and it sets an upper limit to the heritability of the female trait.

The very strength of female mate preferences suggests that there must be considerable fitness benefits to be gained from mate choice. Otherwise, the strength of the preference would diminish rapidly because of the potential costs to females of having a preference. Two main groups of fitness benefits are traditionally contrasted: direct and indirect benefits. Direct benefits are those which accrue directly to the female or her offspring, while indirect benefits are those which accrue indirectly to the female in the next generation. Examples of direct fitness benefits include resources provided by the male, sperm quality, absence of contagious parasites, evasion of costs of mate searching, and avoidance of hybridization. Examples of indirect benefits are increased attractiveness of sons in the next generation as suggested by the Fisher process and increased genetic quality of offspring as proposed in the good genes process. Section 5.4 evaluates these potential fitness benefits of mate preferences in the barn swallow. Finally, in section 5.5 I briefly review different mechanisms for the maintenance of genetic variance in fitness in spite of continuous sexual selection for good genes.

5.2 Theory of mate preferences

Models of sexual selection usually assume that females exert a strong selection pressure on males because of their mate preferences. Mate preferences are the basis for the action of one mode of sexual selection, and models of mate choice assume that there is a genetic basis for the preference. Such a basis has been demonstrated in three insects and one fish species (Crossley 1974; Zouros 1981; Heisler 1984; Majerus *et al.* 1986; Boake 1985; Roelofs *et al.* 1987; Bakker 1993; see also Ritchie 1992). Female mate preferences therefore can be considered a trait subject to processes such as drift, selection and evolution. The preferences are often strong and unanimous, and they form the main selection pressure on male secondary sexual characters. How can such strong mate preferences have arisen and be maintained?

Mate preferences may have arisen for at least three different reasons:

(1) a good genes advantage;

(2) random genetic drift; or

(3) sensory bias in the female sensory system.

Fisher (1915, 1930) was the first to suggest how mate preferences may arise when the main fitness benefit of mate choice is sexual attractiveness of offspring in the subsequent generation. The problem is initially to establish the mate preference if there is no, or hardly any, fitness benefit arising from the preference. An initial good genes advantage therefore had to be invoked as an explanation (Fisher 1915, 1930). This solution may appear counter-intuitive as current models of female preferences with indirect benefits usually contrast the Fisher process with the good genes process, which is also based on an initial good genes advantage. The origin of the initial good genes advantage then has to be explained.

The problem of the origin of female mating preferences was addressed specifically by Fisher (1915, 1930), who suggested that females would favour male traits that were favoured by natural selection. Female choice of such good genes traits was analysed in a quantitative genetic model by Heisler (1984*a*). The initial increase in the female mate preference was governed by a high heritability of the male trait reflecting viability, and an ability by females to discriminate against males deviating from the optimal mate phenotype. The evolutionary history of male traits could therefore determine the extent to which they were close to the optimum phenotype because traits subject to strong directional selection would more often deviate from the natural selection optimum. The recent selection history could thus determine the likelihood of sexual selection affecting a particular trait (Heisler 1984*a*).

A second model of the initial increase in a female mate preference was based on the assumption that the male trait subject to the mate preference may not itself be related to viability, but rather become the target of a female preference because of genetic correlations with other fitness-related traits (Heisler 1985). A female mate preference for such a male trait only indirectly reflecting male viability could under certain conditions result in an exaggeration of both the male trait and the female preference. Selectively arbitrary traits may thus become the target of sexual selection via a female mate preference, if they are relatively immune to random environmental variation and genetically correlated with another trait that affects viability.

Natural selection will favour females that are able to minimize the costs involved in mating. Kirkpatrick (1987) analysed a model of the evolution of a mate preference with a search cost using a similar kind of modelling as Lande (1981) and Kirkpatrick (1985). Search costs were incorporated in the model of an absolute mate preference by assuming that the fitness of females is proportional to the degree of overlap between the phenotype distribution of males and the preference function of females. The actual value of the male trait is thus considered to be unimportant for female fitness. Males are supposed to be polygynous and thus able to handle available females. There is a single equilibrium point for this model which coincides with the optimum phenotype for male survival. Search costs will in this model tend to drive the female's search image towards the mean of the male trait. The male trait is affected by both natural and sexual selection and attains a compromise between the survival and the sexual selection optima. Coevolution will eventually drive both the male trait and the female mate preference to end at the optimum for male survival. If the assumption of the actual value of the male trait being unimportant is relaxed, the equilibium will change either to a point where the costs of the female mate preference are minimized, or to an unstable evolution away from the optimal male phenotype under natural selection. The evolutionary outcome of the process therefore depends to a large degree on the details of how the search costs affect females.

One possibility for how females could be able to tell apart males differing in viability is that the evolution of extravagant secondary sex traits as a response to female choice is related to the presence of fluctuating asymmetry in morphological traits (Møller 1993c). Females may initially prefer males with the largest and at the same time most symmetric morphological traits. Such males will be preferred because they have been able to produce both large and symmetric characters, but also simply because large traits can be perceived at a greater distance than small traits (Møller 1993c). Females would already initially obtain an indirect fitness benefit from their mate choice as a result of this process, and the initial exaggeration of the male character would soon lead to large amounts of fluctuating asymmetry on which further good genes sexual selection may operate (Møller 1993c). Alternatively, the Fisherian mating mechanism may become so relatively important in this second stage of the sexual selection process that it out-competes the effects of the good genes process.

The second possibility for the initiation of the mate preference is random genetic drift (Lande 1981; Kirkpatrick 1982). The preference for a particular male character may initially arise as a result of genetic drift in a small population. An initially rare preference may become relatively common as a consequence of drift. The initial female preference for a specific male phenotype is thus a result of what happens to genotype frequencies for mate preferences in small populations. Given that the female preference is initially of a sufficient magnitude, the exaggeration process of the male character can start.

The third possibility for the initiation of a mate preference is that there was a pre-existing mate preference even before the male secondary sexual character arose (Kirkpatrick 1987; Heisler *et al.* 1987). Such pre-existing biases for particular male characters can be due to bias in the sensory system of the female (Basolo 1990; Ryan 1990; Ryan *et al.* 1990; Ryan and Rand 1990). For example, females may prefer a certain male character because it resembles a common food item, or they may prefer certain male characters simply because they conform with a bias in the sensory system.

Early models of female mate choice such as the first versions of the Fisher process (O'Donald 1962, 1967, 1980a; Lande 1981; Kirkpatrick 1982) assumed that choice was not costly because females were able immediately to pick a male according to their preference. However, recent field studies have shown that females do not pick the first male they encounter, but often incur considerable costs as a result of their exigent mate preferences. Costs of female choice arise in terms of time or energy, increased risks of predation and parasitism during the mate searching process, or lost mating opportunities. These costs are often of such a magnitude that females have to obtain considerable fitness benefits for their preference in order not to have lower fitness than females showing little or no discrimination (Parker 1983*b*; Pomiankowski 1987*b*).

Models of female mate preferences make predictions about the relationship between male secondary sexual characters and the mate preference. The Fisher process assumes coevolution of the female preference and the male character to ever more extreme expressions (Fisher 1915, 1930; Lande 1981; Kirkpatrick 1982). When the female mate preference is costly, as is probably always the case in real organisms, an equilibrium between female preference and the expression of the male sex trait cannot be maintained because female fitness has to be maximized at the equilibrium point (Pomiankowski *et al.* 1991). Biased mutation or a similar process on the male sex trait can, however, lead to establishment of an evolutionarily stable exaggeration of the costly female preference (Pomiankowski *et al.* 1991). The Fisher process thus predicts that the expression of the male secondary sexual character should be positively related to the intensity of the female preference. However, a similar argument can be raised in favour of the handicap process, which also assumes that the male sex trait coevolves with the female preference (Iwasa *et al.* 1991; Houde 1993). In fact, the expression of male sex traits in different populations of guppies *Poecilia reticulata* is positively related to the intensity of the female mate preference (Houde and Endler 1990). The expression of the male trait may be determined by the local balance of natural and sexual selection on the trait, but there could also be an indirect effect of the female mate preference that could be determined by balancing selection pressures (Houde 1993). A study on sticklebacks demonstrated a positive genetic correlation between female mate preferrence and the male secondary sexual character (Bakker 1993). This result is consistent with a Fisherian process of female mate choice.

The theory of the fitness consequences of female mate preferences has been reviewed extensively in Chapter 2.

5.3 Female mate preferences in the barn swallow

Female mate preferences are a central component of current models of sexual selection through female choice. Detailed information about female behaviour and the rules governing female choice is therefore essential for a better understanding of how sexual selection works. This goal is much easier to enunciate than reach because female behaviour in mating systems with 'ordinary' sex roles is often very subtle compared with that of males. I have previously addressed some problems of female choice in Chapters 2, 3, and 4. Here I present data bearing on two important aspects of mate choice raised in section 5.2. The questions I am asking are: (1) is there evidence for female choice being costly, and (2) is there evidence for consistency in mate choice within individuals?

5.3.1 *Costly female choice*

In the following paragraphs I will briefly summarize the evidence for costs of mate choice in the barn swallow.

When female barn swallows arrive at the breeding grounds, there is usually a large number of males already present. The costs of mate choice are therefore relatively small because several unmated males are often present within a colony and available for direct comparison. However, a number of females become members of a solitary pair and therefore have to travel to nearby colonies in order to check whether more attractive males are present.

A number of female barn swallows do not pair up on their date of arrival, and the pre-mating period may last up to a week. This suggests that females sometimes do not immediately find a suitable mate. More often than not, females will visit a number of different displaying males before settling down. Travelling distances may appear short, for example, compared with their world-wide migratory travels, and the time costs of mate choice may be more important in terms of fitness than the expenditure of energy. The probability of successfully rearing a brood decreases continuously throughout the breeding season, and the probability of rearing a second brood drops precipitously in the middle of the summer (Møller 1990*a*). Consequently, females risk a considerable loss of fitness by spending days looking for the right male.

When female barn swallows have established a pair bond with a male, they do not necessarily stay with that male to reproduce. Mate switching as a consequence of females deserting a male in favour of another is a regular phenomenon (see Chapter 11). For example, three out of 19 females in 1984 deserted their mates after a maximum of 16 days and later established a pair bond with another male (Møller 1985). Reproduction will be considerably delayed as a consequence of mate switching, and this may affect both the quantity and the quality of offspring produced.

Female barn swallows may also incur opportunity costs as a result of their mate preferences. As most females establish a pair bond within a few days, those female barn swallows that spend a long time before making their mate choice may find that all the highly preferred males have already been taken by other females.

A relatively large fraction of female barn swallows participate in extra-pair copulations (see Chapter 11). One major cost of this habit is the risk of contagious parasites and venereal diseases. The probability of receiving contagious parasites increases with the number of sexual partners, and a female engaging in extra-pair copulations is more likely to become infected than a female copulating with only its own mate. The most preferred males often copulate with several different females (Chapter 11), and the probability of such a male carrying a contagious disease or parasites therefore should be higher than that of an average male in the population. If extra-pair copulations are considered to be a way for females to adjust their mate choice (Møller 1992a), females may pay a cost of their mate preference in terms of increased risks of infection.

Female barn swallows do not obtain a direct fitness benefit in terms of paternal care as a result of their mate preference. On the contrary, females that mate with the most preferred males make a relatively larger parental investment compared with females mated to average barn swallow males (Chapter 10). There is thus a cost associated with being mated to a highly attractive male.

In conclusion, there is a considerable amount of evidence that suggests that the female preference for long-tailed males carries relatively large fitness costs and thus there must exist direct or indirect fitness benefits to compensate for such costs of mate choice in order for the female mate preference to remain in the population.

5.3.2 *Consistency in mate choice*

Models of female mate choice assume that female mate preferences have a heritable basis. According to the Fisher process, this mate preference coevolves with the male secondary sexual character, while the handicap models similarly assume evolution from an originally low intensity of the mate preference to a higher level at the evolutionary equilibrium. Heritability estimates of a female preference pre-require the ability to test females in different generations under standardized conditions, and this is not feasible in the case of the barn swallow. However, the consistency in female mate preference can be estimated from the repeatability, which quantifies the amount of within-female variance relative to the total within and between female variance in a mate preference (Falconer 1981; Boake 1989). The repeatability also provides an estimate of the upper limit to the heritability.

The repeatability can be estimated for female barn swallows that have more than once chosen a mate, for example in different years. This estimate may be biased because females with surviving males usually re-mate with their previous mate. Male tail length was first used as a measure of the female mate preference, but the relationship between tail length of the first and the second male chosen by a female was not positive, as predicted, if there was consistency in the female preference (Møller 1993d). The estimate of repeatability was small and not statistically significant,[1] and a similarly small repeatability has been reported by Bańbura (1992) for barn swallows.

Moreover, the repeatability may be biased in a number of different ways. First, female mate choices are made in subsequent years, and females thus have a different group of males available each year. Second, tail length of male barn swallows varies with age and year (Møller 1991*a*). Females making a mate choice in some years may thus have at their disposal males with different tail lengths. Female barn swallows choosing a mate for the second time may also have at their disposal males with longer tails because older birds tend to arrive earlier than younger barn swallows.

The most obvious way to adjust for these potential errors in the estimate of repeatability of the female mate preference is to rank a chosen male relative to available unmated males at the breeding site when the female made its choice. This adjustment removes the two biases listed above. It was then found that the relative rank of male mates in terms of tail length was positively correlated in subsequent years (Fig. 5.1). The female mate preference had a statistically significant repeatability of 0.57 (SE = 0.11[2]; Møller 1993*d*), and this clearly indicates that female barn swallows demonstrate consistency in their mate choice among different mate choice events. There was no significant repeatability in female mate choice with respect to other morphological characters of male barn swallows (Møller 1993*d*).

In conclusion, female barn swallows are consistent in their mate choice, which indicates that they follow a specific rule for obtaining a mate.

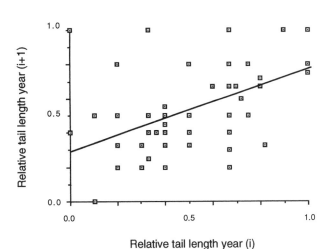

Fig. 5.1 Relationship between relative tail length of male barn swallows chosen by the same female in two different breeding seasons. Relative tail length was the rank for length of the tail of the male chosen relative to the rank of tail lengths available in the breeding site at the time when the female made her mate choice. Adapted from Møller (1993*d*).

5.4 Benefits of mate choice

Traditionally, two major kinds of fitness benefits of mate choice have been postulated, namely, direct and indirect benefits. The potentially great importance of direct fitness benefits has only recently been stressed, and they may be important even for mating systems where females apparently do not obtain anything from males but sperm (Searcy 1982; Reynolds and Gross 1990; Kirkpatrick and Ryan 1991). Even in species with a lek breeding system, females can benefit directly from their mate choice. For example, they may benefit by attending communal displays and thereby avoiding predators during their choice, obtaining mates free from contagious parasites, and acquiring males with superior quality of sperm. It is thus necessary to evaluate carefully all possible fitness benefits of mate choice in order to discriminate between different models of female mate preferences.

5.4.1 *Direct fitness benefits*

The most obvious solution to the question of why female barn swallows choose mates and in that process show a clear preference for the most extravagantly ornamented males is direct fitness benefits. A number of potential direct fitness benefits of mate choice have been suggested:

(1) quality or quantity of sperm;

(2) territory quality;

(3) male parental care;

(4) freedom from contagious parasites;

(5) avoidance of cost of searching for mates; and

(6) avoidance of hybridization.

I shall discuss each of these and finally evaluate their relative importance.

1. *Quality or quantity of sperm.* If males differed even slightly in fertility as a consequence of having reduced numbers of sperm or reduced sperm quality, females would be at a selective advantage if they were able to discriminate between males differing in fertility status (Walker 1980; Gibson and Jewell 1982; Drummond 1984; McKinney *et al.* 1984; Birkhead and Møller 1992). There is very little evidence of total infertility in birds (Birkhead and Møller 1992), but even partial infertility caused by reduced sperm counts could provide a strong selection pressure on female mate preferences. It is perhaps difficult to imagine how male infertility should possibly be reflected in the expression of a male secondary sexual character. A number of different factors including

food deficiency and disease are known to influence male fertility in birds, and these could potentially also influence the expression of male ornaments. Total infertility in the barn swallow is rare as less than 1% of 1289 clutches were completely infertile. This infertility could not directly be partitioned between the two sexes, but one prediction would be that males with long tails only infrequently had totally infertile clutches. There was no relationship between total infertility and male tail length.[3] This does not suggest that females acquire more fertile males by picking a long-tailed mate.

The apparent absence of a negative relationship between complete infertility and male tail length could be caused by extra-pair copulations by mates of short-tailed males masking this effect. This explanation is only likely for barn swallows breeding in colonies, because females of solitarily breeding pairs have never been observed to participate in extra-pair copulations. Yet, the frequency of total infertility was the same among solitary and colonial pairs, and there was no relationship between total infertility and male tail length even among solitarily breeding barn swallows.[4] The effects of extra-pair copulations could thus not account for total infertility being unrelated to male tail length.

Another possibility is that partial rather than total infertility was responsible for the female preference. This, however, does not appear to be the case since partial infertility was unrelated to male tail length among all pairs[5] and among solitarily breeding pairs.[6] In conclusion, there is very little evidence that the female preference for long-tailed males was maintained by selection for fertile males.

2. *Territory quality.* Males of many animal species establish a breeding territory and offer prospecting mates the resources of this territory (Searcy 1982; Alatalo *et al.* 1986). The size of a secondary sexual character could reflect the ability of a male to acquire and defend a breeding territory, and therefore serve as a mate choice cue. This hypothesis assumes that the chief function of breeding territories is male defence of resources essential for successful reproduction. Its validity can be questioned because most intrusions on breeding territories occur during the fertile period of the resident female, and because territory size varies in relation to the fertility status of the resident female (Møller 1987c, 1990b). Barn swallow males establish breeding territories shortly after their arrival. Territories are a few square metres in size and hold no resources but nest sites and perches for roosting. Nest predation is virtually absent and thus is not an important component of territory quality.

If territory quality were important, we should expect settlement sequences of males on the territories to be similar in different years

(Searcy 1979; Yasukawa 1981). The prediction is that arriving males will tend to occupy the same territories first each year if territory quality is important in determining female choice. This prediction rests on the assumption that the quality of territories remains unaltered during the period of study, as seems to be the case in my study area. Correlation analyses revealed small and non-significant relationships between settlement patterns in consecutive years (Møller 1990a). Subsequent analyses based on data from later years gave a similar result. The exclusion of male barn swallows that have already bred once in a study site might be justified, because they usually return to the territory used in a previous year. However, analyses of settlement patterns in sequences of years still gave small and statistically non-significant correlations (Møller 1990a). It is thus unlikely that females prefer particular males because of direct fitness benefits from territory quality.

3.   *Male parental care.* If the expression of the secondary sex trait in males reflects their ability to feed their mate and/or offspring, females would obviously gain a tremendous direct fitness benefit by choosing an effective food provider as a mate (Searcy 1982). This would very much be the case in the barn swallow where males participate both in nest building and in feeding of offspring, although there is no courtship feeding in this species. However, I have assembled a considerable amount of correlational and experimental evidence suggesting that the male contribution to parental care is not positively related to male tail length (Chapter 10). On the contrary, the opposite is the case. Long-tailed male barn swallows contribute relatively less parental care than short-tailed males (see Chapter 10). Females should therefore, if anything, have evolved an aversion to long-tailed males in order to maximize their direct fitness benefits of mate choice in terms of paternal care. It is thus highly unlikely that females gain any direct paternal care benefits from choosing a long-tailed male as a mate. The female preference for long-tailed males during extra-pair copulation attempts also suggests that the male secondary sexual character cannot entirely have evolved because of direct fitness benefits owing to paternal care.

4.   *Contagious parasites.* Ectoparasites and pathogens, such as those causing venereal diseases, which are directly transmitted from one individual to another, probably form a strong direct selection pressure on female mate preferences (Freeland 1976; Hamilton 1990; Clayton 1991). Females that are able to discriminate between infected and uninfected males will be at a considerable selective advantage, if ectoparasites and pathogens constitute a fitness cost for infected individuals. They may also avoid infected males if this reduces the risk of later infection of their offspring (see section 9.5), or if infected males provide less parental care than

uninfected males (see section 9.4). It may be difficult to imagine why females should use the expression of secondary sexual characters as a way of determining infestation by contagious parasites rather than assessing it from direct observations of ectoparasites or from behaviours or activities indicative of male health. Females should always use direct ways of determining the health status of their mate because of inevitable noise and unreliability in indirect assessment.

Female barn swallows seem to estimate the ectoparasite levels of potential mates before making a mate choice since unmated males have more ectoparasites than mated males (Chapters 8 and 9). Thus, ectoparasite load may serve as yet another cue in the mate choice process. Evidence suggests that there is assortative mating with respect to intensities of ectoparasite infestations. The positive correlation between parasite loads of pair members does not arise as a consequence of direct transmission of ectoparasites because (a) the relationships were present from the beginning of the breeding season, and (b) the relationship was present for the bite holes of Mallophaga from the very beginning of the breeding season. These bite holes are produced over long periods (Møller 1991c).

It is possible that female barn swallows gain a direct fitness benefit by avoiding males with high loads of the tropical fowl mite *Ornithonyssus bursa*. Mites are often present on adult barn swallows shortly after arrival at the breeding grounds, and females have to choose among males which differ in their mite loads. The direct fitness consequences of mate choice in terms of effects of mites were demonstrated in an experiment in which mite loads of nests were either reduced by fumigation during egg laying, or increased by mite inoculation during laying, with other nests being kept as controls (Møller 1990c). This experiment thus mimicked the choice of a clean partner (the fumigation treatment), the choice of an infested partner (the mite inoculation treatment), or no choice (the control treatment). There were a number of different fitness consequences of the experimental manipulations, and the annual reproductive success was strongly affected by the treatments. Sprayed nests enjoyed slightly higher reproductive success than controls, which in turn had much higher success than mite inoculated nests (Fig. 5.2). Offspring quality was also affected, and there was an equally large difference in nestling mass between sprayed and control nests as between control and mite inoculated nests (Møller 1990c). This experiment demonstrates that a large fitness gain can accrue as a consequence of choosing the right partner, and that avoidance of infested mates will have much larger fitness consequences than choice of clean partners.

Were infested males also worse parents than uninfested males? There is very little empirical evidence of any relationship between parasite

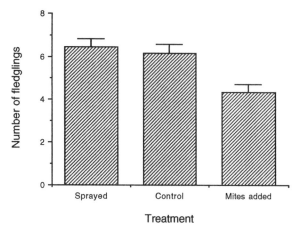

Fig. 5.2 Effect on seasonal reproductive success of experimental manipulation of the loads of tropical fowl mites of barn swallow nests during the laying period of the first clutch. Values are means (SE). Adapted from Møller (1990*d*).

infestation and quality of parental care (Clayton 1991). The role of male barn swallows in providing food for offspring was determined for a number of pairs receiving experimental treatments of mite loads in their first-clutch nest. One assumption of this experiment was that parents as well as nest contents would be infected with mites, and this was verified (Møller 1991*d*). Adult barn swallows with sprayed nests carried fewer mites than birds with control nests, which in turn had fewer mites than ones with nests inoculated with mites. The effects on male parental care of the mite treatment of the nest during the first clutch were determined after statistically controlling for nestling brood size and nestling body mass. These two variables had to be accounted for since the treatment affected the quantity and the quality of offspring. Large broods with heavy nestlings would require more food than small broods with light nestlings. Male feeding rates throughout the nestling period decreased slightly, though significantly, with increasing experimental mite loads (Fig. 5.3), but this was mainly a consequence of reduced brood size and reduced mean nestling mass (Møller 1993*e*). The percentage of all feeding visits by the male did not differ between treatments (Fig. 5.3), neither before nor after controlling for the effects of brood size and mean nestling mass (Møller 1993*e*). The relationship between male provisioning and mite loads of nests was also investigated in nests with natural mite loads. Here it was necessary to control statistically for brood size, mean nestling mass, and breeding date, for the reasons presented above, and determine the relationship between mite loads of nests and absolute and relative male feeding rates. No statistically

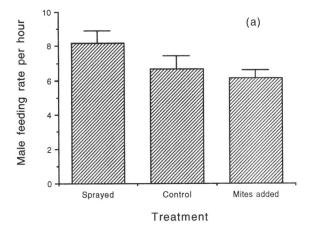

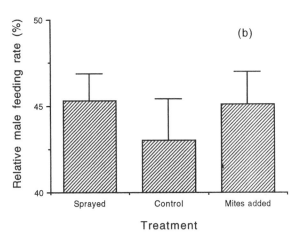

Fig. 5.3 Effect of experimental manipulation of the loads of tropical fowl mites of barn swallow nests during the laying period of the first clutch on (a) absolute male feeding rates of nestlings and (b) relative male feeding rates of nestlings. Adapted from Møller (1993*e*).

significant relationship was found between male provisioning and mite load of nests despite large sample sizes (Møller 1993*e*). Thus tropical fowl mites had little, if any, effect on male parental care in the barn swallow.

In addition, there is some indirect evidence that contagious parasites are unlikely to be particularly important as a selection pressure on female mate preferences. Females should avoid copulating with infested males in order to avoid transmission of contagious parasites and enjoy increased reproductive success. However, this argument can also be turned around as females can transmit contagious parasites to males during copulation. Preferred males are the most likely individuals to be infested with contagious parasites because of their frequent contacts with extra-pair females of generally poor quality. Females searching for direct fitness benefits therefore should not copulate with a preferred male, but rather search for a less preferred male, which is less likely to carry contagious parasites. However, since female barn swallows clearly prefer long-tailed males both as mates and as extra-pair copulation partners (Chapter 4), direct fitness benefits of the mate choice of a parasite-free male are unlikely to be the sole or even an important benefit.

In conclusion, there appears to be some direct fitness to be reaped from choosing mates with low intensities of ectoparasite infestations. These benefits accrue mainly in terms of reproductive success, and to a much smaller extent in terms of paternal care.

5. *Cost of searching for mates.* It can be costly for females to search for mates, and the mate choice may be strongly influenced by this cost (Parker 1983b; Andersson 1986a; Pomiankowski 1987b). Search costs are bound to be high in animal species with low population densities, and if the seasonal conditions for successful reproduction deteriorate rapidly. Female barn swallows, like the females of many other migratory bird species, arrive later than males. They are able to locate males by returning to their previous breeding site or by searching for them in the neighbourhood of their own birth place. Female barn swallows are efficient flyers able to cover large distances in a short time at a low energy cost. Most barn swallows are colonial breeders, and many unpaired male barn swallows often display at the same time in a breeding site. The population density of barn swallows is often high, which suggests that every female independent of its time of arrival, may be able to encounter many potential mates at a low search cost (see also section 5.3.1). Therefore, search costs probably do not play a prominent role in the mate choice process of female barn swallows.

6. *Selection against hybridization.* Female mate preferences may arise as a consequence of hybridization avoidance mechanisms. Females that prefer male phenotypes deviating the most from sympatric closely related species may run little risk of hybridization (Lande 1982; Sanderson 1989). This hypothesis assumes that hybridization will have severe negative fitness consequences for females, and conversely that female fitness may be greatly enhanced as a consequence of having a preference for males with a clearly conspecific phenotype. This assumption is not always fulfilled (Grant and Grant 1992). Barn swallows are known to hybridize with three other hirundine species, namely, house

martin *Delichon urbica*, cliff swallow *Hirundo pyrrhonota*, and cave martin *Hirundo fulva* (Martin and Selander 1975; Martin 1980; Menzel 1984). Hybrids with cliff swallows and cave martins have apparently arisen as a consequence of a recent range expansion of the latter two species caused by the increase in man-made habitats. Hybridization with cave martins and house martins is relatively frequent and may represent a weak but important selection pressure. Hybrids with house martins do not arise as a consequence of a deviant female preference, but because of forced extra-pair copulations (A. P. Møller unpublished), and females thus are hardly able to exert any mate preference in this case. Forced extra-pair copulations may also account for some of the other hirundine hybrids. There is thus little evidence for female mate preferences resulting from selection against hybridization.

In conclusion, a number of direct fitness benefits may potentially arise as a consequence of mate choice in the barn swallow. Only one of these benefits is important, namely, female discrimination against males infested with ectoparasites. Female barn swallows mated to males infested with an haematophagous ectoparasitic mite produced fewer offspring of lower quality in comparison with females mated to uninfested males.

5.4.2 *Indirect fitness benefits*

Indirect fitness benefits of mate choice include:

(1) attractiveness genes of sons;

(2) good genes such as parasite resistance genes and general viability genes;

(3) genetic epistasis and dominance;

(4) social systems; and

(5) mutation pressure on the male trait and the female mate preference.

I shall consider each of these alternatives in an attempt to evaluate their relative importance.

1. *Attractiveness genes.* The Fisher process suggests that the male secondary sexual character and the female preference will coevolve to ever more exaggerated states as soon as the runaway process has been started by genetic drift, an initial good genes advantage, or sensory bias in the sensory systems of the female (Fisher 1915, 1930; Lande 1981; Kirkpatrick 1982; Pomiankowski *et al.* 1991). Contrary to what was previously thought, this coevolutionary process can occur only as a result

of biased mutation or a similar process on the male secondary sexual character (Pomiankowski *et al.* 1991). The indirect fitness benefit of female mate choice is thus the attractiveness of sons to females in the next generation. Similarly, daughters of the high preference genotype may mate to a disproportionate extent with the most extravagantly ornamented males. A second alternative indirect fitness benefit derives from the sexy son hypothesis. This hypothesis suggests that a female mate preference may evolve for attractive males, even if the acquisition of such a mate would result in less assistance in the rearing of offspring and a reduction in fecundity (Weatherhead and Robertson 1979). Females may benefit from choosing attractive mates if the loss in fecundity caused by choice of an attractive male is offset by the advantages of production of more attractive sons. This equilibrium point can be reached in the presence of mutation bias on the male trait because females with stronger preferences acquire mates with larger ornaments and a lowered fecundity caused by the reduction in the contribution by the male (Pomiankowski *et al.* 1991). Both these fitness benefits only apply in the situation where there is biased mutation (or a similar process) on the male secondary sexual character because that is the only condition which allows female choice to be costly at the evolutionary equilibrium (Pomiankowski *et al.* 1991).

How is it possible to evaluate the importance of these fitness benefits in the barn swallow? It is commonly stated that a Fisherian mating advantage is inevitable when there is a genetically based female preference, an attractive secondary sexual character and variance in male mating success (Pomiankowski *et al.* 1991). This should in principle also apply to the barn swallow. However, it is almost impossible to partition the indirect fitness benefits of a female mate preference between advantages of attractiveness genes and good genes.

It is reasonable to assume that female choice is costly in the barn swallow (see section 5.3.1), and the Fisher process can then only exist at an equilibrial state if there is biased mutation (or a similar process) acting on the male secondary sexual character. The majority of mutations are expected to be deleterious and this may particularly be the case for secondary sexual characters that recently have been subject to strong directional selection and thus great exaggeration (Pomiankowski *et al.* 1991). Biased mutation is particularly likely for extravagant ornaments for two reasons (Pomiankowski *et al.* 1991). First, secondary sex traits are greatly exaggerated characters which are likely to be close to their physiological limit of production. It is thus difficult to imagine mutations that improve the quality of such costly displays, but it is very easy to imagine mutations that decrease the quality of the display. Second, extravagant ornaments often demonstrate very complicated structures in terms of design and quality, which may be subject to biased mutation. It is easy to imagine many ways in which a peacock's train can be made less perfect, but it is very difficult to imagine improvements in the design of this elaborate structure. The history of selection on a trait may determine the extent of bias in mutations (Mukai 1964). Traits that have been subject to intense directional selection are more likely to demonstrate mutation bias because any change is likely to result in deterioration of the phenotypic expression. The mean male trait is therefore always displaced from the optimum, and females can continually benefit from exerting a choice of males with the most extreme expression of their sex trait. Processes similar to biased mutation may result in an average deterioration of male secondary sex traits. For example, differential migration of particular phenotypes and fluctuating asymmetry may generate similarly unattractive male sex traits (Seger and Trivers 1986; Møller 1993*f*). Biased mutation is thus likely to lead to less well-ornamented males which are at a selective disadvantage because of their unattractive secondary sex traits.

There is virtually no data on biased mutation in secondary sex traits. I have exploited one field situation for determining whether biased mutations were predominant in secondary sex traits. The radioactive contamination at the Chernobyl nuclear power plant in Ukraine is a potential source of mutation since 1986. Barn swallows are common breeding birds around the power plant as well as in other parts of Ukraine. I investigated the tail ornaments and other morphological characters of barn swallows from two areas near Chernobyl and three uncontaminated areas far from Chernobyl in an attempt to determine whether major changes that potentially could be attributed to mutation had taken place in morphology (Møller 1993*f*). There were two changes in tail morphology; (a) tail deformations, and (b) fluctuating asymmetry in tail length.

First, a number of males had clearly deformed tail feathers in which barbules did not fuse, and such tail feathers clearly looked brush-like (Møller 1993*f*). This abnormal morphology was only recorded in the Chernobyl area, and it was not seen in any of the control areas, among barn swallows in several other parts of Europe, or in various European museum collections. This strongly suggests that this was a novel morphological change associated with the Chernobyl area. The female mate preference can be determined from the relationship between breeding date and male tail length (see section 4.3.2), and the Ukrainian samples showed the common strongly negative relationship between breeding date and male tail length. However, males with brush-like tails were at a selective disadvantage because they bred considerably later than predicted from their tail length (Fig. 5.4; Møller 1993*f*). Therefore,

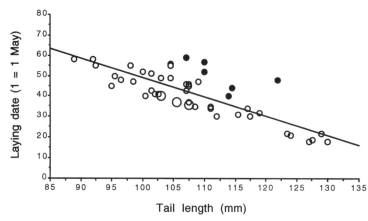

Fig. 5.4 Breeding date of male barn swallows from Chernobyl in relation to their tail length (mm). Males with aberrant brush-like tails (dots) bred significantly later than males with normal tail feathers (open circles). Laying date 1 = 1 May. Adapted from Møller (1993*f*).

some male barn swallows developed aberrant tail morphology in the contaminated area, and this alteration apparently resulted in a delay in reproduction.

The second change was the effect of radiation on fluctuating asymmetry. Fluctuating asymmetry in tail length of barn swallows demonstrates clinal variation, with increasing levels of asymmetry at higher latitudes (Møller 1993*m*). Mean asymmetry in tail length of male barn swallows from the Chernobyl area was more than twice the value predicted from the geographic pattern of clinal variation (Møller 1993*m*). Female tail length did not demonstrate an increased level of fluctuating asymmetry in the Chernobyl area. Females apparently responded to the increase in fluctuating asymmetry in male tail length because males with highly asymmetric tails bred later than symmetric males (Møller 1993*f*).

These changes in morphology of male barn swallows were thus as predicted by the hypothesis of biased mutation. Altered morphology affected the secondary sexual character rather than other morphological traits, the quality of the secondary sexual character deteriorated as a consequence of the change, and females responded by avoiding males with the altered morphological characters. This result suggests that biased mutation may have been responsible for the changes in morphology of male barn swallows, but does not tell us whether the changes in male morphology could have influenced a Fisherian process.

2. *Good genes.* Females may acquire mates with so-called good genes by choosing the most elaborately ornamented males. This mechanism could potentially work in two different ways. First, the male secondary sexual character could be seen as a revealing or a conditional handicap which has evolved to reflect clearly the ability of males to resist the effects of parasites and pathogens (Hamilton and Zuk 1982; see section 9.6). More resistant males will be healthier and in better condition, and this should be reflected in the expression of their sex traits, or males should be better able to develop extreme ornaments or perform elaborate displays at extreme rates in the absence of parasites. Male ornaments could thus directly reflect the possession of genes affecting resistance to parasites and pathogens. This hypothesis is attractive because parasites are ubiquitous and because continuous coevolution between parasites and their hosts may make a female preference for highly ornamented males, and thus indirectly for resistance genes, continuously beneficial. Females with a preference for the most extravagantly ornamented males will tend to mate with resistant males, and the mate preference and the resistance genes will thereby become linked. The male sex trait will subsequently evolve as a result of evolution of more extreme female preferences.

The Hamilton–Zuk hypothesis thus predicts:

(a) a genetic correlation between resistance genes and the female mating preference; and

(b) an evolutionary increase in parasite resistance as a consequence of the female preference (Kirkpatrick and Ryan 1991).

More limited predictions include:

(c) a relationship between the expression of the male sex trait and parasite loads;

(d) female choice of males with fewer parasites as a consequence of their preference for the most elaborately ornamented male; and

(e) resistance genes being related to the expression of the male trait (reviews in Møller 1990*d*; Clayton 1991; Zuk 1991).

These predictions have all been evaluated for the barn swallow in Chapter 9. There is some indirect evidence for predictions (c)–(e). The expression of the male sex trait is related to the parasite load, females acquire mates with fewer ectoparasites as a consequence of their preference for the most elaborately ornamented males, and resistance to parasites is heritable and directly related to the expression of the secondary sexual character. Female barn swallows may also acquire a direct fitness benefit as a result of their choice of parasite-free mates (see the evidence in section 5.4.1), and it is difficult to partition the total fitness benefit from choice of parasite-free males into these direct and indirect components. Even if females acquire a direct fitness benefit from choice of relatively uninfested males, they simultaneously obtain an indirect benefit because males with low intensities of parasite infestations will tend to have resistance genes.

The second possibility of a good genes effect is from choice of males with general viability genes (Trivers 1972; Andersson 1982a, 1986a; Charlesworth 1987; Pomiankowski 1988; Rice 1988; Heywood 1989). Only males in good body condition with genes for high viability may be able to produce the most extravagant secondary sex traits, and a female preference for the most ornamented males will result in acquisition of viability genes for their offspring. This hypothesis thus predicts that (a) male viability is directly correlated with the expression of the secondary sex trait, and (b) offspring viability is directly related to the expression of the ornament of the male parent.

The prediction that male ornament size is a reliable predictor of viability was tested in a population of barn swallows (Møller 1991e). Tail length of males was compared between survivors and non-survivors. There was a highly significant difference between the size of the tail ornament of survivors and non-survivors (Fig. 5.5; Møller 1991e). This difference in tail length also held for unmated males, yearlings, and older males (Fig. 5.5). There was no relationship between male viability and

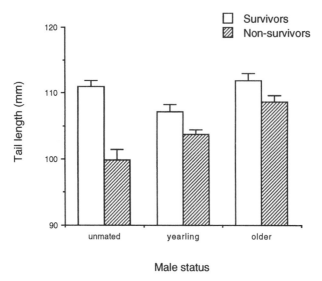

Fig. 5.5 Comparison of tail length (mm) of surviving and non-surviving male barn swallows. Values are means (SE). Adapted from Møller (1991e).

other morphological characters, and the relationship between tail length and survival prospects could thus not be invoked to result from a correlated response to selection on another character. Finally, there was no relationship between tail length and survival prospects in female barn swallows (Møller 1991*e*). Differences in tail ornamentation of male barn swallows therefore reflect their viability in terms of future survival prospects. This may have important indirect fitness consequences for female mate choice as females prefer the most extravagantly ornamented males. Slightly similar, although less detailed, results have been obtained for a few other animals (Watt *et al.* 1986; von Schantz *et al.* 1989; Alatalo *et al.* 1991; Hill 1991; Norris 1993).

Although the correlation between the expression of the tail trait of male barn swallows and their survival prospects superficially may indicate that females could acquire a good genes benefit by mating with long-tailed males, this is not necessarily so. The relationship between male tail length and viability depends on both the direct effects of tail length on viability (which are negative (Møller 1989*a*; Møller and de Lope 1993)) and the positive effects of condition on viability (Zeh and Zeh 1988). A negative relationship between male tail length and viability can easily be obscured by a positive correlation between the male trait and condition. Such masking effects of condition on the costs of a male sex trait can most readily be determined in perturbation experiments where the male trait is displaced from the value chosen by the bird (Møller 1989*a*; Møller and de Lope 1993). The important message is, however, that the expression of the male trait may be positively related to viability even when the female preference has evolved in direct response to the fecundity effects of the male (Price *et al.* 1993). This line of argument is perhaps unlikely to apply in the case of the barn swallow because females have to pay a cost in terms of increased parental effort when mated to the most preferred males (de Lope and Møller 1993).

The second prediction of the good genes hypothesis is that offspring viability is directly reflected by the expression of the secondary sex trait of their father. This prediction was evaluated in the barn swallow by determining the relationship between longevity of sons and the tail length of their putative father. There was a strong positive relationship between offspring viability and tail length of their fathers (Fig. 5.6; Møller 1993*n*). Sons therefore also resembled their fathers with respect to longevity, and there was a strongly positive relationship between longevity of sons and that of their fathers (Fig. 5.7). This relationship could arise as a consequence of male tail length reliably reflecting male viability and male viability having a genetic component. However, a number of maternal or common environment effects could also affect both the expression of male tail ornaments and offspring viability. These

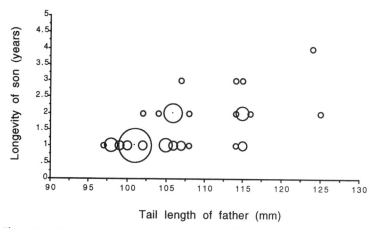

Fig. 5.6 Longevity of male barn swallow offspring (years) in relation to the tail length (mm) of their fathers. Circles of increasing size represent one, two, three, four, five and six observations, respectively. Adapted from Møller (1993*n*).

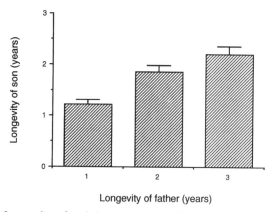

Fig. 5.7 Longevity of male barn swallow offspring (years) in relation to longevity of their fathers (years). Values are means (SE). Adapted from Møller (1993*n*).

factors include a common breeding environment, the amount of parental care, rearing conditions, and the distribution of male phenotypes across colonies of different sizes. None of these most obvious maternal or common environment effects could account for the relationship according to correlation analyses, and the genetic reason for the relationship thus remains a possible candidate (Møller 1993*n*). Several mechanisms could result in parent-offspring resemblance in longevity. One possibility is that small genetic differences among barn swallow nestlings, for example in resistance to parasites, may become amplified into large viability differences among adults, and this could account for the resemblance in longevity between fathers and sons. Only three previous studies have indicated that viality benefits may accrue to choosy females (Partridge 1980; Watt *et al.* 1986; Norris 1993).

3. *Genetic epistasis and dominance.* The exact nature of any genetic basis for female mate preferences is not well known. If the preferences and the male sex traits are quantitative characters with a certain degree of additive genetic variance, then the evolution of the mate preference and the secondary sexual character may follow from the model of Lande (1981) in the case of uncostly mate preferences and the model of Pomiankowski *et al.* (1991) in the case of costly mate preferences. However, the evolution of mate preferences and male secondary sex traits is changed if a small number of major loci and alleles is involved (Gomulkiewicz and Hastings 1990; Barton and Turelli 1991). The evolutionary stability of the exaggeration of the male sex trait depends on the dominance of the female preference, and the evolutionarily stable expression of a male sex trait may change dramatically as the frequency of a female preference gene increases (Gomulkiewicz and Hastings 1990; Heisler and Curtsinger 1990). A population with a preferred male character either fixed or absent would not demonstrate a switch in stability from low expression of the male trait to high expression and then back again as the frequency of a female preference gene is increased (Gomulkiewicz and Hastings 1990). The nature of the mate preference of female barn swallows is unresolved, and the importance of genetic epistasis and dominance for the indirect attractiveness benefits of a female mate preference thus remains unknown.

4. *Social systems.* The evolutionary equilibria of the female preference and the male sex trait under a Fisherian process depend on the behavioural model of female choice (Seger 1985; Pomiankowski *et al.* 1991). The first models of the Fisherian process assumed that females were able to choose from an effectively infinite pool of available males, and that mate choice therefore did not carry any cost to females (O'Donald 1962, 1967, 1980*a*; Lande 1981; Kirkpatrick 1982). The conditions for evolutionary equilibria change dramatically when these highly unlikely assumptions are relaxed. For example, Seger (1985) demonstrated that the evolutionary equilibria become unstable when the pool of available males changes from an infinite number to a small number of males such as those attending a lek. If the equilibria are unstable, the sexual selection process becomes an 'all-or-nothing runaway' exaggeration of the male trait and the female preference (Seger 1985). Instability may, for example, arise as a consequence of female preferences generating frequency-dependent

male fitnesses. For example, if females use a best-of-n-males rule when making their mate choice, positive frequency-dependence will arise when n is small, as is probably always the case in nature. When males without a conspicuous secondary sex trait are very common, only few mating sites will contain a male with a conspicuous ornament, and the average fitness of males without the ornament will be close to that of males with ornaments. The situation is completely different when ornamented males are common because most sites will then contain at least one ornamented male, and males without ornaments will consequently have much lower fitness than ornamented males. The stability and the evolutionary dynamics of the sexual selection process are therefore strongly dependent on the social system under which the mate choice process takes place (Seger 1985). Similar conclusions have been reached for the so-called line of stable evolutionary equilibria claimed to be generated by the runaway process (Pomiankowski *et al.* 1991). As soon as the highly unlikely assumption of uncostly female mate preferences is relaxed, the line of stable equilibria disappears. A stable equilibrium point can only be generated if there is biased mutation on the male sex trait, as previously discussed. Differences in the social system of animals may therefore greatly influence the evolutionary dynamics of the sexual selection process.

The mate choice situation for the barn swallow obviously mimics that envisaged by Seger (1985), particularly during choice of extra-pair copulation partners. However, the social system of the barn swallow will probably result in unstable equilibria of the Fisher process, and the indirect fitness benefits of mate choice in terms of sexual attractiveness of sons will thus only be present during brief bursts of runaway exaggeration interspersed by long periods of no evolutionary change in the male sex trait and the female preference.

5. *Mutation pressure on the male trait.* Models of female mating preferences have assumed that the expression of the male trait is subject to randomly directed mutation (Lande 1981; Kirkpatrick 1982). An extravagant male trait is likely to have a history of directional selection, and the probability that a mutation results in improvements in expression is much less than the probability of deterioration (Mukai 1964). Biased mutation is likely to affect the expression of the secondary sex trait, and the average male will thus always be below its optimal phenotype (Bulmer 1989a; Pomiankowski *et al.* 1991). Costly female mate preferences will be feasible at the evolutionary equilibrium if the male trait is subject to biased mutation (Pomiankowski *et al.* 1991). Biased mutation or a similar process is in fact a necessary condition for maintaining costly preferences (Pomiankowki *et al.* 1991), and such preferences may

be ubiquitous (Parker 1983*b*; Pomiankowski 1987*b*). Mutation pressure on the male trait is thus necessary for the evolution of a female mate preference based on the indirect fitness benefits of attractiveness genes. Possible examples of mutation bias in the secondary sex trait of male barn swallows and their fitness consequences have been described previously in this section.

In conclusion, there is some correlational evidence of genetic benefits arising from female mate choice. Preferred males have genes for resistance to one species of ectoparasite that has severe negative fitness effects on their barn swallow hosts. There is a positive relationship between longevity of fathers and sons and this is related directly to the expression of the preferred secondary sex trait. Some of the hypotheses about indirect fitness benefits of female mate choice are difficult to evaluate and their relative importance awaits further study.

5.4.3 *Other mechanisms*

Three other mechanisms could potentially account for the evolution of female mate preferences. These are:

(1) random genetic drift;

(2) group selection, and

(3) mutation pressure on the female preference.

1. *Random genetic drift.* This hypothesis suggests that the female preference arose as a consequence of drift in small populations (Lande 1981). It is fully feasible that this could be the origin of a mate preference, but it is difficult to imagine that many of the extreme preferences seen today can be accounted for by genetic drift. Extreme female preferences could hardly arise *de novo*, and it is unlikely that the evolutionary increase in female preference could be accounted for by random genetic drift. Mate preferences are often costly to females (Parker 1983*b*; Pomiankowski 1987*b*), and selection against a strong mate preference would result in a reduced cost. Extreme mate preferences are thus unlikely to be maintained by drift alone because of their costs, and direct or indirect fitness benefits of mate choice have to be invoked in order to explain their evolution and particularly their maintenance. Random genetic drift may, however, cause a population to evolve in an unpredictable direction at an accelerated rate (Lande 1981).

2. *Group selection.* The group selection hypothesis states that where genetic or developmental constraints on the expression of a secondary sex trait prevent male and female fitnesses from being maximized simultaneously,

female preferences should evolve to favour males whose expression of the male trait confers low fitness on males but high fitness on females (Seger and Trivers 1986). Migration between partially isolated demes due to selective diffusion of genotypes can move the expression of the male secondary sex trait and the female preference along a line of equilibria in the direction that increases average female fitness while lowering average male fitness (Seger and Trivers 1986). This would result in continuous evolution of a secondary sex trait which promotes fitness. This group selection model suggests a line of equilibria that can only be maintained in the presence of biased mutation on the male trait (or some similar process such as migration, as suggested by Seger and Trivers (1986)). Group selection remains a controversial theme in evolutionary biology, and the relative importance of this mechanism in the evolution of mate preferences remains unknown. This model is not easily evaluated for the specific case of the barn swallow, which does not form demes. Rather the entire barn swallow population can be seen as a large, continuous population with considerable exchange of individuals due to extensive natal dispersal. It is therefore highly unlikely that group selection, as envisaged by Seger and Trivers (1986), can account for the strong mate preference in female barn swallows.

3. *Mutation pressure on the female mate preference.* Lande (1981), Kirk-patrick (1982), and Seger (1985), modelling the Fisher process, have obtained a line of evolutionary equilibria for the male trait and the female preference up and down which the system can drift at random. The line of unstable equilibria arises as a result of the restrictive assumptions of the models, and even small deviations from these assumptions result in the line being reduced to a stable equilibrium (or equilibria). One such common assumption of these models is that mutations on the female preference causes no directional change in the phenotype (for example, Lande 1981). This assumption is highly unlikely because any trait subject to intense directional selection will almost invariably demonstrate biased mutation towards lower expressions of the character (Mukai 1964). Mutation on the female mate preference is therefore likely to be biased (Bulmer 1989a; Pomiankowski *et al.* 1991). A two-locus population genetic model suggests that mutation on the female preference results in the line of unstable evolutionary equilibria being replaced by a stable equilibrium (or equilibria). The female mate preference will probably find its optimum level influenced by mutation, selection and genetic drift, and the male secondary sexual character will be dragged along with the female mating preference along the line of equilibria (Bulmer 1989a).

In conclusion, it is impossible directly to determine the influence of random genetic drift, group selection and mutation on the female mate preference in the evolution or the maintenance of the strong mate preference in the barn swallow. However, both drift and biased mutation are likely to influence the evolution of female mating preferences.

5.5 How can heritability of fitness be maintained?

Female barn swallows exert a strong and continuous selection pressure on the length of male tails, and there is some evidence that females gain a net indirect fitness benefit from this selection. How can this be the case? It has commonly been argued that continuous directional selection on a trait results in depletion of the additive genetic variance, and the time until depletion is determined by the number of genes involved and the intensity of selection. The heritability of a character is thus assumed to be negatively related to the intensity of selection and therefore the relative importance of the character in determining fitness (Fisher's fundamental theorem; Fisher 1930). There is some empirical support for this prediction that the heritability of characters is negatively related to their relative importance in determining fitness (Gustafsson 1986; Mousseau and Roff 1987). Is it possible that female barn swallows acquire indirect fitness benefits from their mate preferences? This is possible because traits even closely associated with fitness usually demonstrate small degrees of heritability (Gustafsson 1986; Mousseau and Roff 1987). There are a number of solutions to the problem of continuous directional selection and heritability of fitness traits, since fitness components can remain heritable in several different ways (review in Charlesworth 1987; see also Price and Schluter 1991). These solutions all potentially apply to the problem of whether females successfully can select mates of high genetic viability, and why barn swallow tails do not become longer even during a short time-scale as a result of this selection process.

How could genetic variation in fitness be maintained despite strong directional selection? There are at least five possibilities:

(1) frequency-dependent selection;

(2) mutation-selection balance;

(3) temporal variation in fitness;

(4) the Red Queen; and

(5) selection on the environmental component of a trait

(Haldane and Jayakar 1963; Gillespie 1973, 1992; Clarke 1976; Felsenstein 1976; Hamilton 1980; Hamilton and Zuk 1982; Turelli 1984; Rice 1988; Bulmer 1989b; Alatalo *et al.* 1990; Price and Schluter 1991; Price *et al.* 1993).

First, genetic variance in fitness can be maintained by frequency-dependent selection (Clarke 1976; Felsenstein 1976; Hamilton 1980, 1986; Hamilton and Zuk 1982). This hypothesis posits that host–parasite interactions may promote the continuous generation of new alleles which determine resistance specificity among hosts and virulence among parasites (Hamilton 1980, 1986; Hamilton and Zuk 1982). Any rare host genotype will be at a selective advantage because it will be more resistant than the average genotype in the population (Haldane 1949). Frequency-dependent selection and mutation may thus lead to single gene substitutions which affect the host–parasite interaction. Genes affecting fitness can remain in a state of continual high heritability if they coevolve with other biotic players such as parasites and pathogens or predators (Hamilton 1986). Additive genetic variance in fitness is thus continuously replenished as a result of aggressive biotic interactions. Frequency-dependent selection due to biotic interactions can be viewed as a special case of genetic variance being maintained by temporal variation in fitness.

Second, genetic variance can be maintained by mutation-selection balance (Mukai *et al.* 1974; Turelli 1984; Kondrashov 1988; Rice 1988; Bulmer 1989b). One of the main sources of heritability of fitness is mutation because all genes affecting functionally significant traits must at least sometimes produce deleterious alleles by mutation. Mutation can affect both the additive and the dominance genetic variance in proportion to the mutation rate, but for mildly deleterious mutations most of the variance will be additive (Mukai *et al.* 1974). The total amount of additive fitness variance arising from mutation-selection balance is small, but significant (Charlesworth 1987; Kondrashov 1988; Rice 1988).

Third, genetic variance can be maintained by temporal variation in fitness (Haldane and Jayakar 1963; Gillespie 1973, 1992). If there is temporal variation in fitness, additive genetic variance will be produced every generation because allele frequencies will never reach equilibrium values (Eshel and Hamilton 1984). Temporal variation in fitness has also been used as an argument by Kodric-Brown and Brown (1984) and Andersson (1986a) as a way of maintaining genetic variation in viability, because most natural populations will probably face changing environments and therefore parallel changes in fitness and selection gradients. The frequency-dependent selection hypothesis in the broadest sense is to some extent a special case of temporal variation in fitness, and the same applies to the Red Queen hypothesis mentioned below.

Fourth, genetic variance in fitness can be maintained by the Red Queen mechanism (Van Valen 1973; Cooke *et al.* 1991). This hypothesis states that all selective improvements in fitness will eventually be nullified because the environment, which includes both biotic and abiotic components, constantly deteriorates from an individual's point of view. Individuals thus have continuously to improve genetically in order just to maintain the current expression of their phenotype. For example, parasites and pathogens continuously develop new means of attacking their hosts, plants develop new anti-herbivory defences, animals evolve new ways of evading attack by their predators, and conspecifics develop new ways of exploiting the same resource as a competitor. Even if there is continuous selection for increased fitness due to a female mate preference, this selective force may never result in a response to selection because the environmental conditions continuously deteriorate. Fitness would be much higher given the degree of heritability and the intensity of selection if it were not for the deterioration of the environment. The mean fitness of a population from one generation to the next does not increase under the Red Queen scenario and a positive correlation in parent and offspring fitness can therefore be maintained.

Fifth, the genetic variance in fitness can be maintained because selection is only or mainly on the environmental component of the trait (Alatalo *et al.* 1990; Price and Schluter 1991; Price *et al.* 1993). Total phenotypic variance of a trait can be partitioned into a genetic, an environmental and a number of other components as stated in section 6.2.1. Phenotypic selection is usually assumed to affect both the genetic and the environmental component of a phenotypic character, but this is not necessarily the case (Alatalo *et al.* 1990; Price and Schluter 1991; Price *et al.* 1993). Let us assume that viability in barn swallows is determined partially by additive genes and a number of environmental components, among these condition. The main determinant of adult viability may be related to the nutritional state of individuals as nestlings, and this could to a large extent be determined by random variation in weather conditions. The environmental component of phenotypic variance in viability could potentially be very large because of variance in weather conditions. Adult barn swallows could vary in viability only for environmental reasons, and variance in male mating success could lead to directional sexual selection on the environmental component of the expression of the viability trait. Directional viability selection would thus be on the environmental component of the viability trait. The genetic variance in viability genes in linkage disequilibrium with the male ornament would remain unaffected by phenotypic selection, and the response to selection would be zero if selection only acted on the environmental component. It is perhaps unlikely that viability is entirely determined by environmental effects, but selection on an additive genetic component would tend to reinforce the effects of selection on the environmental component (Price *et al.* 1993).

A similar line of argument could be developed for the maintenance of additive genetic variance in the male sex trait. Most or all of the phenotypic selection on the male trait could be on the environmental component of the character, and the additive genetic variance in the character would thus only be depleted slowly or not at all. Viability selection during the first months of life would be unrelated to the expression of the tail ornament of male barn swallows, which does not develop until the first moult in the winter quarters.

There is a considerable amount of evidence that currently suggests that there is additive genetic variance in fitness (see review in Charlesworth 1987). This means that the additive genetic variance has not been depleted as a result of continuous directional selection. The different hypotheses for the maintenance of genetic variance in fitness of male barn swallows cannot easily be evaluated. It is straightforward to make arguments in favour of each of these alternatives, but it is much more difficult to carry out specific tests in the field.

5.6 Summary

Female mate preferences in the barn swallow are costly as shown by:

(1) females sometimes spending many days searching for mates, days which could otherwise be used for advancement of breeding date;

(2) female encounters with extra-pair males holding contagious parasites; and

(3) females spending more effort on rearing offspring if their mate is of a preferred phenotype.

These costs have to be balanced by considerable fitness benefits. Female barn swallows have consistent mate preferences from year to year, and this suggests that there could be a genetic basis for the mate preference.

The various potential fitness benefits of the female mate preference for long-tailed male barn swallows were carefully evaluated (summarized in Table 5.1). The only direct benefit of relevance for the maintenance of the female mate preference was avoidance of males with contagious parasites, since the quality and quantity of offspring was considerably reduced by the presence of an haematophagous mite. This direct benefit may, however, have been nullified by a direct fitness cost of the female preference. Female barn swallows mated to the most preferred males invested relatively more parental effort into rearing of offspring than females mated to less preferred males, and this may have resulted in a considerable cost.

Table 5.1. Summary of direct and indirect fitness benefits of mate choice to female barn swallows.

Mechanism	Relative importance
Direct benefits	
Quality or quantity of sperm	Unimportant
Territory quality	Unimportant
Male parental care	Important, but selects against female preference
Contagious parasites	Important
Cost of searching for mates	Unimportant
Selection against hybridization	Unimportant
Indirect benefits	
Attractiveness genes and the Fisher process	Probably important
Good genes	
Resistance genes	Important
General viability genes	Probably important
Genetic epistasis and dominance	Unimportant
Social systems	Unimportant
Other mechanisms	
Random genetic drift	Probably initially important
Group selection	Unimportant

Indirect fitness benefits of the mate preference must have been present because of the strong, non-random extra-pair copulation success of male barn swallows. It is highly unlikely that females obtained any direct fitness benefits by engaging in extra-pair copulations. For example, extra-pair males never provided parental care or territorial resources to extra-pair females, and the evidence for variation in male fertility is at the best very weak. Female barn swallows run a considerable risk in terms of acquisition of contagious parasites and venereal diseases by copulating with extra-pair males, which may have become infected by other females as a consequence of the female preference. Females should thus prefer safe males not engaged in extra-pair copulations and avoid preferred males, if they attempted to increase their direct fitness benefits. Indirect fitness benefits of potential importance include attractiveness genes, good genes for resistance to parasites and general viability genes. Both direct and indirect fitness benefits may therefore play a role for the maintenance of the preference for long-tailed males in female barn swallows (Table 5.1).

1. One-way ANOVA, $F = 1.45$, $df = 38,53$, $P = 0.11$, $R = 0.18$, SE = 0.16.
2. One-way ANOVA, $F = 3.75$, $df = 38,54$, $P < 0.001$.
3. First clutches: $r = 0.05$, $n = 496$, NS, second clutches: $r = 0.01$, $n = 304$, NS.
4. First clutches: $r = 0.00$, $n = 33$, NS, second clutches: $r = 0.00$, $n = 11$, NS.
5. First clutches: $r = 0.04$, $n = 496$, NS, second clutches: $r = 0.01$, $n = 304$, NS.
6. First clutches: $r = 0.02$, $n = 33$, NS, second clutches: $r = -0.01$, $n = 11$, NS.

6

Determinants of tail ornament size

6.1 Introduction

Extravagant secondary sexual characters, such as the tail of male barn swallows, may always be subject to intense directional sexual selection. It would obviously be advantageous for any male to grow an even longer tail, provided that the benefits of doing so exceeded the costs. What determines the size of an ornament of a particular individual? Ornaments, like any other morphological character, are probably determined by the effects of genes, the environment, genetic dominance effects and a number of interaction terms. The relative importance of the various factors generating morphological variation can be determined by comparison of the size of ornaments grown by the same genotypes in different environments and by comparison of the phenotype of relatives, as described in section 6.2.

Ornaments are extremely variable in size compared with other morphological characters which usually vary within very narrow limits. These patterns can be explained in terms of (1) the recent history of selection regimes affecting different morphological characters, and (2) the male trait being a reliable condition indicator. While ornaments have been subject to very intense directional selection, there is a much stronger component of stabilizing selection on ordinary morphological traits. Extreme phenotypic variability also indicates that the development of a specific character is not under rigid genetic control and this opens up for large amounts of another kind of morphological variation, namely, fluctuating asymmetry which is random variation between the two sides of the body. It is generally considered to be difficult to produce a left and a right identical copy of the same trait, but it is particularly difficult to produce two copies of a large and extravagant trait. Ornaments therefore demonstrate higher degrees of fluctuating asymmetry than ordinary morphological characters not subjected to strong directional selection. Alternatively, the evolution of variability in male traits may be directly related to the evolution of condition-dependence

(Grafen 1990*a*; Price *et al.* 1993). When a male sex trait becomes more and more condition-dependent, this should automatically give rise to an increased variability. As the male trait subsequently becomes more variable as a result of condition-dependence, a relatively weaker female mate preference is needed to obtain a mate in good condition. Selection on the male trait and the female mate preference may thus become weaker. These two explanations for the extreme variability in male traits are described in section 6.2.

The extreme degree of variation in ornament size may indicate that it is under great environmental influence. This influence can be partitioned into a number of factors, which affect the expression of the ornament and thereby reflect the overall condition of the individual. The operation of some such environmental factors has been documented by carefully following the same individuals in different years and by comparing the tail length of males subject to different environmental conditions, as analysed in sections 6.3 and 6.4.

How large an ornament should an individual grow? This is basically a problem of optimizing costs and benefits of an ornament, as discussed in section 6.5. If the cost and benefit functions vary among individuals, the result should be variation in optimum ornament size. If the cost function of low quality males increases more steeply than that of high-quality males, individual males of low quality should develop small ornaments while individuals of high quality should develop large ornaments. The presence and the shape of these cost and benefit functions can be determined from experimental manipulations of ornament size and careful measurement of subsequently developed ornaments.

Ornaments have at least two different traits, namely, a mean linear term and a variance term, which consists of variation between the two sides of the body. Variation in the magnitude of both these terms can often easily be assessed visually by humans, and conspecifics, according to experimental evidence, are also able to do so. Males growing an ornament therefore have to optimize both ornament size and symmetry in order to maximize attractiveness and hence mating advantages.

Given that tail ornaments in barn swallows are heritable characters subject to considerable directional selection, why is there no response to selection in ornament size? This question is considered by first checking whether there is a micro-evolutionary increase in ornament size, and then proposing various solutions to the surprising absence of evolutionary response to selection.

6.2 Components of morphological variation

6.2.1 *Theory*

Tail length in barn swallows like many other linear morphological traits is probably a trait with polygenic inheritance. According to standard quantitative genetics, the phenotypic variance in a trait such as tail length can be described by the sum of a number of variance components (Falconer 1981):

$$V_P = V_G + V_E + V_{G \times E},$$

where $V_G = V_A + V_D$.

V_G is the genetical variance, V_E the environmental variance, V_A the additive genetic variance, V_D the dominance variance, and $V_{G \times E}$ the genotype-environment interaction. Additional interaction components can easily be added by substituting the different genetic variance components in the first equation. For practical reasons we are particularly concerned with the variance terms V_G, V_E and $V_{G \times E}$, which account for the genetic variance, the various causes of environmental variance and the genotype-environment interaction.

The phenotypic variation of secondary sexual characters is extremely large compared with that of ordinary morphological traits. This was first realized by Alatalo *et al.* (1988) for feather ornaments in a number of different bird species which showed coefficients of variation sometimes larger than 10%. Ordinary morphological characters not directly subject to strong directional sexual selection usually have coefficients of variation in the order of two or three per cent (for example, Grant and Price 1981; Alatalo *et al.* 1988). The high degree of variation in avian feather ornaments has subsequently been confirmed for a number of other kinds of sex traits (for example, Barnard 1991; Møller and Höglund 1991; Møller 1992*e*). These two kinds of morphological variation beg an evolutionary explanation. Two simple explanations for this bimodal distribution of phenotypic variance among morphological characters are (1) the prevailing selection regimes, and (2) selection for condition-dependence (Møller and Pomiankowski 1993*b*; Andersson 1982*a*; Parker 1983*b*; Grafen 1990*a*; Price *et al.* 1993).

First, ordinary morphological characters such as tarsus length in birds have generally been subject to strong stabilizing selection whereas secondary sexual characters in addition to stabilizing viability selection are also subject to strong directional sexual selection (Fig. 6.1). This is clearly demonstrated by the often strongly marked differences in secondary sexual characters between congeners, in contrast to the similarity in ordinary morphological characters. The prevailing patterns of selection affect the relationship between phenotype and fitness, but also the control of development of the two kinds of characters. Development of characters subject to strong

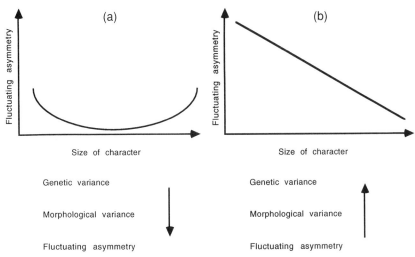

Fig. 6.1 The effects of prevailing selection regime on developmental stability and phenotypic variance of (a) an ordinary morphological character (e.g., tarsus length), and (b) a secondary sexual character. Adapted from Møller and Pomiankowski (1993*a*).

stabilizing selection is strictly controlled, and this is particularly the case for extreme phenotypes which are at a strong selective disadvantage. Any genetic modifier that controls the development of extreme phenotypes will be selectively highly advantageous and contribute to the control of development of extremes (Fig. 6.1).

The situation for secondary sex traits is completely different because of the operation of directional sexual selection. Genetic modifiers, which originally may have controlled the development of the trait before its exaggeration due to sexual selection, have become highly disadvantageous for the largest phenotypes, because males with such modifiers will be unable to develop sex traits as large as males without modifiers. This will affect the sexual selection component of male fitness which will be inversely related to the number of genetic modifiers present. The developmental control of a secondary sexual character will be reduced during its evolution as a consequence of persistent directional sexual selection, because then males will be able to channel extra resources into development of ever larger secondary sexual characters (Fig. 6.1).

There is considerable evidence in favour of this scenario. Directional selection experiments on ordinary morphological characters in a number of different kinds of organisms have generally resulted in an increased level of phenotypic variance (MacArthur 1949; Reeve and Robertson 1953; Falconer 1955; Robertson 1955; Clayton and Robertson 1957; review in Møller and Pomiankowski 1993*b*). This is what one should expect when the prevailing selection regime suddenly changes from strongly stabilizing to strongly directional.

The developmental control of morphological variation has effects on both phenotypic variance and the level of fluctuating asymmetry, which are both directly related to the prevailing selection regime. Fluctuating asymmetry arises as usually small, random deviations from perfect symmetry in a character present on both sides of the body (Ludwig 1932; Parsons 1990). Because both sides of the body develop simultaneously under the control of the same genome, the amount of fluctuating asymmetry reflects the ability of individuals to accomplish perfect symmetry. A number of different genetic (for example, inbreeding, homozygosity, and hybridization) and environmental factors (for example, parasites, nutrient deficiency, and pesticides) are known to result in increased levels of asymmetry (review in Parsons 1990). Fluctuating asymmetry therefore becomes an overall epigenetic measure of the ability of individuals to cope with genetic and environmental stress. The prevailing selection regime affects the degree of developmental control, which is the control of development of the mean and the variance (between the two sides of the body) size of a character (Møller and Pomiankowski 1993*b*). Strong directional sexual selection should result in a weaker degree of developmental control and therefore result both in an increased level of phenotypic variance and an increased level of fluctuating asymmetry. This is exactly what is seen in secondary sexual characters. A diverse array of secondary sexual characters demonstrates considerably larger degrees of fluctuating asymmetry than ordinary morphological characters in the same individuals, and the average asymmetry in secondary sex traits is usually a factor five or more larger than in other morphological traits (Møller and Höglund 1991; Møller 1992*e*). The direct coupling between selection regime and degree of fluctuating asymmetry in secondary sex traits has also been confirmed in selection experiments in fruitflies and rats (Thoday 1958; Reeve 1960; Leamy and Atchley 1985).

Second, the evolution of variability in male traits may be directly related to the evolution of condition-dependence (Andersson 1982*a*; Parker 1983*b*; Grafen 1990*a*; Price *et al.* 1993). The expression of secondary sexual characters may reflect the body condition of males even before any evolution by sexual selection. Sexual selection may, however, result in the evolution of condition-dependence by favouring phenotypic plasticity. Condition-dependence evolves if males in poor body condition suffer higher costs for a certain degree of deviation in the expression of their sex trait from the natural selection optimum. Males will therefore be selected to express a different degree of exaggeration of their sex traits relative to their condition. The effect of sexual selection on the phenotypic variance of the male trait depends on the relationship between condition and exaggeration. When this relationship is monotonically increasing, the variance in the male trait and its relationship with condition will increase with the intensity of the female mate preference. The intensity of natural selection on the male trait will, of course, also influence the variability since intense viability or fecundity selection generally will restrict the magnitude of phenotypic plasticity. When a male sex trait becomes more and more condition-dependent, this should automatically give rise to increased variability. When the male trait subsequently has become variable as a result of condition-dependence, a relatively weaker female mate preference is needed to obtain a mate in good condition. Selection on the male trait and the female mate preference may then become weaker because females may be able to obtain a male in good condition even with a weak preference (Price *et al.* 1993). Natural selection constraints on the variability of male sex traits may therefore lead to a greater degree of exaggeration of a male trait. The reason for this is that the intensity of the female mate preference will remain high in the absence of extreme variability in the male sex trait.

In conclusion, secondary sexual characters are more variable than ordinary morphological characters because of either (1) the strong component of directional sexual selection and the resulting weaker degree of control of development, and (2) direct selection for condition-dependence. In the next two sections (6.2.2 and 6.2.3) I describe the genetic variance and the genotype-environment interaction variance in ornament size. The environmental variance in ornament size is presented in sections 6.3 and 6.4.

6.2.2 *Genetic variance in ornament size*

The additive genetic variance in a character can be determined from the resemblance of the trait in offspring to that in their parents. The heritability, which can be estimated from offspring-parent resemblance, is an estimate of the proportion of the total phenotypic variance that can be attributed to the additive effect of genes (Falconer 1981).

Barn swallows almost always breed away from their natal site, and even a long-term field study will only yield few pairs of measurements of offspring and their parents. Female offspring generally disperse farther than male offspring before settling to reproduce (Davis 1965; Brombach 1977; Jarry 1980; Christensen 1981; de Lope 1983; Glutz von Blotzheim and Bauer 1985), and, as expected, the majority of recruits in my study area were males. The analysis of resemblance between offspring and their parents therefore must be restricted to male offspring and their fathers. This analysis revealed there is, however, a strongly positive relationship between tail length of sons and their fathers (Fig. 6.2; Møller 1991*a*). The estimate of heritability of tail length, which is twice the slope of the linear regression, is $h^2 = 0.59$

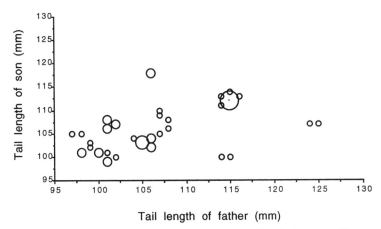

Fig. 6.2 Relationship between tail length (mm) of male barn swallows and tail length of their fathers. Circles of increasing size represent 1, 2, 3, and 5 individuals, respectively.

(SE = 0.10), and this value is about the same as for many other morphological characters.

The estimate of the genetic variance suffers from a number of shortcomings. Offspring may resemble their parents for reasons other than effects of genes because (1) parents and offspring may be reared in similar kinds of environments with similar degrees of favourability, and (2) there may be strong maternal effects if the mates of the most preferred males feed their offspring more lavishly compared with the average reproducing female. On the other hand, if some offspring are fathered by males other than the male attending their nest, this may result in a reduced estimate of heritability. One common method used to circumvent the problems of maternal effects is to make cross-fostering experiments. This is not easily done with barn swallows because offspring do not develop full tail size until they have completed their first moult in the winter quarters. Extensive natal dispersal will prevent most recruits from being checked after having grown their tail ornaments.

I have been able to test for maternal effects in three different ways. First, I captured a number of recruits deriving from two tail manipulation experiments in my Kraghede study area in 1987 and 1991. If maternal effects were common, there should be a strong effect of experimental tail-length manipulation on the tail-length of offspring, while rigid genetic influence should result in a strong effect of natural tail length on the tail length of offspring. The total number of recruits was only 23 males which is a small sample compared with what is usually used in heritability studies. The resemblance between tail length of offspring and experimental tail length of fathers was found to be very weak and not statistically significant,[1] while

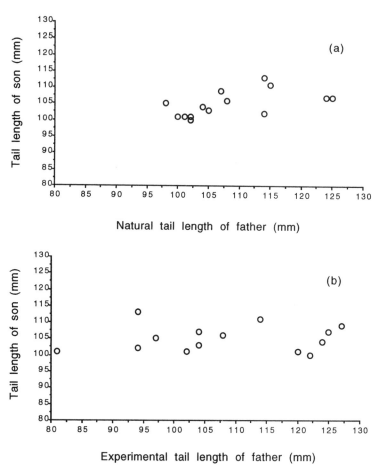

Fig. 6.3 Relationship between tail length (mm) of male barn swallows and (a) natural tail length and (b) experimental tail length of their fathers. Data from tail length manipulations at Kraghede in 1987 and 1991.

there is a significant resemblance between tail length of offspring and natural tail length of fathers[2] (Fig. 6.3; $h^2 = 0.56$ (SE = 0.16)). This suggests that there is a genetic effect on the length of tails in male barn swallows.

Second, if maternal or common environment effects are important, it should be possible to quantify some of these effects. I have tested for the effect of four obvious environmental factors in the larger sample of recruits from all years (Møller 1993*n*). These are:

(1) a common rearing environment of the father and its offspring;

(2) effects of parental care on the expression of the tail ornament;

(3) effects of rearing conditions of offspring on subsequent tail length, and

(4) the distribution of males differing in age and tail length across colonies of different sizes.

(1) *Common rearing environment.* Environmental conditions are likely to be more similar if fathers and sons are reared in the same neighbourhood, and the magnitude of the common environment effect may thus be inversely related to the distance between the breeding site of a father and its son. However, the natal dispersal of sons (and thus the similarity in the breeding environment of the father and the son) was unrelated to their tail length.

(2) *Effects of parental care.* There was no relationship between the total amount of parental care measured in terms of feeding rate per pair and tail length of offspring, and tarsus length and body mass of nestlings were unrelated to tail length of their father.

(3) *Effects of rearing conditions.* Tail length of male swallows was unrelated to their own rearing condition measured in terms of rearing date, number of siblings and the male parent's age.

(4) *Distribution of males.* Male barn swallows were distributed randomly across colony sizes with respect to age and tail length; thus, heterogeneity in the distribution of males varying in tail length or age across colonies can not account for the tail length of sons. In conclusion, there were no obvious environmental factors which affected the tail length similarity between male barn swallows and that of their fathers (Møller 1993*n*).

Third, if there were differences in the quality of parental care provided by parent barn swallows, this could have an effect on the subsequent development of tail ornaments in spring. In the two tail-length experiments on unmated male barn swallows no effects were found of experimental tail length on the quality of offspring measured in terms of tarsus length, which is a skeletal character constant throughout life, and body mass, which is an indicator of rearing conditions (Møller 1988*c*, 1992*d*). de Lope and Møller (1993) experimentally manipulated tail length of male barn swallows after they had acquired a mate, and the distribution of females among males differing in tail length was therefore not affected. Although the sex roles in provisioning were influenced by this experiment, there was no effect on total provisioning by the pair. Offspring quality measured in terms of tarsus length and body mass was unaffected by this experiment. A similar result was obtained by Møller (1993*p*) in another tail manipulation experiment. It is therefore unlikely that the condition of barn swallow nestlings was influenced by female barn swallows mated to males which differed in experimental tail length after pair formation.

In conclusion, there appears to be clear additive genetic effects on the expression of tail length in male barn swallows. A number of different tests failed to demonstrate any clear maternal or common environmental effects on the expression of male tail ornaments.

6.2.3 *Genotype-environment interaction of ornament size*

Individuals with a particular genotype may develop different phenotypes when reared under different environmental conditions. The range of phenotypes developed by a certain genotype under a range or gradient of environmental conditions is said to mirror the reaction norm of the genotype. If different genotypes possess different reaction norms, which differ in slopes and therefore cross, there is a genotype-environment interaction. Most morphological characters demonstrate phenotypic plasticity as a consequence of influences on development of variable environmental conditions. The extent to which secondary sexual characters demonstrate genotype-environment interactions is not well known, although there is considerable evidence for environmental effects on the expression of secondary sex traits of the barn swallow (see section 6.3). However, one kind of genotype-environment interaction might pose a threat to the reliability of secondary sex traits as honest advertisements. If a particular secondary sex trait increases in size along an environmental gradient, this increase could potentially differ between genotypes. Reliable signalling requires that individual genotypes, which develop the smallest ornaments at one end of an environmental gradient, should not be able to develop the largest ornaments at the other end of the gradient (Møller 1993*g*). In other words, there must not be extensive crossing over of the tail length reaction norms in the environments typically encountered. Reliable or honest signalling requires that a display of quality is relatively more costly for low- than for high-quality individuals (Zahavi 1975, 1977, 1987; Grafen 1990*a*, 1990*b*), and this assumption will be violated if the reaction norms for the secondary sex trait were to cross extensively.

Reaction norms are usually studied by determining the expression of a character of interest by the same genotype reared under different environmental conditions. This procedure is not feasible for the tail ornaments of male barn swallows because the genotypes of interest have not been identified. Instead, I have tested for a genotype-environment interaction in barn swallow tail ornaments in two different ways (Møller 1993*g*).

First, I have exploited the fact that individual male barn swallows develop their secondary sex trait anew every year. Environmental conditions measured in terms of precipitation in the winter quarters differ between

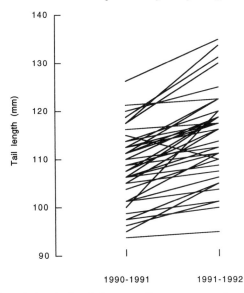

Fig. 6.4 Tail length (mm) of male barn swallows grown under poor and favourable environmental conditions in the winter quarters. Lines connect tail lengths grown by the same male in two subsequent years. Note that most reaction norms had a positive slope. Adapted from Møller (1993*g*).

years, and they have a marked effect on the expression of the tail ornament, probably through effects on availability of food (Møller 1991*a*). Thus, the environmental gradient is the annual variation of environmental conditions in the winter quarters and the trait is tail length developed by the same individual males under these environmental conditions. There was a general increase in tail length of male barn swallows between poor and favourable environmental conditions, and this increase tended to be larger for short-tailed than for long-tailed males (Fig. 6.4). The genotype-environment interaction was statistically significant, but this only means that the slopes for the different genotypes differ, not necessarily that the lines connecting tail length under poor and favourable environmental conditions cross. In fact, there were few cases of extensive crossing between the reaction norms (Fig. 6.4). Male barn swallows with short tails, therefore, do not grow relatively longer tails than long-tailed conspecifics under some environmental conditions. The requirement of reliable signalling that the development and maintenance of a unit of display will be more costly for low-quality males was therefore upheld.

Second, I have exploited the fact that birds grow new feathers if these are plucked. When the outermost tail feathers of male barn swallows are plucked early during the breeding season, their size can be compared with that of feathers grown in the breeding grounds where the environmental

conditions are quite different. The two outermost tail feathers were removed
from a sample of male barn swallows early during the 1989 breeding season,
and the size of the original and the fully re-grown tail feather was measured.
Tail feathers grown in the breeding area were considerably shorter than
those grown in the winter quarters, suggesting that environmental conditions
for feather growth are better at the normal time of moult than during the
breeding season (Møller 1993g). This may seem obvious because barn
swallows growing feathers during the breeding season had to allocate
resources to both feather growth and reproduction, whereas birds moulting
in the winter quarters could allocate all resources to feather growth. More
interestingly, the difference in length of tail feathers grown under the two
different environmental conditions varied with original tail length. Male
barn swallows returning to the breeding area with a long tail also performed
better under the environmental conditions of the breeding season compared
with originally short-tailed males which performed much worse (Fig. 6.5;
Møller 1993g). Although this genotype-environment interaction was statistically significant, there were few cases of extensive crossing over between
the reaction norms. Again, the assumption of the reliable signalling hypo-

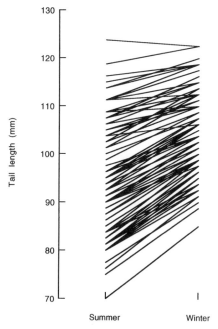

Fig. 6.5 Tail length (mm) of male barn swallows grown in the breeding area and
in the winter quarters. Lines connect tail lengths grown by the same male at two
different times. Note that the degree of crossing over was limited, and that most of
the reaction norms had a positive slope. Adapted from Møller (1993g).

thesis was confirmed; males with the smallest ornaments did not perform better than males with the largest ornaments under any environmental condition.

In conclusion, there were clear indications of genotype-environment interactions for tail length of male barn swallows, but reaction norms rarely cross extensively as required by the reliable signalling hypothesis.

6.3 Ornaments as indicators of condition

The different models for the evolution of secondary sexual characters predict different relationships between the condition of an individual and the expression of its secondary sexual characters (Pomiankowski and Møller 1993a). Condition-dependence is here defined as the development of a more extravagant phenotype by individuals in better body condition. Models of mate choice with a direct fitness benefit assume that the extent of the benefit is directly reflected in the expression of the secondary sexual character. If females choose a mate with an eye on its potential help with raising the offspring, males should not invest so much in the development of a secondary sexual character that their provisioning be hampered by the presence of the ornament. Females should then prefer to mate with males wearing less extreme ornaments which were likely to be better providers.

Two pairs of barn swallows on a wire

With respect to the different models of mate choice with indirect fitness benefits, the good genes and the arbitrary trait models predict different patterns of condition-dependence of secondary sexual characters (Pomiankowski and Møller 1993*a*). It is a common assumption for the different handicap models, irrespective of whether they assume a genetic or a phenotypic basis for the fitness advantage acquired by the female from its mate choice, that the quality and hence the condition of a male will be reflected in the expression of its secondary sexual character (see section 2.5). Females are supposed to be able to assess the quality of the male using the size of its ornament as a cue and thereby indirectly their probable fitness gain from the mate choice. The arbitrary trait models of mate choice do not predict that the male sexual character reflects the condition of the male, because the secondary sex trait is an entirely arbitrary character whose single function is to be attractive. In fact, condition-dependence of secondary sexual characters in the sense that only individuals of high quality are able to develop the most extreme phenotypes indirectly refutes the runaway process of mate choice as the sole explanation for the current expression of a character.

Overall condition of birds is not easily quantified, but I have investigated three environmental effects on the expression of tail ornaments in male barn swallows. These effects are:

(1) age effects on tail length;

(2) effects of environmental conditions in the winter quarters; and

(3) effects of parasites.

First, inevitably the experience of individuals will increase with age, and, as a result, so will skills such as foraging and predator recognition and evasion. The overall condition of individuals therefore can be safely assumed to increase early during their life span (Møller 1992*c*). Ornament size is expected to covary with age-related increases in condition, although it may be unlikely that such age-related effects would override quality differences between individuals. Tail length of male barn swallows increased between their first and second year of life, while increases were much smaller and non-significant in subsequent years (Fig. 6.6; Møller 1991*a*). The effects of age on growth of tail length in females was much weaker than in males, and also in females the increase was significant only between the first and the second year (Fig. 6.6). Such patterns of age-related growth in ornaments agree with the presumed pattern of enhanced body condition during the first part of the life span.

Second, barn swallows grow their feathers in the winter quarters, and environmental conditions during the moulting period may affect the length of tails. Many parts of the winter range have dry climates with distinct dry

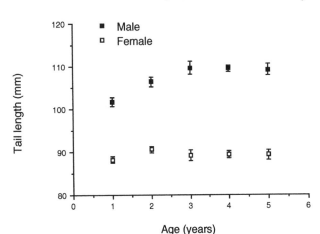

Fig. 6.6 Tail length (mm) of the same individual male and female barn swallows in subsequent years. Values are means (SE). Adapted from Møller (1991a).

and rainy seasons, and aerial insects are known to be much more abundant after than before the start of the rainy season. For example, many ants and termites, which are extremely nutritious, swarm in direct response to precipitation, and the body mass of barn swallows has been found to increase considerably immediately after the start of the rains (Møller *et al.* 1993). The winter mortality of barn swallows is also much higher in dry than in wet years (Møller 1989c). One can therefore expect barn swallow tails to grow longer in years with abundant food. In fact, barn swallow males developed longer tails in wet than in dry winters while female tail lengths were much less affected by the amount of precipitation in the South African winter quarters (Fig. 6.7; Møller 1991a). The conditions in the African winter quarters thus appear to have a direct effect on the expression of male tail length in the barn swallow.

Third, individuals may be able to allocate only limited resources to growth of ornaments because overall resource abundance in the environment is low. Alternatively, resource allocation to ornament growth may be limited because parasites usurp their share of the resources, which as a result cannot be allocated to tail growth. Parasites are but one factor affecting overall body condition, yet if there is a direct relationship between ornament development and levels of parasite infestation, females should be able to assess the abundance of parasites on a potential partner, something that may not easily be assessed directly. Tail length of male barn swallows was negatively correlated with the abundance of two species of ectoparasites, namely the tropical fowl mite *Ornithonyssus bursa* and the feather louse *Hirundoecus malleus* (Møller 1991c). There was also a strong negative correlation between mite loads of male barn swallows on their arrival at the

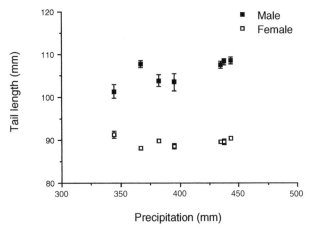

Fig. 6.7 The relationship between tail length (mm) of male and female barn swallows and precipitation (December–March) in the South African winter quarters. Values are means (SE). Adapted from Møller (1991*a*).

breeding grounds and tail growth relative to that of the previous year (Møller 1991*c*). Finally, tail growth from one year to the next was negatively correlated with the abundance of mites in the nests of an individual during the previous breeding season (Møller 1990*e*, 1991*c*). These correlations indicate that there may be a causal relationship between the abundance of these two kinds of ectoparasites and growth of tail ornaments.

A direct test of whether the relationship was causal was provided by an experimental manipulation of mite loads of barn swallow nests. The difference in tail length of male barn swallows from one breeding season to the next was predicted to be negatively related to an experimentally elevated rise in mite abundance if the parasites were directly to affect tail growth. In fact, tail growth decreased with increasing mite loads (Fig. 6.8; Møller 1990*e*), which points to the existence of a direct relationship between parasite numbers and the expression of this particular secondary sex trait. It has also been found that the degree of fluctuating asymmetry will increase with an experimentally increased parasite abundance (Møller 1992*f*).

6.4 Determinants of feather growth

6.4.1 *When to moult and how quickly*

Barn swallows from Northern Europe usually start to moult in the winter quarters after the completion of migration, although individuals from southern populations may start moulting at the breeding grounds and

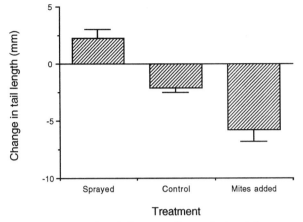

Fig. 6.8 Growth rate of male barn swallow tails (mm) between year (*i*) and year (*i* + 1) in relation to experimental mite treatment of their first clutch nests in year (*i*). Values are means (SE). Adapted from Møller (1990*f*).

suspend moult during migration. Male barn swallows have considerably longer tails than females, and they have to grow these before the start of the breeding season in order to gain the sexual selection benefits of a long tail. Males with long tails more often have mates that lay two clutches per season, and therefore they cannot start autumn migration as early as short-tailed males. Long-tailed male barn swallows also generally arrive earlier at the breeding grounds than short-tailed males (see Chapter 7), and more often have already completed their moult upon arrival (Fig. 3.3). How is it possible for long-tailed male barn swallows to achieve better results and in a shorter time than short-tailed males? There are several possible solutions:

(1) long-tailed males could initiate moulting earlier than short-tailed males;

(2) long-tailed males are in better condition and therefore able to complete moult more rapidly; or

(3) long-tailed males grow their feathers at a higher rate than short-tailed males.

Clearly, it is necessary to study moult in the African winter quarters in order to distinguish between these alternatives.

As long-tailed male barn swallows leave the breeding grounds later than short-tailed males because of more frequent second clutches, they are able to start moulting earlier only if they migrate faster. This could explain both how long-tailed males manage to complete their moult earlier than short-tailed males, and how they manage to arrive at the breeding grounds before other males. However, this possibility is not easily studied because the speed

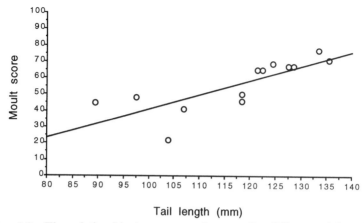

Fig. 6.9 The relationship between moult score for different adult male barn swallows captured at Otjiwarongo, Namibia, January 1991, and their tail length (mm). The moult score is a composite score for primaries, secondaries and rectrices. The line is the model I regression line.

of migration of individual males cannot be measured. An indirect way of determining the start of moult is by catching males in the winter quarters and determining their state of moult, and in fact long-tailed males were in a much more advanced state of moult than short-tailed males (Fig. 6.9; Møller *et al.* 1993). This suggests that long-tailed males either had started moulting earlier, or that they had moulted more rapidly.

The stage of moult of long-tailed male barn swallows appeared to be more advanced than that of short-tailed males, but it is likely that the effect of tail length is simply a reflection of the fact that long-tailed males were in a better body condition. I investigated the independent effects of age, sex, and body condition on stage of moult in an attempt to determine the extent to which each of these factors contributed to the timing of overall moult (Møller *et al.* 1993). The analysis of individual moult scores for tail feathers clearly demonstrated that males were in a more advanced moult stage than females, and that adults were in a more advanced stage than juveniles (Møller *et al.* 1993). However, tail moult score was weakly related to condition estimated either as body mass or residual body mass after the effect of body size had been statistically removed. These effects were virtually unaltered if overall moult score rather than tail moult score was used as the dependent variable.

In conclusion, male barn swallows, particularly old males, were in a more advanced moulting stage during winter than the average barn swallow in the population. The possibility that long-tailed male barn swallows grow their feathers at a faster rate than short-tailed males will be investigated in the next paragraph.

6.4.2 *Feather growth*

The rate of feather growth can be measured directly because daily growth increments are revealed as alternating light and dark growth bars deposited in the feather structure (Michener and Michener 1938; Wood 1950). This pattern of growth bars is a consequence of different growth rates during day and night. Therefore, the size of growth increments can easily be measured and the time required to grow a feather estimated from its length and the mean growth increment. Individual barn swallows vary considerably in mean growth increments, even though there is consistency in the growth rate of feathers from different feather tracts of the same individual (Møller 1993q). Individuals with a fast rate of tail feather growth also grow their wings at a fast rate (Møller 1993q). Birds can be induced to grow new feathers by removal of an old feather. Male barn swallows that had their outermost tail feathers removed at the breeding sites grew new tail feathers which were slightly shorter than their original tail feathers (Møller 1993g). There was a high consistency of feather growth rates between original replacement feathers as the mean length of growth bars under the two different environmental conditions was strongly positively correlated. Thus, some males were always fast in terms of their feather growth while others were slow.

These differences in the size of daily growth increments could result in long-tailed male barn swallows being able to grow their tails more rapidly than the average male in the population. The length of tail growth bars is strongly positively related to tail length in male barn swallows, but not in females (Fig. 6.10; Møller 1993q). This means that long-tailed males grow their tail feathers more rapidly than short-tailed males, whereas that is not the case for females differing in tail length. The duration of the growth period of the outermost tail feather can be estimated by dividing tail length by the daily growth rate. Whereas the growth period is unrelated to tail length in male barn swallows (Fig. 6.10), there is a slight increase in the duration of the growth period with increasing tail length in females (Møller 1993q).

In conclusion, long-tailed male barn swallows are able to shorten the duration of their moult period by growing their outermost tail feathers at a faster rate.

6.4.3 *Order of tail feather growth*

Almost all passerine birds have a uniform pattern of moult of their tail feathers, starting from the centre of the tail and progressing outwards (Stresemann and Stresemann 1966; Palmer 1972; Kasparek 1981). Barn swallows differ by moulting their outermost tail feathers before the second

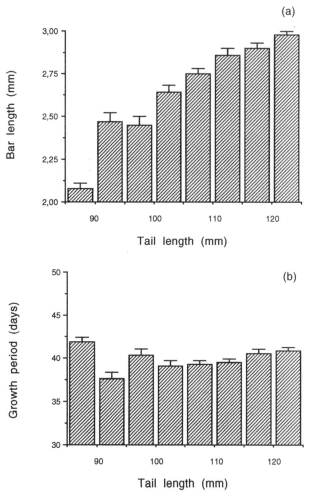

Fig. 6.10 The relationship between (a) the mean length of growth bars (mm) of the outermost tail feathers of male barn swallows and their tail length (mm), and (b) the duration of growth of the outermost tail feathers (days) of male barn swallows and their tail length (mm). Values are means (SE). Adapted from Møller (1993*q*)

or even before the third outermost tail feather has been dropped (Møller *et al.* 1993). There is considerable intraspecific variation in the pattern of tail moult. In particular, males generally start to moult the outermost tail feathers earlier than females (Møller *et al.* 1993). This pattern of moult ensures that the outermost and longest tail feathers of males are fully grown at the start of the breeding season.

Because the sequence of tail moult in the barn swallow does not occur in other well-studied hirundines, its evolutionary origin must be sought among the most recent ancestors of the barn swallow. It is likely that barn swallow ancestors, and particularly males, encountered difficulties completing their moult before the start of the breeding season when a migratory habit arose. The moult in different hirundine species sometimes starts before the autumn migration, but is always suspended during the migration period. Individuals which started to moult their outermost tail feathers relatively early would be at a sexually selective advantage because their tails on average would be longer by the start of the breeding season.

6.4.4 *Quality of feathers*

Feathers produced by individuals of the same bird species may superficially all appear the same. However, feathers may differ in quality measured in terms of size, strength and specific gravity. Differences in feather quality are very important for their perfect functioning during flight. A barn swallow's outermost tail feathers contribute disproportionately to the total lift of the tail, which is concentrated at the leading edges (Thomas 1993). As a result, the outermost tail feathers in the forked tail of a barn swallow are particularly prone to damage. Thomas (1993) argued that square tails with tail feathers of approximately equal length should be the optimal platform for bird species living in habitats such as woodland where damage to the tail feathers is particularly likely. Forked tails should be present in birds foraging in open habitats where the evolution of tail shape to a higher degree is determined by selection acting through aerodynamics. This hypothesis about optimal shape of tails in aerial insectivores assumes that the fully extended tail is triangular. This is not the case in male barn swallows because the long outermost tail feathers extend beyond the optimal triangular tail shape. However, we would expect tail damage to be uncommon in barn swallows because that would strongly disrupt aerodynamics. Here I present evidence for variation in two kinds of feather quality and relate it to male tail length. These are (1) fault bars, and (2) damage to feathers.

Fault bars are transparent light bars in bird feathers developed as a consequence of stressful conditions during moult (Fig. 6.11; Michener and Michener 1938; Wood 1950; Harrison 1985). They are often clearly visible under favourable light conditions, and feathers are particularly prone to break at fault bars. The frequency of fault bars differs remarkably between avian taxa, sexes and age classes. In barn swallows there was no sex difference in whether individuals had any fault bars,[3] or in the number of fault bars in the tail among individuals that had at least one fault bar.[4] However, the relationship between frequency of fault bars and tail length differed clearly between the sexes. While males with short tails had more fault bars than males with long tails[5] (Fig. 6.12), there was no relationship between tail length and frequency of fault bars in females.[6] Long-tailed male

Fig. 6.11 A fault bar in the outermost tail feather of a male barn swallow.

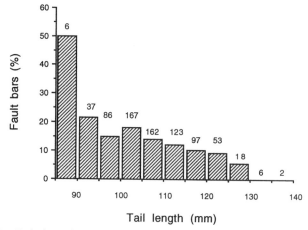

Fig. 6.12 Relationship between the presence of fault bars (% males) and tail length (mm) of male barn swallows. Numbers are sample sizes.

barn swallows therefore less frequently develop fault bars despite the size of their tail ornaments.

The negative relationship between presence of fault bars and tail length in male barn swallows indicates that short-tailed males are unable to produce high-quality feathers even though their tails are relatively short. If there is a direct relationship between the cost of a tail and the occurrence of fault bars, it should be possible to alter experimentally the frequency of fault bars. This was attempted in a tail-manipulation experiment (Møller 1988*c*). When the tail length of male barn swallows was experimentally manipulated in one year, males with elongated tails developed a higher frequency of fault bars in their tail feathers in the subsequent year (Møller 1989*a*). While only 14%

of male barn swallows have fault bars under natural conditions, almost all males with artificially elongated tails developed fault bars during the moult after the manipulation. This suggests that the increased cost of having an elongated tail resulted in higher stress levels during moulting as reflected in the increased frequency of fault bars.

The relationship between presence of fault bars and tail length in male barn swallows might indicate that the frequency of damage to the outermost tail feathers should also decrease with increasing tail length. I recorded tail damage as the frequency of broken outermost tail feathers and the frequency of missing pieces of feather from the vanes. Again, there was no sex difference in whether individuals had damaged outermost tail feathers.[7] However, the frequency of damaged outermost tail feathers decreased with increasing tail length in male barn swallows[8] (Fig. 6.13), but not in females.[9] Male barn swallows therefore suffered damage to their outermost tail feathers more often if their tails were short. The reason for this increased damage to the tail feathers is not obvious. There are at least two not necessarily mutually exclusive possibilities. Either there are differences in the quality of feathers, or males with short tails more often encounter situations that result in feather breakage. The elevated level of damage to the outermost tail feathers in short-tailed male barn swallows should have serious consequences because it will inevitably result in worse flight performance. Feather damage or breakage would resemble individual fluctuating asymmetry in tail length. A forked tail is the optimal solution to maximizing lift, but such a tail is also very sensitive to damage, which greatly distorts the distribution of lift across the tail and introduces rolling and yawing forces (Balmford and Thomas

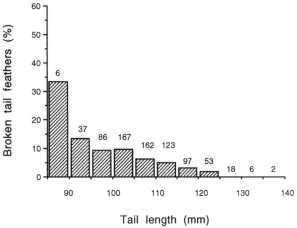

Fig. 6.13 Relationship between the presence of damage to the outermost tail feathers (% males) and tail length (mm) of male barn swallows. Numbers are sample sizes.

1992; Thomas 1993). Damage to the outermost tail feathers thus disturbs air flow over the tail, and this handicap can only partly be overcome if the bird flies with its tail at an angle.

In conclusion, the quality of the outermost tail feathers was inversely related to tail length in male barn swallows, but not in females.

6.5 Why are ornaments not larger?

6.5.1 *Optimal tail length*

If tail length of male barn swallows can be considered a problem of optimizing various costs and benefits that vary in relation to the quality of an individual, it should be possible to determine this optimum by means of optimality models (Sutherland and de Jong 1991). This would require knowledge of the shape of the cost and benefit curves for ornaments of different size developed and maintained by individuals of different quality. The handicap models of sexual selection clearly demonstrate that the single reason for an increase in optimum display rate in relation to individual quality is that low-quality individuals pay a relatively higher cost per unit of display than high-quality individuals (Andersson 1982a; Heywood 1989; Grafen 1990a, 1990b; Iwasa *et al.* 1991; Price *et al.* 1993). The optimality approach at first glance is straightforward because the only requirements are estimates of the cost and benefit functions. However, a closer inspection of the cost and benefit data of long tails reveals that male barn swallows wearing the longest tails nevertheless survive much better than males with short tails (Møller 1991e). This means that a whole suite of fitness components are positively correlated and that there seems to be no trade-off between them. Some males grow very long tails, acquire all the mating advantages, reproduce very successfully and still survive much better than other males which grow short tails and therefore have a very low mating and reproductive success. So why do not all males grow a longer tail?

In fact, there was an indication that males grew longer tails as my study progressed. Mean tail length of males increased significantly during the time series.[10] The increase in mean tail length was not very large as the regression coefficient was only an increase of 0.86mm year^{-1} (SE = 0.17). Barn swallows are very short-lived animals, and average life expectancy is less than two years. The average increase in male tail length between generations is therefore less than 2mm. The predicted increase can be calculated from estimates of the total selection differential on tail length (2.48; Chapter 4.4) and the heritability of the tail trait (0.59), given certain assumptions (Falconer 1981). These are that (1) there is no non-genetic resemblance between offspring and parents, and that (2) fertility and viability are not correlated with the phenotypic value of tail length. Both these assumptions may be violated in the case of the barn swallow. The predicted response to selection in units of standard deviation is 1.46, which equals 12.8mm per generation. This is considerably more than the observed value. This discrepancy can either be as a result of bias in estimates of the parameters, and/or tail length may be determined by phenotype-dependent optimization rather than selection on a quantitative trait.

The standard solution to the problem of individual optimization is experimental pertubation away from the chosen optimum. I used this approach in order to estimate the cost and benefit functions of a long tail ornament in the barn swallow (Møller 1988c, 1989a). The foraging ability of males was impaired by the tail manipulation, apparently because elongation of tails results in a considerably increased drag by the tail streamers (Evans and Thomas 1992; Thomas 1993). As a result, male barn swallows differed in the quality of insect prey items brought to their offspring, and perhaps also in the quality of food eaten by themselves (Møller 1989a). Thus, the combined results of two tail-length experiments in Denmark and one in Spain show that male barn swallows captured significantly larger prey items if their tails had been shortened, and this was particularly the case for males with natural tails shorter than the median tail length (Møller 1989a; Møller and de Lope 1993). However, males with elongated tails captured relatively smaller prey items, and this was particu-larly the case if their natural tail length was below the median tail length. Size of prey captured by male swallows thus decreased with experimentally increased tail length, and this was particularly the case for males with the shortest tails (Fig. 6.14; Møller and de Lope 1993). There was some variation in the three data sets, because the first Danish experiment revealed only a statistically significant change in size of prey among males belonging to the elongated group (Møller 1989a), but then sample sizes were overall quite small.

Impaired foraging ability under stressful conditions such as reproduction, migration and moult should have severe fitness consequences. Therefore, the survival prospects of male barn swallows are likely to be affected by the tail manipulations. It is extremely difficult to demonstrate statistically significant differences in survival rates with small sample sizes because of the large standard errors of survival estimates. However, I was able to capitalize on the unique situation that tail length had been manipulated in four different years. Therefore, the effect of the tail-manipulation experiment on survival prospects could be estimated by calculating the geometric mean survival rate for the experimental groups over four years. This procedure reduces the standard errors considerably and thereby increases the power of the statistical tests. Recapture rates of male barn swallows (which may closely approach survival rates) were negatively related to experimental tail length, as expected from the optimality model (Fig. 6.15; Møller and de Lope 1993). An experimental increase in tail length thus resulted in a decrease in recapture rate.

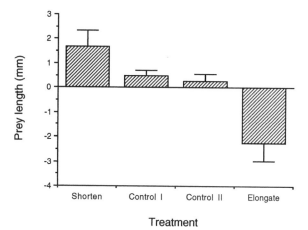

Fig. 6.14 The relationship between prey length (mm) captured by male barn swallows and fed to offspring aged 8–12 days in their first brood in relation to tail-length manipulation. The data have been recalculated as deviations in mean prey size per male from the population mean in each of three experiments. Values shown are means (SE). Adapted from Møller and de Lope (1993).

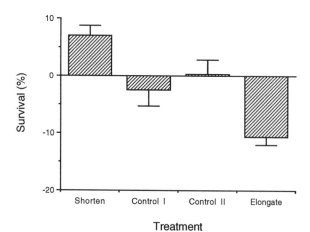

Fig. 6.15 The relationship between survival rate (%) of male barn swallows in relation to tail-length manipulation. Survival rates are deviations in survival rates of experimental groups from the overall estimates for all males involved in each of four experiments. Recapture rates were assumed to approximate survival rates for reasons given in Chapter 3. Values are means (SE). Adapted from Møller and de Lope (1993).

What then is the evidence for individual optimization of tail length? I have previously reported that naturally long-tailed males were better able to pay the extra cost of an elongated tail because survivors of the elongation treatment had longer tails than non-survivors (Møller 1989*a*). This result was replicated in the other three tail manipulation experiments. Naturally long- and short-tailed males differed in their response to tail manipulation. Survival prospects were enhanced among long-tailed males receiving the elongation treatment, but were enhanced among short-tailed males receiving the shortening treatment (Fig. 6.16; Møller and de Lope 1993). The first of these differences in survival prospects suggests that a long tail is costly in terms of reduced survival, but more so in males with short tails (and presumably of low quality) compared with long-tailed ones (presumably of higher quality). The second difference suggests that even the natural tail length is costly, but more so in males with short tails compared with long-tailed males.

Another consequence of the impaired foraging ability of male barn swallows caused by the experimental manipulation was a decrease in tail length among males with elongated tails in the subsequent year (Møller 1989*a*). As the size of the tail ornament is of major importance for the fitness of male barn swallows, a reduction in tail length will affect mating advantages and seasonal reproductive success. Males belonging to the elongation group had a reduced tail length in the year after the experiment, with the result that (1) the pre-mating period lasted longer, and (2) the seasonal reproductive success fell (Møller 1989*a*).

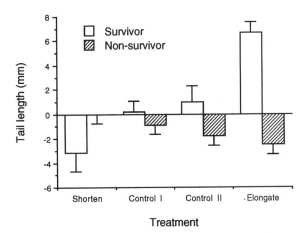

Fig. 6.16 Tail length of male barn swallow survivors and non-survivors in relation to tail-length manipulation. Values are tail lengths relative to the mean of all males involved in each of four experiments. Values are means (SE). Adapted from Møller and de Lope (1993).

In conclusion, male barn swallows with optimal tail length enjoy maximal fitness. The optimal solution depends on the quality of an individual; low-quality individuals have lower optima than high-quality individuals. This optimization process is ultimately controlled by the phenotype-dependent cost and benefit functions. An individual male that grows too long a tail is severely penalized in terms of natural selection, while an individual that grows too short a tail is severely penalized in terms of sexual selection.

6.5.2 *The optimal degree of fluctuating asymmetry*

Length is but one of many features of the tail ornament of male barn swallows that may be subject to sexual selection. Like many other structures, tails have two sides which are not necessarily equally long. Random deviations from symmetry in otherwise symmetrical traits are referred to as fluctuating asymmetry, which arises as a consequence of an inability of individuals to cope with genetic and environmental stress (Ludwig 1932; Parsons 1990). Secondary sexual characters display much larger degrees of asymmetry than ordinary morphological traits as a consequence of strong directional selection caused by female mating preferences (Møller and Pomiankowski 1993*a*). The degree of ornament asymmetry is frequently so large that it can be easily perceived by a human observer, and therefore also probably by the animals themselves. The main reasons why asymmetry is so easily perceived is that almost all extravagant secondary sex traits are paired structures placed next to each other, and that interactions with prey and predators make everyday use of symmetry a prerequisite for survival.

Fluctuating asymmetry in ornaments basically serves as a health certificate of the individual, and conspecifics able to perceive the degree of asymmetry may assess the quality of potential mates or competitors. When this is the case, both size and asymmetry become targets for an optimization process (Møller and Pomiankowski 1993*b*; Pomiankowski and Møller 1993*b*). A male has to optimize tail length in order to maximize its mating success, but the tail should not be so long that the male runs into difficulties during the development of the trait because then the male would be penalized in terms of increased fluctuating asymmetry in the tail trait. Rather than having a single handicap the male will then have to cope with two handicaps, namely, tail length and tail asymmetry. Asymmetry can be seen as a device which prevents cheating and deceptive signalling; males cannot grow tails above their individual abilities because that would be revealed immediately by an increased level of asymmetry. The optimal solution to the asymmetry problem is to develop an optimal tail length for a given quality. The higher level of asymmetry frequently seen in males with small ornaments highlights the fact that they have to work harder in order to develop a unit of

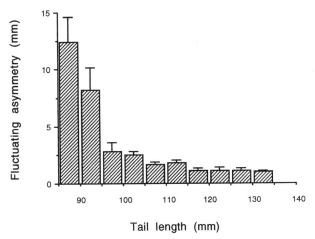

Fig. 6.17 Relationship between individual fluctuating asymmetry in tail length (mm) in male barn swallows and the length of their tails (mm). Values are means (SE).

ornamentation[11] (Fig. 6.17). This is exactly the standard requirement for reliable signalling in sexual and other contexts.

Fluctuating asymmetry in ornament size differs from that of ordinary characters because it is very difficult to make a perfectly symmetric trait in the absence of strong developmental control, and developmental control is reduced under a regime of intense directional sexual selection (Møller and Pomiankowski 1993b). However, there is no upper limit to the extent of elaboration of ornaments because other features of secondary sexual characters revealing male quality may similarly become the target of a female mating preference.

6.6 Summary

Like many other morphological characters, ornament size is probably a polygenic trait, the expression of which is determined by genetic, environmental and genotype-environment effects. The heritability of tail length in male barn swallows is relatively high, notwithstanding a number of potential maternal effects. Several environmental factors (male age, weather in the winter quarters, and parasite loads) have been found to affect the expression of the tail ornament. The size of a tail developed by the same individual in a sequence of years was used as a way of estimating the genotype-environment interaction. The reaction norms for tail length of male barn swallows only rarely crossed each other.

Long-tailed male barn swallows started moulting earlier and grew their tail feathers at a faster rate than short-tailed males. The outermost and longest feathers start growing before the second outermost or even the third outermost tail feather, and moult of the longest tail feathers is thus usually completed before the start of the breeding season. Feather quality is positively related to tail length as shown by the frequency of fault bars and feather damage.

Ornament size of male barn swallows was optimized as evidenced by manipulation of tail length. Males suffered foraging, survival and mating costs in the subsequent season as a result of tail elongation, and these costs were larger for males with naturally short tails than for males with long tails.

STATISTICS

1. Linear regression: $F = 0.001$, $df = 1,21$, $P = 0.99$.
2. Linear regression: $F = 11.97$, $df = 1,21$, $P = 0.0023$.
3. Males: 14.3%, $n = 809$, females: 14.3%, $n = 869$.
4. Males: 1.45 (SE = 0.09), $n = 116$, females: 1.47 (SE = 0.09), $n = 124$.
5. Linear regression: $F = 7.99$, $df = 1,806$, $P < 0.0001$.
6. Linear regression: $F = 0.44$, $df = 1,866$, $P > 0.20$.
7. Males: 6.7%, $n = 377$, females: 10.1%, $n = 437$.
8. Linear regression: $F = 3.56$, $df = 1,380$, $P < 0.01$.
9. Linear regression: $F = 2.27$, $df = 1,430$, $P = 0.13$.
10. Linear regression: $F = 26.79$, $df = 1,8$, $P = 0.0008$.
11. Linear regression: $F = 30.02$, $df = 1,370$, $P < 0.0001$.

7

Advantages of early arrival

7.1 Introduction

From everyday experience we all know that there are optimal times for doing different things. One should not arrive too early at a party, because there might be costs associated with this behaviour. For example, the hosts may suggest that you help prepare the meal. Late arrival may also be costly; perhaps all the food is gone. Animals face similar kinds of problems, and the question becomes how to solve them in order to maximize net benefit.

The question of optimal arrival time at the breeding sites from the winter quarters for males and females may be answered by means of optimality models. A number of such models exist for the frequent phenomenon of male insects emerging earlier than females (protandry) (section 7.2). The optimal arrival time for individuals of the two sexes will depend on the costs and benefits of arriving at a particular time as discussed in section 7.3. The costs of early arrival include high maintenance costs due to bad weather, whereas the benefits of early arrival include acquisition of a high-quality territory and mate and a high seasonal reproductive success. The costs of late arrival include acquisition of a low-quality mate and territory, or no mate at all, and a low seasonal reproductive success, while the benefits include low maintenance costs due to avoidance of inclement environmental conditions.

There is considerable annual variation in arrival date of barn swallows, some of which may be accounted for by environmental conditions. This possibility is considered by determining the relationship between weather conditions and arrival date in different years (section 7.4).

The cost and benefit functions of arrival date may vary relative to the quality of individuals. High-quality male barn swallows may have lower costs of early arrival than low-quality individuals, and they may also be able to derive higher benefits from an early arrival if females discriminate against low-quality males. Tail length of male barn swallows may be a general measure of individual quality, and long-tailed individuals are predicted to arrive earliest at the breeding grounds. I shall use male tail length as a measure of male quality and determine whether differences in arrival date are associated with differences in tail length in section 7.5.

Male barn swallows with long tails may arrive early for a number of reasons. For example, they may be of high quality, they may depart from the winter quarters early, or they may migrate at a higher speed. These alternative explanations for early arrival of male barn swallows will be investigated in section 7.6 using data on timing of moult and aerodynamic theory.

7.2 Models of optimal arrival

The arrival time can be predicted once the cost and benefit functions for different arrival times are known. There is an extensive literature on optimal emergence times for insects that spend the winter dormant in the ground based on optimality and game theory models (Wiklund and Fagerström 1977; Fagerström and Wiklund 1982; Bulmer 1983a, 1983b; Iwasa et al. 1983; Parker and Courtney 1983). The most common pattern of emergence in insects is protandry, with males arriving at the mating sites before females. The problem facing males is to arrive at a time in relation to arrival of females and environmental conditions that maximizes their fitness.

The optimality models developed by Wiklund and Fagerström were based on the assumption that the emergence time of males was related to that of females. While males are selected to emerge at a time when they can encounter mates, females are selected to emerge both when mates are available, but also at a time when conditions for successful reproduction are optimal. The temporal emergence curves were supposed to be fixed and under genetic control. All female insects mate once on the day of emergence. This is likely to be the case whenever the costs of delaying reproduction are high. The duration of the pre-mating period can therefore be minimized by emerging at the time when the maximum number of males are available. If the variance in male emergence is given, it is possible to determine whether males emerging at the peak emergence time will gain a larger number of mates than males emerging at other times. It turns out that the optimal degree to which males emerge before females is approximately equal for similar lengths of the emergence periods of the sexes. This is the case irrespective of the life-expectancy of the males. If the duration of the emergence periods of the two sexes are markedly different, there will be an evolutionary conflict of interest between the sexes concerning when males should emerge relative to females.

An alternative solution to this optimality problem of emergence times is the evolutionarily stable strategy (ESS) approach adopted by Bulmer (1983*a*, 1983*b*), Iwasa *et al.* (1983), and Parker and Courtney (1983). Iwasa *et al.* determined the optimal time of emergence for males so that males emerging on different days enjoy equal fitness. Their approach was thus to predict the position and the shape of the male emergence curve for any given female emergence schedule, and these two properties of the male emergence curve were thus assumed to be under natural selection. The number of males present on a given day is determined by the number of males emerging on that day and the number emerging on previous days given their pre-emergence and post-emergence mortality. Females emerge every day and they mate with males that are present on that day. The fitness of a male emerging on a given day should be the product of survivorship until emergence and the number of matings after emergence. Because of competition for mates, the mating success of a male emerging on a given date decreases with the number of other males emerging around that date. The evolutionary equilibrium curve of male emergence has a width similar to that of the female emergence curve, so that males emerging at different dates enjoy the same fitness. Iwasa *et al.* (1983) obtained two results from their game theory model. First, if the female emergence curve is smooth and unimodal, the reproductive season can be divided into an early phase when males emerge, and a late phase with no males emerging. The male emergence curve is truncated at the boundary between these two phases, which is determined by the difference in male pre- and post-emergence mortality. The larger the difference in the two male mortalities the later the truncation point occurs during the season. The second result was that the sex ratio (the number of males to virgin females) decreases with time during the period when some males emerge every day. Female availability per male thus increases with time as long as males emerge every day. Bulmer's (1983*a*, 1983*b*) models can be viewed as a special case of the model by Iwasa *et al.* (1983).

Parker and Courtney (1983) also used a game theory approach to determine the mixed evolutionarily stable strategy for male emergence time so that male fitness is equal for all individuals irrespective of their emergence times. The variation in the timing of male emergence is thus supposed to be adaptive, because fitness gains by each individual at a given time in the season are inversely related to population density. The ESS-model was based on the assumption that female availability varies seasonally. The number of males is determined by the rate of male emergence and from their mortality rates. Environmental effects must cause some degree of variation in emergence times in a given cohort of males genetically programmed to have the same emergence date. However, it is possible that most of the observable variation in male emergence date is strategic. Parker and Courtney (1983) assumed that the fitness gain to an individual on a given day is equal to the number of females available divided by the number of males. When the mortality rate of males is high, the emergence curve of males should closely track the seasonal distribution of females (Fig. 7.1b). When the mortality rate is low, the peak of male emergence should occur well before the peak of female availability, and the male emergence curve should be much tighter than that of females (Fig. 7.1d). There are three effects on male emergence when the mortality rate decreases:

(1) the variance in male emergence time decreases;

(2) the peak in emergence time is shifted towards an earlier date; and

(3) the skewness of male emergence increases despite a normal distribution of female availability (Fig. 7.1). These three effects are also increased when there are hierarchies of male competitors as in the case of males differing in attractiveness (Fig. 7.1c).

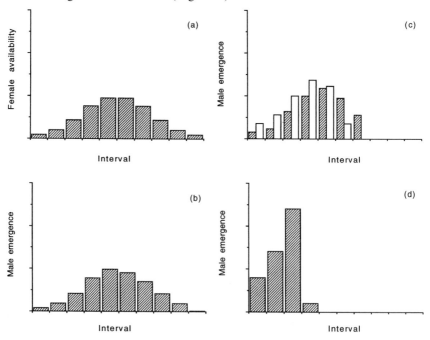

Fig. 7.1 An optimality model of arrival date. (a) Female availability at different times during the season. (b)–(d) Stable frequencies for male arrival strategies. When inter-interval survival is low (b), male arrival closely tracks the availability of females. When survival is high (d), peak male arrival anticipates peak female availability and the distribution of male arrival becomes strongly skewed. Intermediate survival (c, shaded histogram) yields an intermediate result. When there are hierarchies of male qualities, the arrival of high-quality individuals is earlier (c, open histogram) than that of low-quality individuals (c, shaded histogram). Adapted from Parker and Courtney (1983).

In conclusion, these different models of optimal male emergence time suggest that the shape and the peak of male emergence can be determined by factors such as the mortality rate of males, and the quality of competitors. The patterns of arrival of male and female barn swallows are analysed in detail in the following sections testing the qualitative predictions on the relationship between individual quality and optimal time of arrival.

7.3 Costs and benefits of early arrival

The costs and benefits of early arrival are likely to determine the optimal time of arrival. The major cost of early arrival for swallows is probably poor environmental conditions and the major benefit is various factors related to mating success. These benefits include territory quality, mating success, breeding date, which influences seasonal reproductive success and the probability of recruitment by offspring, and the quality of mates. Is there any evidence for these costs and benefits used in the models of optimal arrival date? The single most obvious cost of early arrival in a migratory bird species is that environmental conditions for survival are at the lowest at the beginning of the arrival season, but improve as the season progresses (Møller 1993*h*). This is particularly important for insectivorous birds, which rely on the seasonal burst of insect food for reproduction. Thus, the probability of encountering poor weather conditions decreases dramatically as the season progresses. This is demonstrated by the frequency of frosty nights, which decreases rapidly during spring, and daily mean temperatures, which increase rapidly during spring (Fig. 7.2).

Migrating swallows

Extreme weather conditions sometimes kill large numbers of migratory birds. For example, on 13 May 1867, a sudden snowstorm in Denmark resulted in a dramatic drop in temperatures at a date when many migratory

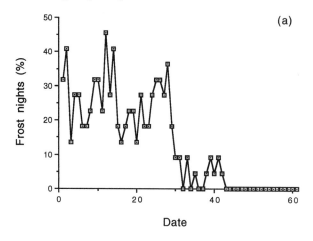

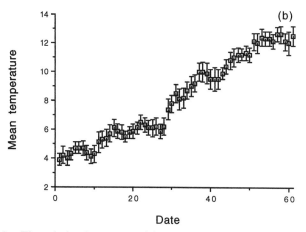

Fig. 7.2 The relative frequency of frosty nights (a) and mean daily temperatures during spring (b) in my Kraghede study area 1971–1992. Temperatures (°C) are mean (SE). Dates are days since 31 March. Barn swallows start arriving on date 20 with peak arrival on date 40–50.

birds already had arrived (Fischer 1869). Large numbers of insectivorous migrants, among them many barn swallows, were found dead or dying on the ground. Less severe weather conditions during April and May have sometimes resulted in mortality of early-arriving barn swallows. More than twenty male barn swallows were reported dead by farmers in my Kraghede study area during a cold spell early in the 1987 breeding season, and two of these males weighed only 15g which is a quarter less than the normal body mass. Less than 10% of the males and no females had arrived at this early stage of the breeding cycle. Most important, the tail of males which survived the cold spell was significantly longer than that of males found dead (Møller 1993*h*). This suggests that susceptibility to adverse weather conditions is directly related to the length of tails in male barn swallows. It is thus costly to arrive too early because of the increased risks of bad weather and starvation.

At least five types of benefit of early male arrival can be suggested:

(1) mating success;

(2) seasonal reproductive success;

(3) recruitment of offspring;

(4) mate quality; and

(5) territory quality (Møller 1993*h*).

Since there is an excess of males in the barn swallow population (Chapter 8), a fraction of males have to remain bachelors. If male mating success depended entirely on their time of arrival, only the latest arriving males would be unable to acquire a mate. However, male arrival date is not the only determinant of mating success since male tail length is also important. Analysis of the independent effect of male tail length and male arrival date suggests that both play a role in determining mating success (Møller 1990*a*). For example, experimental manipulation of male tail length has been found to affect the time it takes a male barn swallow to acquire a mate, such that short-tailed males acquire a mate more slowly, irrespective of their time of arrival (Møller 1988*c*). However, arrival date contributed independently to male mating success. There was a negative relationship between male mating success and arrival date[1] (Fig. 7.3), and this effect was found to remain even after controlling statistically for male tail length.[2] Early-arriving males thus have a higher probability of acquiring a mate than late-arriving male barn swallows.

Barn swallows are seasonal breeders, and there is only a limited time available for reproduction each year. A major determinant of reproductive success of mated male barn swallows is the number of broods reared per season rather than the number of nestlings per brood. Most barn swallows rear two broods per season, but second broods are never initiated after 10 August. This means that some late-breeding pairs will be unable to produce two broods in a season. The probability of raising two broods per season decreases as the season progresses (Fig. 7.4). Whereas all early-breeding barn swallows attempt to rear two broods, the frequency of second broods drops to zero for the late breeders.[3] Late arrival therefore confers a cost in terms of reduced or lost opportunities for raising a second brood.

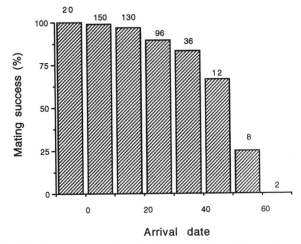

Fig. 7.3 Mating success of male barn swallows in relation to their arrival date. Dates are days since 30 April. Numbers are number of males. Adapted from Møller (1993*h*).

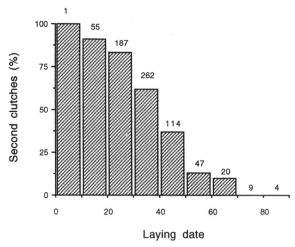

Fig. 7.4 The relative frequency of second clutches (% of pairs having two clutches) in relation to start of laying of the first clutch for barn swallows. Numbers are number of pairs.

As the probability of raising two broods decreases during the breeding season, so does the conditions for successful reproduction. Late-reproducing pairs are likely to encounter bad weather when raising their second brood, and nestling and fledgling mortality will therefore increase. The decreasing seasonal favourability for reproduction is reflected in the decreasing mean body mass and tarsus length of nestlings as the breeding season progresses.[4]

For a number of different bird species the probability of recruiting into the breeding population is known to depend on nestling body mass (Magrath 1991). Late-fledging barn swallows had lower recruitment rates than early ones (Fig. 4.3). In conclusion, late arrival confers a cost in terms of lower seasonal reproductive success and lower probability of offspring recruitment.

Early-arriving male barn swallows will be present before any females. They therefore have the opportunity to acquire females of the highest quality, whereas later-arriving males have to mate with the unmated females left. A reliable measure of the quality of a female is survival probability. If early-arriving males acquire mates of higher quality, their mates would be expected to have a higher rate of survival than mates of later-arriving males. This is a very conservative test because the mates of early-arriving male barn swallows more often lay two clutches per season and for that reason reproduce at a higher rate than the mates of late-arriving male barn swallows. Nevertheless, there was a highly significant effect of both male arrival date and male pairing date on female survival prospects (Møller 1993*h*). Females mated to long-tailed male barn swallows therefore appear to be of higher phenotypic quality since their survival rate increases with male tail length (Møller 1991*b*). It is therefore likely that early-arriving males acquired mates of higher quality.

Finally, early-arriving males may be able to acquire territories of superior quality because their choice is unconstrained by the choice of previously established males. Barn swallow territories are small, often only a few square metres, and the distance to the nearest neighbour is usually 4–8m. The only resources in the territory are nest sites and perches used for roosting and copulations. Nest predation was only rarely recorded during the whole study period, and thus safety from nest predators was not an important cue for males when choosing a territory. However, other properties of the territory, such as the quality of nest sites or the safety of perches, could be important. If females were to employ territory quality as a cue in their mate choice, male barn swallows should use a similar cue when choosing territories. Territories of high quality in one year would also be predicted to be of high quality in subsequent years if there were no obvious changes in their appearance (Møller 1990*a*). A territory was assumed to be occupied in two years if territories in those years overlapped partially. Territories were mapped carefully during observations of interactions between territory owners and intruders. The order of occupation of the same territory differed between years.[5] This suggests that territory quality was not an important factor affecting the order of settlement by males. Consequently, territory quality was not an important benefit accruing to early-arriving males.

In conclusion, early-arriving male barn swallows experienced a number of different costs and benefits of early arrival. Males, particularly those with short tails, sometimes died during adverse weather conditions early in the season. Early-arriving male barn swallows were more likely to acquire a mate of high quality and to reproduce early and thus enjoy high seasonal reproductive success, and the recruitment rate of their offspring was also higher than that of late arriving males.

7.4 Annual variation in arrival time

It is a common misconception that migratory birds arrive during a short time interval. Spring arrival by migrants demonstrates considerable variability like any other costly activity. Barn swallows arrive during a prolonged period in spring (Fig. 7.5). Males arrive between 23 April and 2 July, on average 16 May (SE = 0.48, n = 697), females arrive slightly later, between 26 April and 1 July, on average 20 May (SE = 0.80, n = 211).

There is considerable annual variation in average male arrival date, which may vary from 10 to 24 May. The annual variation in arrival date is large and highly significant, accounting for a small proportion of the variation.[6] These annual differences in arrival may be due to environmental conditions in the winter quarters, during migration, or in the breeding areas. I investigated whether this was the case by determining the relationship between arrival date and weather conditions in the African winter quarters during spring migration and in the breeding areas during spring arrival.

The departure from the African winter quarters and the speed of migration could be influenced by the condition of barn swallows and thus indirectly by the weather conditions in the winter quarters. Precipitation during winter and just before departure is known to influence overwinter

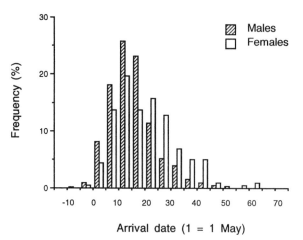

Fig. 7.5 Arrival dates for male and female barn swallows. Sample sizes are 697 and 211, respectively. Adapted from Møller (1993*h*).

survival in the barn swallow (Møller 1989c), and I therefore analysed whether male spring arrival at the breeding grounds was related to precipitation in the African winter quarters. The relationship between mean arrival date for male barn swallows and precipitation was statistically non-significant during the period 1978–1992 (Møller 1993h). However, the variation in arrival date of males measured as the coefficient of variation was positively related to precipitation in the South African winter quarters in different years.[7] High levels of precipitation just before spring departure thus appeared to increase variation in arrival date of male barn swallows. The only evidence for an effect of winter weather conditions on time of arrival at the breeding grounds was that variation in arrival was increased by precipitation in the winter quarters just before spring departure from Southern Africa.

Weather conditions in the breeding areas could also affect the time of arrival if barn swallows, for example, could anticipate weather in the breeding areas (Møller 1993h). This could be the case if weather conditions were similar on a large geographical scale. This is often the case because temperatures in large parts of Northern Europe are determined by regular low pressures passing from west to east. Mean arrival date of male barn swallows at the breeding sites was negatively related to mean minimum temperatures in May in different years,[8] whereas mean temperatures were unimportant for mean arrival date. Males therefore on average arrived later during cold springs. Weather conditions at the breeding grounds affected the timing of spring arrival by male barn swallows.

7.5 Individual quality and early arrival

The cost and benefit functions of arrival date presented previously (see section 7.2) may not be similar for all individuals. On the contrary, it is highly likely that individuals differing in quality have different cost and benefit functions (Møller 1993h). The cost function may be higher for low-than for high-quality individuals, if high-quality individuals have larger energy reserves, are better able to capture food, or to sustain periods of food shortage. This hypothesis was supported by field data because long-tailed male barn swallows were less susceptible to mortality during periods of adverse weather in early spring. The benefit function of high-quality individuals may also be higher than for low-quality individuals. For example, only high-quality males may enjoy the advantages of acquiring a high-quality mate if females discriminate against low-quality males, as suggested in Chapter 4. The optimality model for individuals differing in quality predicts that high-quality individuals should arrive earlier than low-quality individuals (Fig. 7.6).

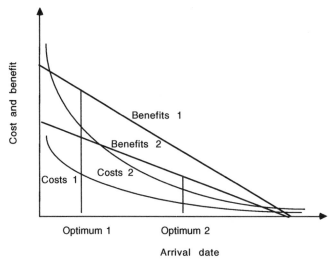

Fig. 7.6 An optimality model of arrival date for individuals differing in quality. Individuals are either of high (costs 1, benefits 1, optimum 1) or low (costs 2, benefits 2, optimum 2) quality.

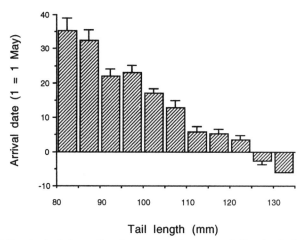

Fig. 7.7 Arrival date of male barn swallows in relation to their tail length (mm). Values are means (SE). Adapted from Møller (1993*h*).

One measure of male quality in barn swallows is tail length. The optimality model predicts that high-quality individuals should arrive earlier than individuals of low quality, and arrival date of male barn swallows should thus be negatively related to their tail length. Arrival date was actually related to male tail length (Fig. 7.7). Early-arriving male barn swallows had longer tails than late arriving males, with tail length explaining 35% of the variance in arrival date.[9] Older males are known to have slightly longer tails than young ones (Møller 1991*a*), and the relationship between arrival date and tail length thus could be due to long-tailed males being older and therefore more experienced. However, this did not appear to be the case since only tail length explained a significant proportion of the variation in arrival date[10] (Møller 1993*h*). Male barn swallows with long tails therefore arrived earlier than short-tailed males regardless of age.

If environmental conditions differ among years, one should expect the relationship between male arrival date and tail length to differ relative to the net benefits of early arrival (Møller 1993*h*). Males should arrive late only when conditions for spring arrival are poor. Poor environmental conditions may differentially affect individuals of low phenotypic quality, and low-quality male barn swallows should therefore arrive later than high-quality males in years when good weather comes late. Environmental conditions for arrival are better in years when the good weather comes early, and even males of poor quality should be able to arrive relatively early under such conditions. The negative relationship between male quality and arrival date should thus have been more pronounced under poor environmental conditions. This was in fact the case. Both male tail length, years, and their interaction affected male arrival date.[11] The effect of tail length on arrival date was much stronger in years when males arrived late than in early years (Fig. 7.8). The cost of early arrival should be relatively higher in years with a late average arrival date, and quality differences should affect arrival more markedly under such environmental conditions. Long-tailed males therefore arrived relatively earlier in years with poor environmental conditions.

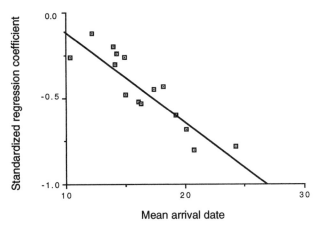

Fig. 7.8 Relationship between dependence of arrival date on male tail length (standardized regression coefficients) and mean arrival date. Data points are different years. Dates are days since 30 April. Adapted from Møller (1993*h*).

7.6 Aerodynamics and early arrival

Individual variation in spring arrival among male barn swallows can be explained to some extent by variation in male quality and responsiveness to environmental conditions in the winter and the breeding ranges. Long-distance migration requires much effort, particularly for a small bird weighing the same as an ordinary letter. Functional morphologists and theoreticians in aerodynamics have described the relationship between speed of migration and morphology. For example, the speed at which birds fly is directly proportional to their wingspan (Norberg 1990). Similarly, fluctuating asymmetry in tail (or wing) length could affect the speed of migration. A high proportion of the lift is generated by the outermost tail feathers of a barn swallow, and asymmetry in the length of the outermost tail feathers will greatly reduce lift (Balmford and Thomas 1992; Thomas 1993). Furthermore, the distribution of lift across the tail will be affected by asymmetry, which introduces large rolling and yawing forces. These can be partly overcome by flying with the tail at an angle. Manoeuvrability of male barn swallows will thus be affected by tail asymmetry. Could some of the individual variation in time of arrival at the breeding grounds be explained simply by variation in morphology?

The speed of migration was predicted to increase with increasing wing span. Wing span of barn swallows increases with their tail length, and that is the case in both males and females.[12] Long-tailed individuals should thus be able to migrate more quickly than short-tailed barn swallows. However, the predicted wing span for a male barn swallow with the shortest tail of 85mm is 324mm and that for a male with the longest tail of 137mm is 335mm. A difference in wing span of 9mm would only account for a difference in time of arrival by 3.4%, or approximately one day for a 30-day trip from South Africa to Denmark. Similarly, female barn swallows have tail lengths varying between 73 and 106mm, and predicted wing spans from 317 to 330mm. The difference in wing span of 13mm accounts for a difference in arrival time of 4.1% or 1.23 days. These differences in the duration of migration are minute compared with the arrival dates for male barn swallows of different tail length, as these vary by more than 40 days (Fig. 7.7). It would seem that individual differences in wing span do not account for individual differences in arrival times.

Fluctuating asymmetry in morphology could potentially affect the flight of barn swallows and thus the timing of their arrival, because asymmetries in wing and tail morphology may affect flight or manoeuvrability directly. Asymmetry in wing length and tail length was not positively correlated,[13] and either measure of fluctuating asymmetry could thus potentially affect speed of migration and timing of arrival independently. Arrival date of male barn swallows was not influenced by the degree of asymmetry in wing length,[14] whereas increasing levels of tail asymmetry apparently delayed the arrival of males.[15] This suggests that asymmetry in morphology may have some influence on the flight performance of barn swallows and thus their time of arrival at the breeding grounds. However, this conclusion rests on the untested assumption that the time of departure from the wintering grounds is unrelated to asymmetry in morphology.

Tail length was used as an index of male quality in previous analyses of arrival date. Fluctuating asymmetry in tail length was negatively related to tail length in males, but not in females.[16] Tail length and tail asymmetry could therefore independently affect the time of arrival at the breeding grounds. These effects could potentially be determined in a multiple regression analysis of the relationship between date of arrival by male swallows and four morphological variables: wing length, wing asymmetry, tail length, and tail asymmetry. There was indeed a highly significant relationship between arrival date and morphology, which explained a large proportion of the variation.[17] However, only male tail length accounted for a statistically significant proportion of the variance in arrival date.[18] This suggests that tail length *per se*, or properties related to tail length, strongly influence time of arrival, whereas wing length and wing and tail asymmetry are much less important. However, this conclusion needs to be confirmed by an experiment.

7.7 Summary

Arrival times for migratory animals can be seen as the result of an optimization process of costs and benefits of early arrival. The major cost of early arrival in the barn swallow is poor environmental conditions, which sometimes kill early-arriving individuals. The major benefits of early arrival are a higher mating success, enhanced reproductive success, improved recruitment rates for offspring, and an enhanced mate quality. Annual variation in arrival date is related to weather conditions at the breeding grounds, but also to some extent to weather conditions in the African winter quarters.

Individual variation in arrival time can be explained by phenotype-dependent cost and benefit functions of early arrival. Long-tailed male barn swallows arrived earlier than short-tailed males. The costs of early arrival should be particularly high under poor environmental conditions, and this prediction was confirmed by the stronger effect of male tail length on date of arrival in years when arrival was relatively late because of poor weather. Individual variation in male wing and tail morphology accounted for a large proportion of the variation in their date of spring arrival.

STATISTICS

1. Kendall $r = -0.31$, $z = 6.20$, $n = 181$, $P < 0.001$.

2. Kendall partial $r = -0.29$, $z = 5.85$, $n = 181$, $P < 0.001$.

3. Kendall $r = -0.39$, $z = 16.46$, $n = 801$, $P < 0.0001$.

4. First brood: body mass: $r = -0.33$, $t = 4.89$, $n = 198$, $P < 0.0001$; tarsus length: $r = -0.35$, $t = 5.00$, $n = 181$, $P < 0.0001$; second brood: body mass: $r = -0.29$, $t = 2.62$, $n = 77$, $P < 0.02$; tarsus length: $r = -0.25$, $t = 2.00$, $n = 62$, $P < 0.05$.

5. Spearman $r = -0.22$ to 0.26, $n = 10$–73 territories per year, $n = 9$ pairs of years, NS.

6. One-way ANOVA: $F = 6.28$, $df = 14,786$, $P < 0.0001$.

7. Linear regression: Mean precipitation in March: $F = 5.15$, $df = 1,6$, $\beta = 0.71$, $P < 0.05$.

8. Linear regression: $F = 5.50$, $df = 1,12$, $\beta = -0.56$, $P = 0.037$.

9. Linear regression: $F = 126.60$, $df = 1,231$, $P < 0.0001$.

10. Multiple linear regression: $F = 67.24$, $df = 2,208$, $P < 0.0001$, β (tail length) $= -0.63$, $P < 0.0001$, β (age) $= 0.006$, $P = 0.92$.

11. One-way ANCOVA: $F = 16.02$, $df = 15,217$, $P < 0.0001$; tail length: $F = 55.69$, $P < 0.0001$, year: $F = 7.68$, $P < 0.0001$, tail length $\times$ year: $F = 6.85$, $P < 0.0001$.

12. Linear regression: Males: $F = 34.13$, $df = 1,561$, $\beta = 0.24$, $P < 0.0001$; females: $F = 72.22$, $df = 1,614$, $\beta = 0.33$, $P < 0.0001$.

13. Spearman correlation coefficient: males: $r_S = 0.07$, $n = 372$, NS; female: $r_S = 0.002$, $n = 399$, NS.

14. Linear regression: $F = 0.23$, $df = 1,122$, NS.

15. Linear regression: $F = 6.97$, $df = 1,115$, $\beta = 0.24$, $P = 0.009$.

16. Linear regression: males: $F = 30.02$, $df = 1,370$, $\beta = -0.27$, $P < 0.0001$; females: $F = 0.05$, $df = 1,395$, NS.

17. Multiple linear regression: $F = 35.91$, $df = 4,112$, $R^2 = 0.56$, $P < 0.0001$.

18. Partial regression coefficient: $\beta = -0.74$, $P < 0.0001$.

8

Options for unmated males

8.1 Introduction

Even though barn swallows have a socially monogamous mating system, because of a male-biased sex ratio not all males are able to acquire a mate. The causes of variation in the proportion of unmated males in different years will be discussed in section 8.2. Why do some males remain unmated while others readily acquire a mate? This question will be analysed in section 8.3 by determining morphological, behavioural and parasitological differences between mated and unmated individuals.

There are two fundamentally different views about the reproductive success of unmated males. Either they are a permanently doomed group of low phenotypic quality with no or little reproductive success, or they are adopting an alternative reproductive strategy with a similar fitness pay-off to the mated male strategy. Given that male barn swallows do not acquire a mate, what are their reproductive options? There are at least three different, but not mutually exclusive ones: a commonly considered possibility is that unmated males have higher survival prospects than mated ones if it is more costly for males to reproduce than skip a year of reproduction. Enhanced survival prospects relative to mated males is therefore an option (section 8.4). The second possibility is a switch from mate acquisition effort to engagement in extra-pair copulations (section 8.5). This second option is based on the idea that males can allocate their mating effort to acquisition of mates or copulation partners. An unmated male barn swallow may not be worse off in terms of reproductive success if it is able to acquire extra-pair copulations. The net cost of reproduction for a 'professional' extra-pair copulation male would be low because it would not have to provide any costly paternal care.

The third reproductive option for unmated barn swallow males is to acquire a mate by inducing divorce in an already mated pair (section 8.6). Females of a large number of bird species frequently divorce their mates as a consequence of reproductive failure (Rowley 1983; Rohwer 1986). Unmated males may exploit this propensity to their own advantage by killing the offspring of breeding pairs. If females divorce their mates because of an enforced reproductive failure, the unmated male may subsequently have an opportunity to take over the divorced female, and may be able to raise at least one brood of young, provided that it accomplished the infanticide early in the season (Crook and Shields 1985). Obviously, mated barn swallows should attempt to avoid the risks of infanticide by guarding their nest and chasing off all potential infanticide perpetrators.

The fitness of unmated male swallows thus results from the combination of the reproductive success in future breeding seasons, success in extra-pair fertilizations, and success in committing infanticide. Unmated males can only be considered to engage in a true mixed reproductive strategy if their fitness equals that of mated males, whereas their behaviour can be considered a best-of-a-bad-job strategy if their lifetime reproductive success is lower than that of mated males (Smith and Arcese 1989). This chapter evaluates these alternatives.

8.2 The sex ratio

The sex ratio of animals is usually assumed to be equal because any pair investing in production of offspring of only one sex will be at a selective disadvantage. Production of broods with a biased sex ratio will soon make it highly advantageous for other pairs to produce broods with an opposite bias. The evolutionarily stable equilibrium therefore is an equal investment in offspring of the two sexes (Fisher 1930).

Sex ratios can be measured at different stages of the life cycle. The primary sex ratio is the ratio at conception, the secondary sex ratio is that at emergence of offspring, and the tertiary sex ratio that at the start of reproduction. Barn swallows have a male biased tertiary sex ratio, and the average excess of males in my study population was 12.5% (SE = 1.6, $n = 16$ years) during the period 1977–1992. The annual variation in the excess of male barn swallows could be due to higher female mortality rates in years with poor environmental conditions. Barn swallows suffer by far the most of their annual mortality outside the breeding season (Møller 1989c). This mortality is independent of density and related to environmental conditions in the winter quarters. Winter mortality is higher during relatively dry years, apparently due to scarcity of invertebrate food. If the excess of male barn swallows in the population were due to excess female mortality during poor environmental conditions, the excess of males in the breeding population should have been positively related to the overall winter mortality. This turned out to be the case. There was a strong positive relationship between overwinter mortality and the percentage of unmated males in the breeding population in the following breeding season.[1] There were relatively more unmated males in years when mortality had been high during the preceding winter.

The excess of adult male barn swallows could be caused by:

(1) a biased primary sex ratio;

(2) a biased secondary sex ratio; or

(3) a higher mortality rate among adult females.

I have no data on the primary sex ratio of barn swallows. It is impossible to sex barn swallows from external features until they become adults. However, it is possible to obtain an estimate of the secondary sex ratio from sexed dead juveniles. This method is based on the assumption that dead individuals comprise a random sample of birds. Of 69 roadkills in my study area during the years 1988–1992, 52.2% were males, which does not deviate from the null expectation of an equal sex ratio.

The hypothesis that the annual mortality rate was higher among females than among males could be tested because of the high breeding fidelity of barn swallows of both sexes. The annual survival rate of adult males was 28.4% (SE = 0.18, n = 9 years) which was significantly higher than the annual survival rate of adult females[2] (25.5%, SE = 0.9). The male bias in the breeding population can be calculated directly from the ratio of the estimates of sex-specific survival rates. This calculation predicts 11.7% unmated males, which is not far from the observed value of 12.5%. This sex difference in adult mortality rate is therefore sufficient to explain the excess of male barn swallows in the adult population.

8.3 Characteristics of unmated males

When choosing their mates female barn swallows may use as cues the male's appearance, its parasite loads, and it display activity. How did unmated males differ from mated males in terms of these three variables?

8.3.1 *Morphological differences*

Unmated males differed from mated barn swallows in age, the former being significantly younger than the latter.[3] A total of 46 males were unmated during one year, four during two years and one during three years. As most unmated males were yearlings, I have compared the morphology of unmated and mated yearling male barn swallows (Table 8.1). Unmated males differed from mated ones in six out of 12 variables. They had shorter tails and wings

Table 8.1. Morphology of mated and unmated yearling male barn swallows. Values are means (SE).

Morphological variable	Unmated male	Mated male	t	P
Beak length (mm)	7.69 (0.05)	7.64 (0.02)	0.73	0.46
Beak width (mm)	12.08 (0.09)	11.96 (0.03)	1.33	0.18
Beak depth (mm)	2.86 (0.04)	2.93 (0.01)	1.63	0.10
Tarsus length (mm)	11.51 (0.11)	11.42 (0.04)	0.75	0.45
Keel length (mm)	21.69 (0.33)	21.75 (0.07)	0.21	0.83
Wing length (mm)	125.02 (0.45)	126.72 (0.14)	3.73	0.0002
Wing asymmetry (mm)	0.88 (0.12)	0.56 (0.03)	2.60	0.009
Tail length (mm)	99.02 (1.05)	106.57 (0.40)	5.99	0.0001
Tail asymmetry (mm)	15.87 (1.71)	2.33 (0.21)	2.05	0.014
Short tail length (mm)	43.20 (0.31)	44.23 (0.12)	2.46	0.014
Badge area (mm^2)	212.82 (3.84)	201.30 (1.26)	2.41	0.017
Body mass (g)	19.00 (0.17)	19.15 (0.06)	0.85	0.40
n	47	439		

than mated males, but whereas the difference in tail length was 7.6%, the difference in wing length was only 1.4%. They also tended to be lighter in mass than mated males. There were clear differences in the degree of individual fluctuating asymmetry in both tail and wing length between unmated and mated males (Table 8.1), the former generally being more asymmetric than the latter. These differences in relation to mating status were not independent because morphological characters are correlated with each other. When these six morphological variables were used as predictors of male mating success in a stepwise logistic regression analysis, only tail length and individual tail asymmetry entered as a statistically significant variable, with unmated males having shorter and more asymmetric tails than mated male barn swallows.[4] Thus, tail length and tail asymmetry were the most important morphological differences between males of different mating status.

8.3.2 *Arrival date*

The arrival date of male barn swallows is strongly negatively related to their tail length, with long-tailed males arriving earlier than short-tailed males (Fig. 7.7). Males of different mating status also differed in their arrival date. Whereas mated males on average arrived on 15 May ((SE = 0.8 days), $n = 209$), unmated males arrived on average more than two weeks later (1 June (SE = 4.0 days), $n = 18$). This difference is statistically highly significant.[5] Arrival date also differed if the analysis was restricted to yearling males.[6] Arrival date and tail length contributed independently to male mating success as determined from a stepwise logistic regression analysis.[7] This was also the case when only yearling males differing in mating status were compared with each other.[8] The result of this analysis is also consistent with the results of tail-length manipulation experiments in which males were randomly assigned to treatments. Male barn swallows in different treatment groups therefore on average arrived at the same time, but still differed in the speed at which they acquired mates as determined by their experimentally manipulated tail length (Møller 1988c).

In conclusion, early-arriving male barn swallows are more likely to become mated than late-arriving males.

8.3.3 *Parasite loads*

One reason why some male barn swallows remain unmated could be that they carry larger parasite loads than the average male in the population. Females may be able to discriminate against males with high levels of parasite infestation by direct inspection or by the use of morphological or behavioural displays reflecting the parasite burden. There is a considerable amount of data suggesting that males indirectly reveal their parasite burdens in displays (see Chapter 9).

8.3.4 *Mate acquisition behaviour by unmated males*

Male barn swallows establish a small breeding territory shortly after arrival at the breeding grounds. The mating success of all males is still not determined at this early stage of the breeding cycle, because the proportion of unmated males depends on the degree of male bias in the sex ratio of adults returning to the breeding grounds.

All male barn swallows attempt to attract a mate by singing, and displaying their tails in flight and while perched. However, it is still interesting to ask whether males that remain unmated for the entire breeding season differ from mated males in their display activity. Males that remained unmated throughout the breeding season did not sing less than males that acquired a mate.[9] However, the rate of male flight displays of their tail feathers shortly after arrival differed; male barn swallows which subsequently became mated displayed at a higher rate than permanently unmated males.[10]

In conclusion, male barn swallows that subsequently achieved different mating status differed in their initial display rates; unmated males displayed their tail ornaments less frequently than mated males.

8.4 Survival prospects of unmated males

The survival prospects of male barn swallows can be estimated from the proportion of mated and unmated males returning to the breeding area. Survival rates calculated from return rates assume that almost all individuals are recaptured and that there is no relationship between dispersal and mating status of males. The first assumption can be tested by recording the number of individuals not recaptured in one year, but recaptured in a subsequent year. The number of more than one-year-old male barn swallows captured in one year, but not captured in the previous year was very low (6 of 401 males). This suggests that almost all ringed males are recaptured if alive. The second assumption, that dispersal distances do not differ between mated and unmated males, can be tested by comparing dispersal distances. Almost all males established territories in the same site in subsequent years. Only 0.7% ($n = 401$) of males moved to another site during one season or between seasons, and the largest distance moved by any male was 600m, a short distance compared with the size of my Kraghede study site.

The survival rate of unmated male barn swallows was only 18.2% (SE = 3.6, $n = 9$ years), while that of mated males was considerably higher with 29.1% (SE = 0.7, $n = 9$ years) survival. Most of the unmated male barn swallows were yearlings, and therefore, a more meaningful comparison of survival rates is between unmated and mated yearling males, respectively. The survival rate of unmated yearling males was 14.0% (SE = 3.1, $n = 9$ years) which is significantly smaller than the 29.4% (SE = 0.8, $n = 9$ years) survival rate of mated yearling males.[11] There is therefore no reason to believe that unmated male barn swallows survived better than mated males. On the contrary, unmated males suffered a higher mortality rate than mated male barn swallows.

8.5 Extra-pair copulations

Unmated males could allocate their mating effort to acquisition of extra-pair copulations rather than acquisition of a single mate. A 'professional' extra-pair copulation strategy would theoretically be highly profitable in terms of reproductive success because such males could glean all the benefits without paying any of the costs of raising offspring. Females could potentially also benefit from engaging in extra-pair copulations with un-mated males, if such copulations prevented these males from committing infanticide (Hrdy 1977, 1979; Altmann *et al.* 1978; Shields and Crook 1985). Is there any evidence that unmated male barn swallows were more successful in acquiring extra-pair copulations than mated males?

Both unmated and mated male barn swallows pursue extra-pair copulations by visiting fertile females in their territories, singing and displaying their tails and attempting to copulate with them. I recorded a total of 34 extra-pair copulation attempts by unmated male barn swallows, but none of these resulted in cloacal contact. In almost all these cases males did not succeed because females behaved aggressively towards the male or flew away. None of the 16 unmated male barn swallows which were followed throughout a breeding season was successful in acquiring an extra-pair copulation. The situation for mated male barn swallows was quite different. The total number of extra-pair copulation attempts involving mated male barn swallows was 474, and 26.9% of these resulted in what appeared to be successful copulations with cloacal contact. The proportion of mated males that succeeded in acquiring some extra-pair copulations was 33% ($n = 164$), which is considerably larger than the total absence of copulation success among unmated males.[12] Thus, unmated male barn swallows were much less successful in achieving extra-pair copulations than their mated conspecifics.

8.6 Infanticide behaviour

A farmer once told me that he had repeatedly put small barn swallow nestlings back into a nest, but that they kept falling out! Barn swallow nests are placed high above ground level, and there is a premium for nestlings to remain in their nest; their survival prospects are then better! What the farmer did not realize was that he had witnessed a case of infanticide by an unmated male barn swallow.

Unmated male barn swallows visit neighbouring territories particularly during two periods of the reproductive cycle, namely, the fertile period of the female, and the first days of the nestling period (Fig. 8.1). A number of different nests are visited by each male during the early nestling period, ranging from 0 to 5 with a mean of 2.1 nests (SE = 0.2, $n = 10$ males) (Møller 1988a). Unmated males will invariably be chased away by the nest owners if present. Visits sometimes become concentrated at a specific nest where live nestlings are subsequently removed by the unmated male.

I have witnessed infanticide on nestling barn swallows on four different occasions, and farmers have reported two additional cases of nestlings being thrown out of their nest. The unmated male flew to the unattended nest, picked up a nestling by its head, flew away and dropped it to the ground. The male usually then returned immediately, picked up another nestling and dropped it, continuing until the nest was empty. In 15 nests, all nestlings disappeared during the first few days after hatching and as nest predation was virtually absent, the cause of mortality of these broods was probably

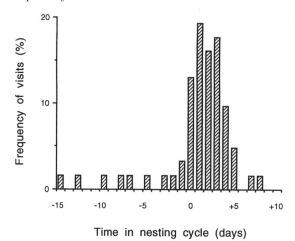

Time in nesting cycle (days)

Fig. 8.1 Visits by unmated male barn swallows at neighbouring nests in relation to time in the breeding cycle (day 0 is the first day of the nestling period) during the 1985 breeding season. Adapted from Møller (1988*a*).

Infanticide male

also infanticide. Nestlings were 1 to 4 days old, mean 2.3 days (SE = 0.3, $n = 19$ nests), when they disappeared.

In 15 of the 19 cases reported here, the male barn swallow nest owner disappeared before infanticide took place. Death was probably the cause of disappearance as was in fact verified in 6 cases. The unmated male that committed infanticide later mated with the widowed female. In the remaining 4 cases, the female was still mated when infanticide occurred, but subsequently mated with the killer of its previous brood. Renesting during the same season took place in 16 cases, and a new nest was initiated in the other three cases. All the renestings subsequently resulted in fledglings. Infanticide was the major cause of nestling mortality in the barn swallow population, as it accounted for 32.1% of 131 dead nestlings in 308 broods from the period 1982–1986 (Møller 1988a).

Male barn swallows should only attempt infanticide as long as female victims can renest soon afterwards. Infanticide is therefore predicted to occur only in first broods. Infanticide took place in 2.7% (19 of 701) of first broods, but never in any of 420 second broods during which renesting is not feasible. This differs markedly from the null expectation of equally many infanticides in first and second broods.[13] Unmated male barn swallows left their territories on average 9 July, range 18 June to 22 August ($n = 8$), much earlier than any other adult barn swallows.

Infanticide nests were not a random sample of all barn swallow nests in the population. In nests with infanticide egg laying had started later than in non-infanticide nests.[14] The tail length of infanticidal males was longer than that of unmated males that did not succeed in committing infanticide (Fig. 4.13). This suggests that infanticidal male barn swallows were not a random sample of the unmated male population.

Nearly one quarter of unmated males acquired a mate by committing infanticide as 23% of the 81 unmated males were successfully infanticidal.

8.6.3 *Nest guarding as anti-infanticidal behaviour*

I have never seen unmated male barn swallows fly up to a neighbouring nest in the presence of the nest owners. Barn swallows always attack or display aggressively to intruders approaching the nest. Infanticide should therefore be virtually impossible in the presence of at least one nest owner within the small territory. Intruding unmated males were regularly attacked if they approached an occupied nest, and all intrusions within a distance of 2m of the nest invariably resulted in an attack (Møller 1988a). A nest could therefore be considered to be guarded if a nest owner was present within that distance.

If effective nest guarding prevents infanticide, infanticide nests should be guarded less carefully than nests not suffering from infanticide. Nest guarding was estimated as the proportion of time spent within 2m of the nest by at least one pair member from daily one-hour observation periods during the first five days of the nestling period. Infanticide nests were guarded less closely than non-infanticidal nests.[15] This relationship was

Nest-guarding swallow

verified as being causal in a detention experiment in which males of colonial and solitary pairs were detained temporarily during a single morning of the early nestling period (Møller 1988a). Removal of the male caused a decrease in nest-guarding intensity from 85.6% in colonial control pairs to 69.2% in colonial experimental pairs, and from 77.2% in solitary control pairs to 50% in solitary experimental pairs. Nestling barn swallows were removed from 4 out of 10 experimental colonial nests but from none of the 10 control colonial nests. They were not removed from any of the 5 experimental solitary nor the 5 control solitary nests. Experimental colonial nests suffering from infanticide were guarded less intensely than other experimental colonial nests.[16] Nest guarding during the early nestling period was therefore an efficient way of preventing infanticide. The main reason for the success of infanticidal males was that nest owners were unable to guard their nest properly against potential infanticidal perpetrators. Newly hatched nestlings have to be brooded by the female, but they also need to be fed regularly. Male barn swallows provide a majority of the food for nestlings during the first days of the nestling period. Widowed females cannot brood (and thus guard their nest) and provide food at the same time. Therefore, their nests become exposed to unmated males because widowed females are forced to spend some time foraging away from the nest.

Both males and females contributed to overall nest guarding, although the female's contribution (87% of the time) was larger than that of the male (35.8%) (Møller 1988a). Pair members spend some time together at the nest, but this should be minimized in order to increase overall nest-guarding intensity, given that one parent is just as efficient a nest guarder as two parents. Barn swallow pairs were not all equally efficient nest guarders. The intensity of nest guarding during the first five days of the nestling period increased with male tail length and colony size.[17] The higher intensity of nest guarding by long-tailed male barn swallows was due to a higher intensity of guarding by the female rather than the male pair member.[18] Males with long tails therefore appeared to have acquired mates of higher quality, and these mates were more efficient nest guarders than the average female in the population.

In conclusion, nest guarding was a highly efficient way of preventing infanticide. Widowed females were unable to guard their nests intensely and at the same time provide their nestlings with food, and their nests were therefore exposed to infanticidal males. Barn swallow pairs with a long-tailed male were more efficient nest guarders than others, apparently because long-tailed males were paired with efficient nest guarders.

8.7 Summary

The tertiary sex ratio in barn swallows is usually male biased as a consequence of higher annual mortality rates among females compared with males. Unmated male barn swallows differed from mated conspecifics with respect to tail length, tail asymmetry, arrival date and parasite loads, and mated males performed better in every respect than unmated males. Male barn swallows that remained unmated during the breeding season displayed less than mated males. Unmated males did not achieve fitness equal to their mated male conspecifics, because their survival prospects were lower and they were less successful in acquiring extra-pair copulations than mated males.

A large proportion of unmated male barn swallows acquired a mate late during the breeding season by killing the nestlings of broods that were not properly guarded by their parents. Unmated males that were successful in killing nestlings invariably mated with the mother of the brood, because the female had either previously been widowed or subsequently divorced its mate. Several pairs that formed as a result of infanticide reproduced successfully later during the same breeding season. Barn swallow pairs attempted to guard their nestlings against potentially infanticidal male swallows during the first days of the nestling period, and nest guarding was an efficient way of preventing infanticide. The mates of long-tailed males were more efficient nest guarders than those of short-tailed males.

Unmated male barn swallows performed less well than mated males with respect to every fitness component, and they probably had lower lifetime reproductive success than mated males. Unmated males therefore apparently adopted a best-of-a-bad-job strategy.

STATISTICS

1. Spearman correlation: $r = 0.85$, $n = 9$, $z = 2.40$, $P = 0.016$.

2. Paired t-test, $t = 3.88$, $df = 8$, $P = 0.005$.

3. Unmated males: 1.04 years (SE = 0.03), $n = 51$; mated males: 1.50 (SE = 0.03), $n = 674$; Mann-Whitney U-test, $z = 4.42$, $P < 0.0001$.

4. $F = 18.78$, $df = 1$, 484, $P < 0.001$; tail length: partial $\beta = -0.27$, $P < 0.0001$; tail asymmetry: partial $\beta = -0.15$, $P < 0.01$.

5. Two-sample t-test: $t = 5.77$, $df = 225$, $P < 0.0001$.

6. Mated males: 17 May (SE = 1.0 days), $n = 143$; unmated males: 1 June (SE = 4.3 days), $n = 17$. Two-sample t-test: $t = 3.40$, $df = 158$, $P < 0.001$.

7. $F = 16.43$, $df = 2$, 209, $P < 0.0001$; tail length: partial $\beta = 0.12$, $P = 0.002$, arrival date: partial $\beta = 0.28$, $P = 0.0008$.

8. $F = 12.56$, $df = 2$, 156, $P < 0.0001$; tail length: partial $\beta = 0.10$, $P = 0.003$, arrival date: partial $\beta = 0.30$, $P = 0.002$.

9. Unmated males: 4.9 songs per hour (SE = 0.5), $n = 8$; mated males: 5.0 songs per hour (SE = 0.05), $n = 46$; Mann-Whitney U-test, $U = 65$, NS.

10. Unmated males: 0.7 displays per hour (SE = 0.2), $n = 8$; mated males: 1.2 displays per hour (SE = 0.3), $n = 46$; Mann-Whitney U-test, $U = 38$, $P < 0.05$.

11. Paired t-test, $t = 2.89$, $df = 8$, $P = 0.020$.

12. Fisher exact probability test, $P = 0.005$.

13. Wilcoxon matched-pairs signed-ranks test, $z = 3.72$, $P < 0.001$.

14. Infanticide nests: 27 June (SE = 3.1, $n = 19$); non-infanticide nests: 21 June (SE = 0.8, $n = 462$); two-sample t-test, $t = 2.00$, $P < 0.05$.

15. Infanticide nests: 42.7% (SE = 6.1), $n = 5$; non-infanticide nests: 81.4% (SE = 1.6), $n = 32$; two-sample t-test: $t = 4.92$, $P < 0.001$.

16. Infanticide nests: 51.3% (SE = 8.6), $n = 4$; non-infanticide nests: 81.2% (SE = 3.8), $n = 6$; Mann-Whitney U-test, $U = 0.5$, $P < 0.01$.

17. Multiple linear regression: $F = 3.33$, $df = 2$, 128, $P = 0.04$; tail

length: partial $\beta = 0.18$, $P < 0.05$; colony size: partial $\beta = 0.12$, $P < 0.05$.

18. Pearson product moment correlation: female: $r = 0.32$, $t = 3.84$, $P < 0.01$; male: $r = 0.09$, $t = 1.03$, NS.

9

Parasites and sexual selection

9.1 Introduction

Parasites constitute more than half of all organisms of the world, and individuals of most organisms are infected with parasites at least at some time during their life span (Price 1980). Parasites therefore become a potentially important force during natural and sexual selection. Parasites exploit the resources of the live bodies of other species, on which individual parasites spend most of their lives, causing some or even a great deal of damage (Price 1980). Parasites are often subdivided into microparasites (viruses, bacteria, fungi, protozoa) and macroparasites (helminths, arthropods). Microparasites are characterized by small size, short generation times, and high rates of direct reproduction within their hosts. The duration of an infection is relatively short, and these parasites induce either lifelong immunity, chronic infections, or death of the host. Macroparasites have longer generation times than microparasites and complete some part of their life cycle outside the host. The immune responses of the host generally depend on the number of macroparasites present, and are of relatively short duration, which makes the host susceptible to continual reinfection.

The conventional wisdom amongst parasitologists is that parasites usually do not severely reduce the fitness of their hosts. However, much experimental evidence suggests that parasites often constitute an important selective force as a result of their detrimental effects on fitness components of infected hosts (for example, Elton 1927; Anderson and May 1982; Møller *et al.* 1990).

Parasites continuously respond to the evolution of host resistance by adapting to the new host environments, and they may enter coevolutionary relationships with infected host populations, with changes in host resistance being followed by changes in virulence of parasites (Futuyma and Slatkin 1983). Coevolution of parasites and their hosts may continuously generate new additive genetic variance in the resistance to parasites and in the ability of parasites to exploit their hosts (Clarke 1979; Hamilton and Zuk 1982). Parasites therefore thrive better on very common host genotypes, and hosts may be better able to defend themselves against very common parasite genotypes (Haldane 1949). Frequency-dependent selection on hosts will reduce the rate of dispersal of parasites and the amount of damage they cause.

Considering the omnipresence of parasites, it is remarkable that their potential role in sexual selection was ignored until the influential paper by Hamilton and Zuk (1982). Their idea was that secondary sexual characters and sexual displays evolved to reveal the parasite status of hosts, and that females chose resistance genes as a result of their choice of the most extravagantly ornamented males. The Hamilton and Zuk hypothesis on parasites and sexual selection simultaneously provided an explanation for the strong female mate preferences and for the maintenance of genetic variance in viability. Virtually no empirical studies of the effect of parasites on host sexual selection were published for several years. Not until the publication of a series of papers from a symposium on parasites and host sexual selection in American Zoologist in 1990 and of the volume *Bird–parasite interactions* edited by J. Loye and M. Zuk was the topic properly addressed. It soon became clear that females also might benefit directly from their choice of parasite-free males. Direct fitness benefits could accrue either as a result of parasite-free males being more efficient parents, or of a lower probability of transmission of contagious parasites (review in Clayton 1991). These three different hypotheses probably will be very difficult to disentangle in real host–parasite relationships. The three hypotheses, the revealing handicap, the parental care, and the contagious parasite hypothesis, and their assumptions and predictions are described in section 9.2.

To date, relatively few attempts have been made to test the three current hypotheses on parasite-mediated sexual selection. Parasites often have a very skewed frequency distribution among hosts, with a minority of all hosts being infected and the variance in the intensity of infection being high because a few individuals suffer from high intensities of infection. Hosts may be uninfected either because of resistance to a particular parasite or because of lack of transmission. It is difficult to use natural variation in parasite intensities among hosts to test the hypotheses of parasite-mediated sexual selection because heavily infested host individuals may be poor-quality individuals for other reasons and therefore avoided by females prospecting for a mate. Experimental manipulation of parasite intensities therefore becomes essential for testing the predictions of the various hypotheses. I have used the barn swallow and its ectoparasites for this purpose. The different kinds of parasites, their abundance and their life cycles will be described in section 9.3. In the subsequent three sections (9.4 to 9.6) the barn swallow and its parasites will be used to discriminate between some of the predictions of the three hypotheses of parasite-mediated sexual selection.

9.2 Models of parasite-mediated sexual selection

Parasites potentially play a role in sexual selection of their hosts in at least three different ways. These are:

(1) female avoidance of infected males with poor parenting abilities;

(2) female avoidance of males with contagious parasites; and

(3) female choice of males with a genetically based resistance to parasites along the lines suggested by Hamilton and Zuk (1982).

These three alternatives are briefly discussed in the following paragraphs.

9.2.1 *Parasites and parenting ability*

Parasites exploit resources which otherwise might be used by the host for maintenance or reproduction. Heavily parasitized individuals are often young and of poor phenotypic quality, and the detrimental effects of parasites on low-quality hosts may be particularly severe. In the case of limited availability of resources, for example due to the detrimental effects of parasites, individual hosts are expected to reduce their resource allocation to reproduction at the expense of maintenance in an attempt to increase their residual reproductive value. Parasitized hosts should thus be less efficient parents than unparasitized hosts. If male hosts play a role in reproductive activities, females may simply discriminate against parasitized males because a male's contribution to reproduction would be expected to be inversely related to its degree of parasitism. Females may benefit in ways other than courtship feeding or male participation in parental care. For example, males may defend a breeding territory with resources essential for successful reproduction, and parasitized males may be less able to do so than parasite-free males. Consequently, females would acquire a direct fitness benefit from their choice of a parasite-free male. The negative effect of parasites on parenting ability should be even more important in females, which usually provide a larger contribution to parental duties than males. The effects of parasites on parenting ability should thus be relatively more important for male mate choice than for female choice.

Parenting ability is thus only important as a reason for female discrimination against parasitized males when males defend food resources and nest sites, or when they contribute to care of nestlings. All types of parasites may affect parenting ability. The fitness benefit of choosing a parasite-free male in this case is a direct one.

9.2.2 *Avoidance of contagious parasites*

Contagious parasites may affect sexual selection if individual hosts are able to assess the state of parasite infestation at a distance and thereby avoid contact with infected conspecifics (Freeland 1976). Individuals prospecting for a mate may avoid becoming infected themselves by keeping their distance from conspecifics which are or seem to be infected. Alternatively, contagious parasites of a partner may infest current or future offspring, and hence avoidance of infected individuals of the opposite sex could result in a direct fitness benefit. A special kind of contagious parasite generates venereal diseases, which are associated with the reproductive system and transferred during copulation. Avoidance of venereal diseases may have direct fitness benefits because of the effects of the disease on viability or functioning of the reproductive system. Contagious parasites are transmitted directly and therefore include all ectoparasites such as mites, ticks, louseflies, fleas, and feather lice, and certain species of endoparasites living in the reproductive tract such as venereal disease bacteria of the genus *Mycoplasma*. Sexual selection mediated by contagious parasites involves ectoparasites and venereal diseases, and the fitness benefits of mate choice by avoiding infected conspecifics are direct. The effect of contagious parasites on mate choice should apply to both sexes, unless there is a sex bias in prevalence and intensity of parasite infections. The effects of contagious parasites on parasite-mediated sexual selection should thus be as important for male mate choice as for female choice.

The critical problem for hosts of contagious parasites is how to assess the presence of parasites and their abundance on conspecifics. Contagion may be assessed at a distance:

(1) by direct observation of parasites;

(2) indirectly from the behaviour of hosts; and

(3) indirectly from the appearance of hosts.

The feasibility of direct assessment depends on crypticity of the parasites and thus also their size. Direct assessment by observation of parasites at a distance should be possible for some of the large taxa, such as ticks, whereas bacterial venereal disease probably is more difficult to assess directly. Ectoparasites, which leave bite wounds in bare skin patches, can be assessed indirectly by visual inspection.

The behaviour and appearance of contagious parasites have been modified by the anti-parasite reactions of hosts through evolution, and many ectoparasites are cryptically coloured as a way of avoiding attack by the host (Marshall 1981). The behaviour of ectoparasites has also been modified by host behaviour, and many ectoparasites are active primarily at times when hosts are less able to interfere with their resource acquisition (Marshall 1981). If the cryptic appearance and behaviour of ectoparasites have evolved primarily as a response to anti-parasite strategies of already infested hosts, it is easy to imagine that the presence of these parasites is even more difficult to assess at a distance.

The second way in which individuals prospecting for a mate may be able to assess infestations of ectoparasites is by means of host behaviour, such as various kinds of comfort behaviour and sexual displays. Hosts use a number of different comfort behaviours to eliminate or reduce parasite infestations. Among birds these may include scratching, preening of feathers, dust bathing, and the application of formic acid to feathers, so-called anting behaviour. Sexual displays and other energy-expensive activities may also be used by conspecifics to assess parasite infection levels.

Scratching swallow

Finally, individuals prospecting for a mate may use the appearance of conspecifics as a way of determining their parasite status. Parasites by definition exploit the resources of hosts and they may inflict damage if their resource acquisition severely interferes with that of the host. Sapping of the host's resources by parasites may result in secondary sexual characters and sexual displays being less extravagant than if the host was parasite-free. This is particularly likely to be the case among young, inexperienced hosts of poor body condition and naive immune defences. Females may be able indirectly to assess from the expression of sexual displays whether males are infested with serious contagious parasites and the extent of the infection.

Contagious parasites live on individual hosts, which constitute a patchy environment with a limited carrying capacity. After a certain population growth reproductive success can be achieved only by dispersal to new hosts. It is therefore of essential importance for parasites that infested and uninfested host individuals meet. While it is relatively important for hosts to mate with uninfested conspecifics, it is paramount for contagious parasites to achieve dispersal. Parasites therefore are at conflict with their hosts, and the parasites have a major advantage in this conflict because of their small size and short generation time. The entire fitness of the parasites depends on their successful exploitation of a host while host fitness usually is only partly determined by parasitism. The selection pressure on parasites is therefore stronger than that on hosts. It is in the interest of contagious parasites to hide their presence in their host because otherwise transmission may become impeded or impossible.

There is a second reason why it may be difficult for host individuals to avoid infestation with contagious parasites during mate choice. The sex roles during courtship usually differ dramatically. Whereas individuals of the choosy sex are coy, individuals of the chosen sex are eager to establish brief, but efficient sexual relationships with many individuals of the opposite sex. Sales resistance thus often characterizes female behaviour while males perform sexual salesmanship. These sex roles during courtship lead to physical encounters being initiated by individuals of the chosen sex and avoided by individuals of the choosy sex. How should females be able to keep their physical distance from potential mates during sexual display so as to avoid infestation with contagious ectoparasites? The answer is obviously that it is usually not possible, and that many ectoparasites, which are able to move rapidly, easily can transfer from one individual to another during courtship situations. The only contagious parasites which may be completely avoided by individuals prospecting for mates are those generating venereal diseases.

9.2.3 Parasites and the revealing handicap

Parasites withdraw essential resources from their hosts, and hosts therefore continuously evolve anti-parasite defences which are then overcome by the parasites. Parasite–host interactions may lead to a coevolutionary arms race in which the small, short-lived parasites with short generation time appear to be one step ahead. A major enigma in evolutionary biology is the consistent mate choice in many animal species, even when there appears to be no direct fitness benefits, as for example in animal leks. This so-called lek paradox of consistent choice without any apparent direct fitness benefits can be resolved if in fact females choose mates on the basis of their advertised resistance to parasites (Hamilton and Zuk 1982). The idea is based on the revealing handicap, which evolves independently of intensity of infection. It is only the expression of the male sex trait which depends on parasite abundance. A revealing handicap has therefore evolved as the most efficient way of revealing the level of parasite infection. Alternatively, a male sex trait may be a condition-dependent handicap with the level of exaggeration depending on condition. Parasites are likely to affect the body condition of their hosts negatively and thus indirectly affect the expression of their secondary sex traits. Condition-dependent male sex traits have, however, not specifically evolved to reveal the level of parasite infection. By choosing the most extravagantly ornamented males, females will acquire mates with few parasites if only parasite-free males are in a sufficiently good condition to be able to produce an extreme secondary sexual character. The secondary sexual character can be viewed as a revealing handicap which has evolved to facilitate discrimination among males with different levels of parasitism. To mate with the most ornamented males means to mate with resistant males, and as a consequence a genetic correlation becomes established between the female preference and the signal for resistance to parasites. An evolutionary improvement in parasite resistance will also cause the evolution of a more extreme mate preference, which will promote the evolution of ever more extravagant secondary sex traits. The ornament and the preference are supposed never to reach an evolutionary equilibrium because host genes for resistance change continuously as a result of the coevolutionary arms race between hosts and parasites.

The parasite hypothesis of Hamilton and Zuk (1982) makes several predictions, two of which are particularly important. First, the hypothesis assumes a genetic correlation between resistance genes and female mate preference, and, second, an evolutionary increase in resistance among hosts as a result of the female preference (Kirkpatrick and Ryan 1991). These two predictions are not readily tested in free-living animals because of the need for extensive genetic information across generations. Therefore, a number of field studies have investigated more limited predictions which, as we shall see, may be open to alternative interpretations (Møller 1990*d*; Clayton 1991). These predictions are that:

(1) parasites reduce host fitness;

(2) there is heritable variation in host resistance;

(3) the expression of secondary sexual characters is affected by the intensity of the parasite infestation;

(4) females get mates with few parasites as a consequence of mate choice based on the secondary sexual character; and

(5) there is assortative mating with respect to parasite loads.

It has been claimed that revealing or condition-dependent handicaps are unable to evolve in a monogamous mating system because there is no force for sexual dimorphism in the absence of an initial asymmetry in the expression of the characters in males and females (Heisler *et al.* 1987). For example, mate choice could occur in both sexes as a result of a parasite-dependent expression of a morphological character. An initial sexual asymmetry could, however, arise in a number of different ways. For example, males and females always differ in their reproductive roles, even in monogamous species with shared parental duties. Different roles in reproduction might render individuals of one sex more susceptible to parasitism because of the immunosuppressive effects of circulating hormones (Grossman 1985; Folstad and Karter 1992). Sex differences in reproductive roles could also affect the costs of sexual dimorphism, even in the initial stages of the process of sexual selection. Alternatively, different sex roles during reproduction might result in biases in sex differences in the rates of recombination and ultimately the benefits of mate choice (Trivers 1988). Female choice has usually been assumed to favour traits in males, such as extravagant tail ornaments, which are useful for their sons' mating success. However, sexual selection and female choice may often be adaptive with a bias towards the genetic interests of daughters (Seger and Trivers 1986). The heterogametic sex usually shows less recombination across its autosomes than the homogametic sex. Males may have been selected to link their genes more tightly in order to preserve their more intensely selected beneficial gene combinations.

9.3 Description of barn swallow parasites

Barn swallows, like most other organisms, are infected by a number of different parasite taxa. A total of 12 different species regularly occur in my Kraghede population, and several microparasites probably have gone undetected. The most common barn swallow parasites are depicted in Fig. 9.1, and information on their life cycle, behaviour, prevalence, intensity and damage to the host is given in the following sections. A number of other parasites have been found on barn swallows in other areas. These include 11 species of blood parasites (Protozoa), 16 species of helminths (eight Cestoda, five Trematoda, three Nematoda), eight species of mites (Acari), and among the insects three species of louseflies (Hippoboscidae), seven species of fleas (Siphonaptera), and one species of feather louse (Mallophaga).

The tropical fowl mite *Ornithonyssus bursa* is a gamasid mite with a body length of 0.7mm (Fig. 9.1). It is a generalist cosmopolitan species identified from a number of different bird species, but only quite rarely from mammals (Micherdzinski 1980; Gjelstrup and Møller 1986). The life cycle includes one larval and two nymphal stages known as proto- and deutonymphs (Sikes

Ornithonyssus bursa

Stenepteryx hirundinis

Ceratophyllus hirundinis

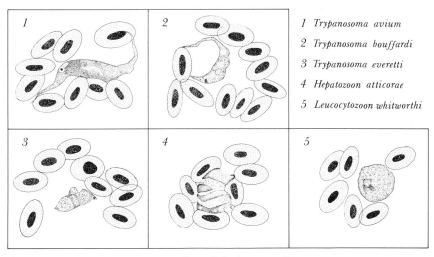

1 Trypanosoma avium

2 Trypanosoma bouffardi

3 Trypanosoma everetti

4 Hepatozoon atticorae

5 Leucocytozoon whitworthi

Fig. 9.1 The most common parasites of the barn swallow.

and Chamberlain 1954). Each female mite produces one to seven, most frequently two to five eggs per clutch, and clutches are produced with brief intervals. The eggs hatch within 1.5–2 days, and the larvae, which do not feed, moult into the protonymph stage after less than a day. The protonymph needs two blood meals to moult into the deutonymph stage, which takes place after a period of 2–3 days. The deutonymph moults without feeding within less than a day. The entire life cycle is 5–7 days, and one barn swallow reproductive cycle thus equals 8–10 mite generations. Numbers of mites build up rapidly; the maximum number extracted from one barn swallow nest was about 14,000 (Gjelstrup and Møller 1986). Adult mites are continuous parasites, requiring frequent blood meals throughout their life. They need at least two blood meals before being able to reproduce, and whereas female mites remove on average 0.077mg of blood per meal, males remove only 0.025mg. Each bout of blood-sucking from barn swallow nestlings only takes a few minutes.

The behaviour of mites changes during the nestling period of their barn swallow hosts (Powlesland 1978; Petersen 1979; own observations). When nestlings are small with an undeveloped plumage, mites are photophobic, sucking blood from the very thin skin on the feet and spending most of their time in the nest material. When the skin of the feet hardens, mites transfer to the developing plumage, especially on the wings and the head. Many mites are probably transported from the nests as nestlings fledge. A day after fledging, the mites left in the nest become phototactic and aggregate on the rim of the nest where they may infest visiting barn swallows.

Mites live on adult barn swallows and in their nests throughout the year. Nests harbour up to 500 mites before the spring arrival of the barn swallow hosts, and this is the case even for nests subjected to frost during winter. Mites have also been found on recently arrived barn swallows at sites where all old nests had been removed. Mite abundance in nests after host reproduction is positively related to that of adult barn swallows upon arrival at the breeding grounds (Møller 1990c). Mite populations in nests thus primarily originate from propagules inoculated by the adult barn swallows. Approximately one third of the newly arrived adult barn swallows carry mites; 38 is the maximum number recorded so far (Table 9.1). Most mites are found on the head of adult hosts because that is the site where hosts have most difficulty removing parasites. Approximately one third of the barn swallow nests are infested with the tropical fowl mite, slightly more in second than in first clutches (Table 9.1).

Birds infested with the tropical fowl mite become anaemic, droopy and progressively emaciated because of the loss of blood, the many bite wounds, and a state of immune shock (Furman 1963). Mites may also act as vectors for diseases and are known to transmit, for example, the western equine virus (Sulkin and Izumi 1947).

The feather lice *Myrsidea rustica* and *Hirundoecus malleus* are abundant on adult and nestling barn swallows. *M. rustica* is a *c.* 1.5 mm feather louse only known from the barn swallow. Adults chew small holes in feathers and other hard tissues. They are also said to suck blood and are therefore potential vectors for microparasites. Their presence is readily visible owing to the small holes they make in the feathers, and adult *M. rustica* sometimes move around between the feathers of their hosts when the plumage is disturbed during capture and handling. Their damage consists only of minor destruction of the plumage and probably insignificant blood loss. The small holes in the feathers may affect the energy cost of flight and the costs of thermoregulation. The prevalence and intensity of *M. rustica* are high in both nestling and adult barn swallows (Table 9.1).

The feather louse *Hirundoecus malleus* is only known from the barn swallow. The species is small (*c.* 1 mm) and individuals, which live between the barbules of feathers, can be seen as small dark spots when the feathers are held against a light source. Adult *H. malleus* feed on feathers and blood and the only damage they cause is their consumption of parts of the feather, which potentially may affect its strength, the energy cost of flight and the costs of thermoregulation. The prevalence and intensity of *H. malleus* are high in both nestling and adult barn swallows, and individual hosts sometimes harbour hundreds of feather lice (Table 9.1).

Table 9.1. Prevalence and intensity of infection of barn swallows with parasites. Intensities for the Haematozoa are reported as numbers per microscope field of a blood smear, and intensities of *Myrsidea rustica* as number of holes in the tail of the barn swallow host.

Parasite species	Prevalence	Intensity	*n*
Haemoproteus prognei	7.7	0.7	535
Trypanosoma avium	0.2	1.0	535
Trypanosoma bouffardi	0.2	1.0	535
Splendidofilaria mavis	0.4	1.5	535
Leucocytozoon whitworthi	0.6	1.3	535
Ornithonyssus bursa			
Adults	27.9	7.0	1567
Nests	43.0	73.0	833
Hirundoecus malleus	84.3	82.9	1567
Myrsidea rustica	78.1	15.6	1567
Stenepteryx hirundinis			
Adults	0.4	1.0	1567
Nests	2.2	2.7	833
Ceratophyllus hirundinis			
Adults	0.3	1.0	1567
Nests	1.0	1.7	833

The swallow louse fly *Stenepteryx hirundinis* is a large parasitic fly *c*. 8mm long (Fig. 9.1). The adult flies are barn swallow specialists and live between the feathers of adult and particularly nestling barn swallows, where they feed on blood. Swallow louseflies are usually scarce, but somewhat more prevalent on nestlings compared with adult birds (Table 9.1). Louseflies act as vectors for Haematozoa (Marshall 1981).

The swallow flea *Ceratophyllus hirundinis* is a parasite *c*. 1.5mm in size and known only from the nests of barn swallows (Fig. 9.1). The adult fleas regularly suck blood from their hosts while the larvae live from debris in the nest material. Prevalence and intensity of the swallow flea are very low (Table 9.1).

The blood parasites include *Haemoproteus prognei*, *Trypanosoma avium*, *Trypanosoma bouffardi*, *Splendidofilaria mavis*, and *Leucocytozoon whitworthi* (Fig. 9.1). *H. prognei* is transmitted by ornithophilic knots of the genus *Culicoides*, but there has been no life history studies of this parasite. Haematozoa of the genus *Haemoproteus* tend to be host-family specific. Common vectors are louseflies or biting midges. The parasite is ingested by a vector in which it undergoes a series of developmental changes before transmission to another host via a bite. In this definitive bird host the infestive sporozoite disappears from the blood and multiplies in the lungs and reticuloendothelial system. Gametocytes then appear in the blood where they infect up to 90% of the erythrocytes with high intensity infections lasting many months. Infestations may last more than a year with maximum intensities during reproduction and other periods of stress. Immature bird hosts usually have higher prevalences and intensities of infestation than adults. Few and weak symptoms and pathological effects of hosts have been reported for *Haemoproteus* (Bennett 1987; Threlfall and Bennett 1989).

The life cycles of the two species of *Trypanosoma* have not been described. Trypanosomes are intercellular organisms living in the blood stream of their hosts. They are generally not host specific. The intensities of infestation are often low, making detection difficult. Vectors of trypanosomes are biting Diptera, which ingest infestive life stages in a blood meal. Binary fission occurs in the stomach of the vector, and infective stages of the parasite are deposited on the host in faeces. The trypanosomes subsequently penetrate the skin, possibly through the bite wound produced by the vector. No symptoms or clinical effects have been reported for hosts from Haematozoa of the genus *Trypanosoma* (Bennett 1987; Threlfall and Bennett 1989).

L. whitworthi is transmitted by ornithophilic simuliids, but the life cycle has not been described. Haematozoa of the genus *Leucocytozoon* are host-family specific with usually only one species being found in each family. Infective stages develop in a vector, which becomes infested during feeding. Oocytes give rise to sporozoites, which migrate to the salivary glands of the vector. After transmission to the final bird host, prepatent periods occur in the liver and various other organs. Gametocytes then appear in the erythrocytes. Infections may last for more than a year, with maximum prevalences and intensities of infection occurring during reproduction and other periods of stress in the host. The symptoms of infections in hosts are anaemia and in severe cases mortality, sometimes mass mortality (Bennett 1987; Threlfall and Bennett 1989).

The life cycle and vectors of *S. mavis* have not been studied, but other avian Filaroidea have a life cycle of at least a year, with biting Diptera acting as vectors.

The prevalences and intensities of all the blood parasites in barn swallows were quite low (Table 9.1). This may be due to the aerial behaviour of barn swallows and their nesting sites, which are not particularly favourable for biting dipteran vectors.

9.4 Parasites and parental care

Male barn swallows defend a small nest-site territory and participate in nest building, incubation (the North American subspecies only), and feeding of nestlings. All males regardless of their parasite load start defending a small territory upon arrival. There was no apparent preference for particular territories, since the tail length of male barn swallows occupying the same territory in subsequent years was not positively correlated (see section 5.4.1). The ectoparasite loads of males occupying the same breeding territory in two subsequent years were not positively correlated.[1] Male tail length and their ectoparasite loads were therefore not associated with occupation of a particular breeding territory.

Female barn swallows may benefit directly by choosing a parasite-free male if the extent of male parental care is associated with parasite load. This hypothesis was tested experimentally by manipulation of the intensity of infection of first clutch nests with the tropical fowl mite. Nests were either sprayed with a pesticide, kept as controls, or inoculated with *c.* 50 mites. This manipulation affected the intensity of infestation of both nestling and adult barn swallows (Møller 1991*d*). The relationship between male feeding rate and experimental treatment was investigated for the treatment alone, and after controlling for brood size and mean nestling mass (Møller 1993*e*). Brood size and mean nestling mass had to be controlled statistically because large broods of heavy nestlings need higher feeding rates than small broods of light nestlings. Male feeding rate was determined as the absolute mean feeding rate during one-hour daily observation periods throughout the nestling period and as the percentage of feeding visits by the male.

There was a weak and statistically non-significant relationship between absolute male feeding rate and mite treatment (Fig. 5.3a). This relationship disappeared completely when brood size and brood mass were controlled statistically. The relative feeding rate by the male parent was unrelated to mite treatment (Fig. 5.3b). This was also the case after controlling for the effects of brood size and mean nestling mass (Møller 1993*e*). The feeding rate of unmanipulated barn swallow broods was also unrelated to the intensity of mite infestation in previous years both before and after the effects of brood size, mean nestling mass, and breeding date were controlled statistically (Møller 1993*e*). Although this experimental test of the parenting ability hypothesis cannot reject the null hypothesis of no relationship between the extent of male parental care and the intensity of mite infestations, this may not justify acceptance of the null hypothesis. Sample sizes of the experiment were relatively large, and as the alternative hypothesis clearly was directional, a one-tailed test could be used. The power of the statistical tests thus exceeds 80%, which justifies acceptance of the null hypothesis of no relationship between mite abundance and parenting ability.

Debilitating parasites should affect the parenting ability of male barn swallows as well as females. The effect on females might be expected to be stronger than that on males because females invest relatively more in reproduction than males. The parenting ability hypothesis would thus predict that males should choose parasite-free mates because of their superior parenting ability. However, female barn swallows that had nests experimentally infested with many tropical fowl mites did not feed their offspring less often than females which had their nests sprayed with a pesticide (Møller 1993*e*).

In conclusion, there is little evidence of a direct fitness benefit accruing to females in terms of increased paternal care by parasite-free mates.

9.5 Avoidance of contagious parasites

Contagious parasites may play a major role in parasite-mediated sexual selection. Often when I handle adult barn swallows after capture, several species of ectoparasites, such as tropical fowl mites, swallow louseflies, and feather lice of the species *Myrsidea rustica*, move across to my hand from the bird. Some of these cause quite irritating or painful experiences, and I therefore always do my best to avoid transfer of contagious parasites. Barn swallows, too, may suffer from these parasites, and may likewise attempt to avoid being infested.

Which signs could female barn swallows potentially use to assess whether males are infested with contagious parasites? The intensity of comfort behaviour could be used if it reflected the infestation rates by parasites. This hypothesis was tested experimentally by manipulating parasite loads of barn swallow nests. The intensity of infestation by the tropical fowl mite of barn swallow first clutch nests was manipulated by either spraying nests with an insecticide, keeping them as controls, or increasing mite loads by inoculating nests with *c*. 50 mites. This manipulation affected the mite intensities of nests and adult and nestling barn swallows (Møller 1991*d*). Barn swallows regularly preen their feathers, and one of the functions of this activity may be removal of ectoparasites. Male barn swallows have been found to preen more than their mates, and the preening activities of pair members were positively correlated (Møller 1991*d*). This is consistent with the observation that the mite intensity of male barn swallows is positively correlated with that of their mates (Møller 1991*c*; see also Fig. 9.10). However, the preening activity of barn swallows was unrelated to the experimental treatment of mite loads. Adult barn swallows which had mites added to their nests did not preen more than controls. The controls did not preen more than adults which had their first clutch nest sprayed with a pesticide (Møller 1991*d*). Barn swallows may preen in order to remove other ectoparasites, but preening activity of adults was not related to intensities of feather louse infections. It was therefore impossible to assess mite infestations of adults from their preening activity.

Preening swallow

Nestling barn swallows were apparently more susceptible to the effects of the tropical fowl mite as their preening activity was clearly affected by the manipulation of mite loads (Møller 1991*d*). Nestlings from sprayed nests preened much less than controls which preened less than nestlings from mite-inoculated nests. Tropical fowl mites were therefore able to affect the comfort behaviour of nestling barn swallows, although adults were unaffected.

A second possibility for female barn swallows to assess male infestations by contagious parasites is by the use of song and other sexual displays. The tropical fowl mite sucks blood from its host, and haematophagy may result in a reduced blood volume as well as an immune response by the host. The haematocrit value of adult male barn swallows was affected by the mite manipulation experiment, with males from sprayed nests having a mean haematocrit of 63% and males from mite-inoculated nests one of 50% (Møller 1991*f*). Energy-demanding activities like sexual display will probably be affected by a change in the metabolic efficiency caused by a reduction in blood volume. Male barn swallows sing at a maximum rate during the fertile period of their mate, during both first and second clutches. Song output by male barn swallows during the fertile period was clearly negatively related to the abundance of tropical fowl mites in their nests (Møller 1991*f*). The causality behind this relationship was tested experimentally by manipulating parasite loads of first-clutch nests during the laying period. There was no difference in song output between the three groups of male barn swallows during the first fertile period, as parasite manipulations had not been effective at that stage, and because males were randomly assigned to treatments (Fig. 9.2). However, the singing activity of male barn swallows was affected during the second fertile period because males which had mites added to their first-clutch nest sang considerably less than males receiving the other two treatments (Fig. 9.2). Males with mite-inoculated nests also sang less during the second fertile period than they did during the first fertile period. Females (and males) therefore potentially could assess the intensity of infestation of conspecific males by the tropical fowl mite by listening to song activity. Male barn swallows also perform a sexual display of exposing

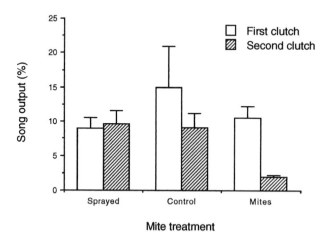

Fig. 9.2 Song output (% scans every second minute with song) of male barn swallows during their first and second clutches in relation to experimental treatment of mite loads of first-clutch nests during the egg-laying period. Values are means (SE). Adapted from Møller (1991*f*).

their long tail feathers, and their display intensity may potentially also be affected by mite infestation. The size of the outermost tail feathers in male barn swallows is also influenced by the presence of tropical fowl mites, as we shall see in section 9.6.

Even if contagious ectoparasites are transmitted between infested males and their mates or offspring, this may not necessarily be particularly costly measured in terms of fitness. For example, some parasites do little damage in terms of reducing offspring growth or survival, and perhaps a minor extra parental investment may overcome such effects (Møller *et al.* 1990). The fitness consequences of contagion were investigated by simulating a good and a poor mate choice. First-clutch nests of barn swallows were randomly assigned to one of three treatments:

(1) the nest was sprayed with a pesticide that killed all parasites;

(2) the nest was kept as a control; or

(3) the nest was inoculated with *c.* 50 tropical fowl mites.

The spraying treatment would mimic the choice of a parasite-free partner while the mite-addition treatment would mimic the choice of a heavily infested partner because mite loads of nests are determined by the mite infestations of nest attendants (Møller 1990*c*). Fifty mites were close to the maximum number estimated on an adult barn swallow.

The tropical fowl mite had several effects on the reproduction of its barn swallow host. Most prominently, overall breeding success from egg laying until independence of fledglings was strongly negatively affected by the ectoparasites. A comparison of overall breeding success revealed that sprayed nests had approximately 30% higher success than mite-infested nests, in both first and second clutches (Møller 1990*c*). Seasonal production of independent offspring was therefore strongly affected by the presence of the tropical fowl mite (Fig. 5.2). The reproductive success of parents was reduced by the parasite treatment for several different reasons. First, adult barn swallows discriminated between old nests with and without mites, and the start of reproduction was delayed if only infested nests were available (Møller 1990*c*). Second, nestling barn swallows fledged prematurely in the mite addition treatment apparently in an attempt to escape the biting mites (Møller 1990*c*). Third, reproduction was delayed in the second clutch if the first clutch was infested with mites, because barn swallows then usually built a new nest for the second clutch (Møller 1990*c*). Fourth, a smaller proportion of barn swallows reared a second clutch if their first clutch nest was infested with mites (Møller 1990*c*). Fifth, adult barn swallows alleviated the negative effects of mites on offspring growth and survival not by adjustment of their optimal clutch size during laying (Møller 1991*g*), but by adjustment through brood reduction (Møller 1993*i*). Sixth, the tropical fowl mite increased the cost of reproduction of their barn swallow hosts because reproductive activities such as production of a second clutch at all, timing of the second clutch, and the size of the second clutch were affected by both reproductive decisions and mite intensities during the first clutch (Møller 1993*i*). Finally, offspring weighed less if heavily infested with mites. Their survival prospects would therefore be reduced after fledging (Møller 1990*c*). These fitness consequences of ectoparasitism in the barn swallow are not unique because a large number of experimental studies of birds have also revealed important reductions in host fitness (review in Møller *et al.* 1990).

Contagious parasites can obviously be transmitted between all infested and uninfested hosts. The risk of transmission of contagious parasites should therefore affect the mate-choice decisions of both male and female barn swallows. The effect on females might, however, be expected to be stronger than that on males because males more often have bodily contact with conspecifics during extra-pair copulation attempts. The mates of the most attractive males should therefore run higher risks of becoming infested with contagious parasites, and this should clearly result in a female mate preference for less attractive males.

In conclusion, the tropical fowl mite is a serious contagious ectoparasite with severe fitness consequences for its barn swallow host. Avoidance of potential mates infested with tropical fowl mites therefore may play an important role in sexual selection.

9.6 Indirect benefits of choice of parasite-free mates

The Hamilton–Zuk hypothesis of sexual selection makes a number of predictions, two of which would most directly support the sexual selection process for resistance genes namely (1) a genetic correlation between resistance genes and the female mate preference, and (2) an evolutionary increase in resistance among hosts. These two predictions would require detailed genetic analyses of host populations during a period sufficiently long to allow an evolutionary response to selection in the host. As the duration of the host-parasite cycles are unknown, a direct test of these predictions would lie outside the possibilities of most field workers.

Previous tests of the Hamilton–Zuk hypothesis have therefore investigated a number of more indirect predictions (reviews in Møller 1990*d*; Clayton 1991; Zuk 1991). This hypothesis predicts that:

(1) parasites reduce host fitness;

(2) there is heritable variation in host resistance;

(3) the expression of secondary sexual characters of the host is affected by the intensity of the parasite infestation;

(4) females acquire mates with few parasites as a consequence of mate choice based on the expression of secondary sexual characters; and

(5) there is assortative mating with respect to parasite loads.

These predictions were tested for the barn swallow, as will be described in the following.

1. *Parasites reduce host fitness.* Parasites must be detrimental to host fitness for the costly female preference of parasite-free mates to evolve and be maintained. Benign parasites would not fulfil this requirement. There is considerable evidence for the negative effects of the tropical fowl mite on various fitness components of the barn swallow (see section 9.5; Møller 1990c, 1990e, 1993i). Prevalence or intensity of infection by the 11 other parasites are only weakly related or unrelated to survival prospects, or any of the measures of reproductive success recorded for the barn swallow during the period 1983–1992. It is thus unlikely that these parasite species currently play an important role in sexual selection of their barn swallow host. The prediction that parasites reduce host fitness is not exclusive for the Hamilton-Zuk hypothesis as the parenting ability and the contagious parasite hypotheses also assume a detrimental effect of parasites on host fitness.

2. *There is heritable variation in host resistance.* Demonstration of heritable variation in resistance requires estimates of the resemblance between relatives in parasite intensities. An upper limit to the heritability of resistance is the repeatability of parasite numbers on individual hosts at sequential sampling events (Falconer 1981). Ectoparasite loads of barn swallows are highly repeatable both within and between years (Møller 1991c). The differences between within-year and between-year repeatabilities are usually small, which indicates that individual barn swallows with many ectoparasites on one occasion are very likely to have many ectoparasites also on subsequent sampling occasions independent of whether they are in the same or a different year (Møller 1991c). Parasites often affect young individuals more than older ones because of an acquired immunity, and parasite intensities may therefore drop in older age classes (Wakelin 1978; Wikel 1982; Baron and Weintraub 1987; Wakelin and Blackwell 1988). However, standardization of ectoparasite intensities of barn swallows to a mean of zero and a variance of unity in each sample did not affect the repeatability estimates (Møller 1991c).

Ectoparasites are contagious and move readily between individuals that have bodily contact with each other. A significant resemblance in parasite intensities of adult and nestling barn swallows might therefore suggest that nestlings were infected from their parents, and their parasite intensity might thus resemble that of the parents. This problem was solved in a partial cross-fostering experiment. Adult barn swallows were sampled for tropical fowl mites at capture upon arrival. The mite intensity of nestlings was manipulated by adding *c*. 50 mites to the first-clutch nest during the egg-laying period because the natural prevalence of mites in nests is only 43%. When the barn swallow eggs hatched, half the nestlings were transferred between nests with similar hatching dates. The mite intensity of nestling barn swallows was assessed when they were seven days old (Møller 1990*e*).

Barn swallow nestlings which were raised in their nest of hatching resembled their parents in mite intensity (Fig. 9.3). More interestingly, the nestlings that were cross-fostered in another barn swallow nest resembled their true parents, but not their foster parents in mite intensity (Fig. 9.3). This strongly indicates that nestling barn swallows resemble their parents in respect of intensity of infestation by the tropical fowl mite because of similarity in genetics, and not because of similarity in environmental conditions (Møller 1990*e*). The resemblance between offspring and parent barn swallows is not a true estimate of heritability because nestlings were sampled at a different age from their parents, but it is unlikely that offspring will cease resembling their parents in respect of parasite intensities.

If mite intensities were to be partially controlled by heritable resistance among barn swallow hosts, and genes for resistance affected the exaggeration of secondary sexual characters (see the next paragraph), then there should have been a negative phenotypic correlation between the size of the secondary sexual character of the male parent and the average experimentally manipulated mite load of offspring (Møller 1990*e*). There was in fact a strong negative correlation between intensities of tropical fowl mites on barn swallow offspring reared in their nest of hatching or in a foster nest and tail length of their male parent (Fig. 9.4). This correlation disappeared when mite intensities of offspring reared in a foster nest were related to the tail length of the male foster parent (Fig. 9.4). These correlations suggest that genes for resistance to the tropical fowl mite are associated with tail length of male barn swallows. This result is therefore consistent with the first, direct prediction of the Hamilton–Zuk hypothesis that there should be a genetic correlation between resistance genes among hosts and genes for the female mate preference.

The prediction of heritable variation in host resistance is not necessarily exclusive for the Hamilton–Zuk hypothesis because genetic variance in host resistance to parasites may even be present according to the hypotheses on contagious parasites or parasites that affect the parenting ability of male barn swallows. However, females should particularly choose the direct benefits of a parasite-free male or an efficient male parent rather than a resistant male if the parenting ability or the contagious parasites hypotheses accounted for the female preference.

3. *The expression of secondary sexual characters is affected by the intensity of the parasite infestation.* This prediction can be tested by determining the relationship between ornament size of hosts and their parasite abundance. Male barn swallows with long tails had fewer tropical fowl mites and feather lice of the species *Myrsidea rustica* than short-tailed males, which suggests that the expression of male tail ornaments may reflect parasite abundance (Møller 1991c). A slightly stronger test can be made by determining the relationship between parasite abundance and the change in the size of tail ornaments from one year to another. Male tail length increases on average from one year to another, but there is considerable variation in annual growth increments (Møller 1991a). The growth rate of male tail size was in fact strongly negatively related to the intensity of the tropical fowl mite in the nest during the previous year (Fig. 9.5). This relationship was directly caused by the haematophagous mite because experimental manipulation of mite intensities resulted in a negative relationship between mite abundance and the change in male tail length from one year to another (Møller 1990e; see Fig. 6.8). Because tail length of male barn swallows is a highly repeatable morphological trait independent of annual growth increments (Møller 1991a), it is possible that the expression of male tail length is influenced by the presence of mites after the first complete moult.

The expression of the tail ornament in male barn swallows should be particularly responsive to the presence of detrimental parasites if the trait were a revealing handicap. In other words, tail length should be affected by the presence of debilitating parasites, while other morphological traits such as wing length or the length of the short central tail feathers should be unaffected. This was in fact the case because neither wing length nor short tail length were affected by the experimental manipulation of the abundance of tropical fowl mites.

Another feature of secondary sexual characters is their high degree of fluctuating asymmetry, that is, random deviations from perfect symmetry between the two sides of the body (Møller and Pomiankowski 1993b). Asymmetry develops as a result of a number of different

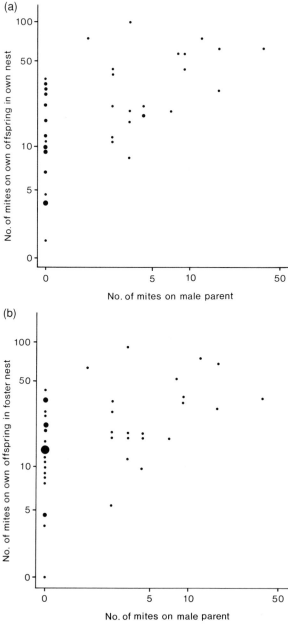

Fig. 9.3 Mean number of tropical fowl mites on seven-day-old barn swallow nestlings in relation to estimated number of mites on male parents upon arrival in spring. Nests were inoculated with *c*. 50 mites during the egg-laying period. (a) Mite intensities of own offspring reared in own nests and of male parents. (b) Mite

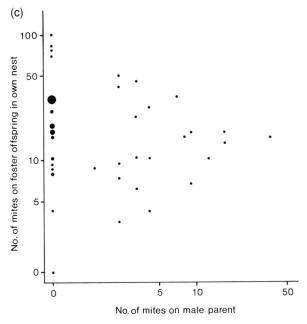

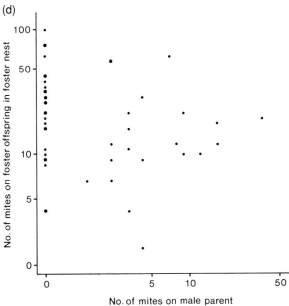

intensities of own offspring reared in other nests and of male parents. (c) Mite intensities of other offspring reared in own nests and of male parents. (d) Mite intensities of other offspring reared in other nests and of male parents. Symbols of increasing size represent from one to five individuals. Adapted from Møller (1990*e*).

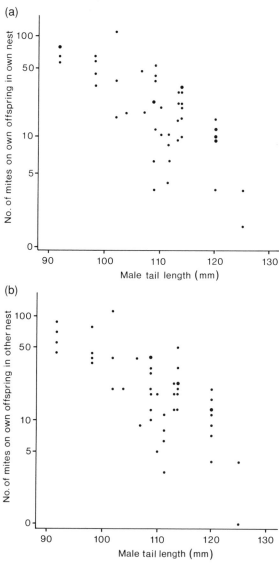

Fig. 9.4 Mean number of tropical fowl mites on seven-day-old barn swallow nestlings in relation to tail length (mm) of the attending adult male. Nests were inoculated with *c.* 50 mites during the egg-laying period. (a) Mite intensities of own offspring reared in own nests and tail length (mm) of male parent. (b) Mite intensities of own offspring reared in other nests and tail length (mm) of male parent. (c) Mite

environmental and genetic stressors. Fluctuating asymmetry of sexual ornaments should be particularly responsive to the presence of important indicators of viability if the secondary sexual character were a

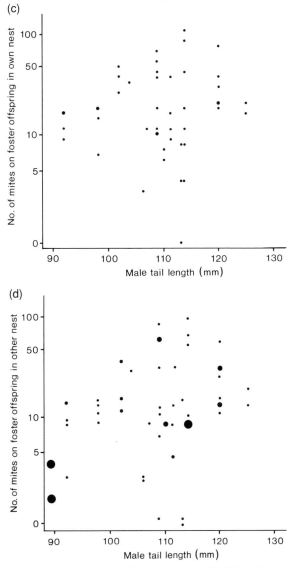

intensities of other offspring reared in own nests and tail length (mm) of male parent. (d) Mite intensities of other offspring reared in other nests and tail length (mm) of male parent. Symbols of increasing size represent from one to five individuals. Adapted from Møller (1990*e*).

revealing handicap. Parasites should therefore cause particular increase in fluctuating asymmetry in tail length, but a much smaller increase in the asymmetry of other morphological characters (Møller 1992*f*). This

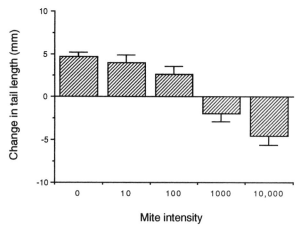

Fig. 9.5 Change in tail length (mm) of male barn swallows from year (*i*) to year (*i* + 1) in relation to intensities of mite infestations of first clutch nests in year (*i*). Values are means (SE).

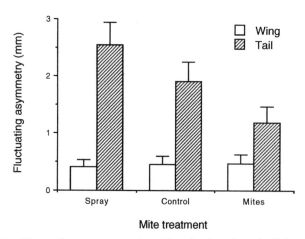

Fig. 9.6 Fluctuating asymmetry (mm) in wing length and tail length of male barn swallows in year (*i* + 1) in relation to mite treatment of their first clutch nests in year (*i*). Values are means (SE). Adapted from Møller (1992*f*).

prediction was in fact fulfilled, since fluctuating asymmetry in tail length, particularly in males, was affected by experimental manipulation of the abundance of the tropical fowl mite (Fig. 9.6). Asymmetry in wing length and the length of the short central tail feathers were virtually unaffected by the experimental manipulations (Fig. 9.6; Møller 1992*f*). One species of ectoparasite thus differentially affected the expression of the barn swallow tail ornament, namely in its degree of fluctuating asymmetry.

The prediction that the expression of the secondary sexual character is affected particularly strongly by parasites is not necessarily exclusive to the Hamilton–Zuk hypothesis. It is difficult to imagine that contagious parasites should affect the expression of ornaments because that would directly prevent or reduce their own dispersal and reproductive success. The probability of transmission to other host individuals and thus dispersal will supposedly be inversely related to the ability of host conspecifics to assess parasite prevalence and intensity. Contagious parasites should therefore not affect the expression of a secondary sex trait because that would interfere with their efficient transmission. The parenting ability hypothesis may predict a negative relationship between the expression of a secondary sexual character and parasite abundance if parasites exploit host resources used for both development of the secondary sex trait and provisioning of offspring. This is not particularly likely for the barn swallow where male parenting is *inversely* related to the expression of the tail ornament, and not directly as predicted by the hypothesis. In fact, female barn swallows would obtain relatively more help with feeding their offspring by mating with short-tailed males with many parasites.

4. *Females get mates with few parasites as a consequence of basing their mate choice on the secondary sexual character.* This prediction can be tested by comparing the parasite abundance of mated and unmated male barn swallows. It can also be tested by determining the relationship between parasite loads of males and the duration of their pre-mating period because heavily infected male barn swallows should take longer to acquire a mate than parasite-free males. An even stronger test is the comparison of the parasite abundance of male barn swallows that are successful and unsuccessful, respectively, in acquiring extra-pair copulations, because this test will exclude any effects of male parenting ability.

The mating status of male barn swallows was found to be related to both the prevalence and the intensity of the tropical fowl mite and the feather louse *Myrsidea rustica* (Møller 1991*c*). A smaller proportion of mated males were infested with ectoparasites when compared with unmated males (Fig. 9.7a), and the intensities of infestation were also significantly smaller among mated males (Møller 1991*c*). Thus, this result is consistent with the prediction.

If unmated male barn swallows have more ectoparasites than do mated males, this suggests that it is relatively more difficult for a male with many parasites to acquire a mate. In other words, it takes longer for male barn swallows with many parasites to acquire a mate. The duration of the pre-mating period, which is a measure of the ability of

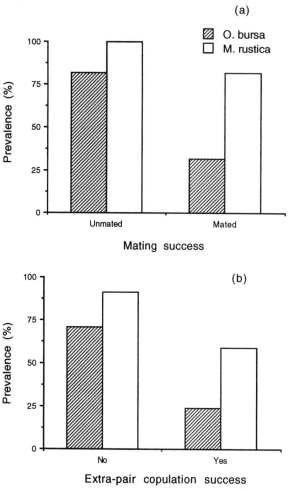

Fig. 9.7 Prevalence of ectoparasite infestations of male barn swallows in relation to their mating success. (a) Mating success. (b) Extra-pair copulation success. Adapted from Møller (1991*c*).

males to attract a mate, was positively related to the intensity of infestation by the tropical fowl mite and *Myrsidea rustica* (Møller 1991*c*). It is likely that male barn swallows with many ectoparasites took longer to acquire a mate because females used male tail length to assess their parasite abundance, as described in the previous section.

It is possible that the reason for female barn swallows to prefer long-tailed parasite-free males is that such males have a superior parenting ability. This possibility can be excluded by determining the parasite abundance of male barn swallows in relation to their extra-pair copulation success (Møller 1991c). Male barn swallows that were successful in acquiring extra-pair copulations had lower prevalences and intensities of infestation by the two species of ectoparasites than unsuccessful males (Fig. 9.7b; Møller 1991b). Again females may have assessed the parasite abundance of extra-pair males indirectly from male tail length (see the previous section).

The prediction that female barn swallows acquire mates with fewer parasites as a result of their preference for long-tailed males is not a critical test of the Hamilton–Zuk hypothesis. Females may also avoid short-tailed, heavily parasitized males in order to acquire mates with superior parenting ability. Alternatively, female barn swallows may avoid short-tailed parasitized males in order not to be infested by contagious parasites. However, the low parasite intensities of male barn swallows with high extra-pair copulation success cannot be explained by the parenting ability hypothesis because extra-pair males never provide any parental care.

5. *There is assortative mating with respect to parasite intensities.* If both male and female ornaments reflect resistance to parasites, and if there is mate choice by both sexes, there should be assortative mating with respect to parasite intensities. The main reason for assortative mating is that individual hosts should be unwilling to mate with any conspecific with a parasite intensity considerably higher than their own. Barn swallows were in fact found to mate assortatively in regard to parasite intensities (Møller 1991c). For example, males with many tropical fowl mites mated with females with many mites (Fig. 9.8). There was a similar, albeit weaker, relationship for the feather louse *Myrsidea rustica* (Møller 1991c). Barn swallows also mate assortatively with respect to tail length (Fig. 12.4), and assortment concerning parasite intensities could be a direct consequence of mate choice based on the size of secondary sexual characters. It could be argued that barn swallow pair members resemble each other in ectoparasite intensities because of already completed transfer of contagious parasites. This, however, is probably not the case. The abundance of tropical fowl mites was estimated on barn swallows captured upon arrival, and males on average arrive several days before females. There had thus been virtually no opportunity for transfer of mites from females to their mates when parasite loads were checked. The abundance of the feather louse *M. rustica* was estimated from the number of small holes chewed in the tail and wing feathers of barn swallows. It takes several weeks to make large numbers of holes in the feathers, and male and female barn swallows therefore could not resemble each other in abundance of feather lice as

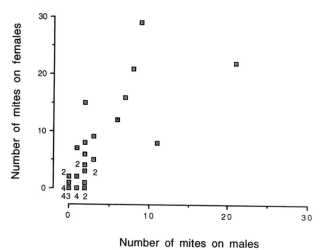

Fig. 9.8 Assortative mating in the barn swallow with respect to the intensity of infestation by the tropical fowl mite. Numbers refer to number of barn swallow pairs. Adapted from Møller (1991c).

a result of rapid transfer of parasites at the very beginning of the breeding season.

The prediction on assortative mating, again, is not exclusive to the Hamilton–Zuk hypothesis since mate choice by both sexes in respect of parasite intensities may also be expected under the parenting ability hypothesis and the contagious parasite hypothesis.

In conclusion, my data do not refute the Hamilton-Zuk hypothesis. On the contrary, corroborative evidence exists for the predictions that (1) there should be heritable variation in host resistance to one species of parasite, and that resistance is related to the expression of the major secondary sexual character in the males, and that (2) the expression of the tail ornament should be affected by the abundance of the same ectoparasite. However, most of the predictions are open to alternative interpretations.

9.7 Summary

The barn swallow is a host of a number of different parasite species. The three hypotheses of parasite-mediated sexual selection (parenting ability, contagious parasites, revealing handicap) were tested using observational and experimental studies of the relationship between the barn swallow and its parasites. There was virtually no direct evidence for the parenting ability hypothesis. There was some evidence consistent with the contagious parasites hypothesis. The revealing handicap hypothesis of parasite-mediated sexual selection was supported because there was heritable variation in host resistance to the tropical fowl mite, and this resistance was inversely related to the expression of the major secondary sexual character.

In conclusion, it is likely that female barn swallows experience a combination of direct and indirect fitness benefits related to parasites because of their preference for long-tailed males.

STATISTICS

1. *Ornithonyssus bursa*: nine correlations between pairs of years 1983–1984 to 1991–1992: $r = -0.17$ to 0.26, $n = 10$–67, all NS; *Myrsidea rustica*: nine correlations between pairs of years 1983–1984 to 1991–1992: $r = -0.21$ to 0.22, $n = 10$–67, all NS; *Hirundoecus malleus*: nine correlations between pairs of years 1983–1984 to 1991–1992: $r = -0.18$ to 0.25, $n = 10$–67, all NS.

10

Paternal care and male ornamentation

10.1 Introduction

Measurement of indirect fitness benefits through sexual selection are simplified in the absence of obvious direct fitness benefits such as parental care (see Chapter 2). Analysis of the relative contribution of direct and indirect fitness benefits through mate choice quickly becomes difficult when both components are present. The expression of a secondary sexual character may reflect indirect fitness benefits such as male attractiveness or superior genetic constitution, but the extent of paternal care may also be reflected in the expression of a male sex trait. Therefore, the problem is to determine whether the expression of the secondary sex trait reflects direct or indirect fitness benefits to the female.

At least two models have addressed the relationship between male parental care and degree of ornamentation (section 10.2). The good parent model is based on the assumption that the expression of the secondary sexual character reflects parenting ability, such that females are actually choosing future male parenting ability by assessing the elaboration of the male secondary sexual character (Kirkpatrick 1985; Heywood 1989; Hoelzer 1989; Grafen 1990a; Price et al. 1993). In other words, females are interested in male quality in terms of paternal care as a direct fitness benefit. The differential allocation hypothesis is based on the assumption that females are after indirect fitness benefits reflected in the expression of secondary sexual characters (Burley 1986). Male sex traits in this case are assumed to reflect properties of male quality not related to parenting ability, and females have to pay the price of acquiring a male of otherwise high quality by undertaking themselves a disproportionately large share of parental duties.

Male barn swallows participate considerably in reproduction, as do males of most bird species (see section 10.3). Nest building or refurbishing of an old nest is a duty undertaken by both pair members, and may last several weeks. Females of all subspecies of the barn swallow, with the exception of the North American *erythrogaster*, incubate the eggs on their own for two weeks. All barn swallow females also brood the nestlings during their first week of life. Both male and female barn swallows provide food for their young; nestlings are fed for approximately three weeks and fledglings for another week or two. Finally, males and females defend their offspring against potential predators during the nestling and the early fledgling period by giving alarm calls and direct attacks, an activity that is risky and sometimes fatal to the parent.

The four kinds of male parental care are all costly in terms of time, energy and/or mortality risks. How does the male contribution to these activities relate to its degree of ornamentation (section 10.4)? The good parent model of sexual selection predicts a positive relationship whereas the indirect fitness models predict a negative relationship. I shall use my extensive data on barn swallows to examine the relationship between degree of ornamentation and male relative contribution to parental duties. Data from a study in Canada of the North American subspecies will be used to determine the relationship between male incubation and degree of ornamentation.

The differential allocation hypothesis predicts that females have to pay for the acquisition of a relatively attractive mate by providing a disproportionate share of parental duties. This hypothesis was tested experimentally by manipulating the tail length of male swallows after mate acquisition. If females invest differentially in reproduction relative to the attractiveness of their mates, female reproductive effort should be directly related to the experimental level of ornamentation (section 10.5).

10.2 Models of sexual selection and parental care

There are several different models of sexual selection and parental care, namely, the good parent model (Kirkpatrick 1985; Heywood 1989; Hoelzer 1989; Grafen 1990*a*; Price *et al.* 1993) and the differential allocation model (Burley 1986). These models differ primarily in the kind of fitness benefits signalled by the male secondary sexual character. Females may gain a number of different direct benefits during mate choice, such as a high-quality territory, a parasite-free male, or an efficient father for the offspring, but these benefits should be assessed directly whenever possible. The reason for this is that indirect assessment, for example through the expression of a secondary sexual character, is bound to be less reliable than direct assessment. Paternal care is the only kind of direct fitness benefit that cannot be assessed directly, because mate choice precedes paternal activities. In this case, indirect assessment is the only possibility.

10.2.1 *The good parent model*

The simplest possible situation for the evolution of a female mate preference and a male trait is when the female benefits directly from its mate preference. The good parent model assumes that the expression of the secondary sexual character reflects overall male quality and that females obtain a benefit by using the expression of the male sex trait in assessment of male quality. Males which are good at developing an extravagant secondary sexual character are also assumed to be good at providing male parental care.

Kirkpatrick (1985) developed an explicit genetic model of direct fitness benefits from a uncostly mate preference in order to investigate the sexy son hypothesis (see section 2.4.4). The male secondary sexual character was supposed directly to affect their fecundity. The single stable equilibrium for the male trait occurs at the phenotype with the maximum fecundity per mating. Female fecundity is therefore maximized at the evolutionary equilibrium. The male secondary sexual character may, however, evolve to extreme expressions which seriously impair male survival. The alternative outcome is an unstable evolutionary equilibrium from where the male trait and the female preference can evolve to ever more extreme expressions by the runaway process of self-reinforcing feedback. Female fecundity is then continuously impaired as a result of exaggeration of the male trait.

Heywood (1989) modelled the evolution of a female mate preference and a male handicapping trait when there is non-heritable variation in paternal investment. Under certain conditions the male trait and the female preference evolve to either of two equilibria: (1) a complete loss or fixation of the male trait, or (2) polymorphic equilibria of the male trait and the female preference along a line. If the male trait is a condition-dependent handicap, it generates a very strong association between the male trait and paternal investment. This enhances the exaggeration of the male trait and the female preference even with a small variance in paternal investment. Furthermore, males remain phenotypically variable after the male trait has gone to fixation, and the handicap mechanism thus continues to favour the female mate preference until it is also fixed in the population.

Hoelzer (1989) developed a simulation model of the good parent process where the male trait was supposed to advertise a non-heritable component of parental quality. The male trait was able to increase in frequency because of a temporal association between parental quality and presence of the male trait following sexual selection in each generation. The evolution of the male trait is dependent on a linear trade-off between the benefit that male care confers on offspring viability and the viability cost of the male trait. Extreme exaggeration of the male trait would thus impair further evolution because of its negative effects on the quality of male parental care. Phenotypic plasticity in the male trait would potentially allow low-quality males not to reveal their quality. However, phenotypic plasticity further enhances the evolution of a good parent trait because when the cost of the expression of the male trait to low-quality males is small, the mean fitness of males with the sex trait relative to males without the sex trait is greater at small values for the male parental care trait.

Grafen (1990*a*, 1990*b*) analysed the evolution of a male sex trait that reliably reflected male non-genetic quality by determining the phenotype-dependent evolutionarily stable strategies. Females were assumed to have higher fecundity if they mated with a male of higher quality. The advertisement and preference rules where no alternative strategy could invade were the non-signalling equilibrium at which males either advertise at the lowest possible level and at which females mate independent of male phenotype, or male advertising is a continuously increasing function of male quality and the female preference for higher quality males is costly. Males as well as females may pay a cost at this second equilibrium because the male trait reduces male viability, and females may experience reduced fitness if male survival is reduced to such an extent that females are less able to mate at the best time of the season.

Price *et al.* (1993) analysed the joint evolution of a condition-dependent male trait reflecting direct fitness benefits to the female and a female mate preference. The number of offspring produced by a pair was assumed to decrease with the expression of the male trait and increase with male condition. Sexual selection favours evolution of condition-dependence of the male trait, and a female mate preference for condition-dependent traits evolves even if the male trait has negative effects on male viability and female fecundity. The mean fecundity of females is not maximized at the equilibrium and may continuously decline as the male trait and the female preference evolve because of the condition-dependent expression of the male trait.

Secondary sexual characters reflecting parenting ability are unlikely to be extraordinarily extravagant because very large secondary sex traits would interfere with efficient parenting. Reliable indicators of parenting ability are therefore supposed to reflect the ability to forage efficiently or find rare, but highly nutritious food items. A typical example is male coloration, which requires acquisition of carotenoids from food (Milinski and Bakker 1990; Hill 1991). Efficient foragers with particular foraging skills are likely to acquire more carotenoids or acquire them more rapidly, and for that reason develop the most extravagant coloration. Bright coloration *per se* will hardly impose a cost on foraging activity.

10.2.2 *The differential allocation model*

The differential allocation model assumes that the male secondary sexual character reflects properties of its genetic quality, and that females choose indirect fitness benefits by choosing the most ornamented males (Burley 1986). Males may play an important parental role in species with biparental care. The relative contribution by the two sexes is usually assumed to reflect a compromise in the sexual conflict over parental effort (Houston and Davies 1985). The differential allocation hypothesis rests on the assumption that both the compromise on parental investment between the sexes and the relative quality of a partner play a role in determining the relative magnitude of female effort.

The indirect fitness benefits signalled by the expression of the male secondary sexual character could either be an arbitrary attractiveness or a good genes benefit. Females may be willing to work relatively harder during rearing of offspring if they have acquired a mate of relatively high quality because they would gain an indirect benefit in terms of the mating advantage of their surviving offspring. The expression of the secondary sexual characters of offspring would resemble those of the father, and females with a relatively attractive mate would raise more sons with a Fisherian mating advantage. This is virtually equivalent to the sexy son process of female mate preferences.

However, this process could work equally well for secondary sexual characters reflecting genetic quality of males (Burley 1986). Males with secondary sexual characters reliably signalling their condition could acquire mates that were willing to provide parental care in relation to the relative quality of the male. Females that acquired a male of relatively high quality would provide relatively more parental care if they as a result gained an enhanced indirect fitness benefit. Sons of males with the most extravagant ornaments are likely to share viability genes with their fathers, and the expression of secondary sexual characters will be positively correlated between subsequent generations. The cost of the relatively high parental effort of females may therefore be offset by benefits of their male offspring in terms of mating success and by benefits of all their offspring in terms of viability. A condition-dependent secondary sexual character is by definition costly, and this could interfere with male parenting ability. Females mated to the most ornamented males could be forced to compensate for the inability of males to provide extensive parental care owing to the presence of an ornament. However, ornaments are supposed to be relatively more costly to low- than to high-quality males, and males with small secondary sex traits should therefore provide relatively more parental care than males with large sex traits. This should result in relatively higher female parental effort if a female is mated to a low-quality male. If females mated to males with the largest ornaments provide relatively less parental effort, this could only be explained by differential allocation of parental effort and not by compensation for the cost of a secondary sexual character. The Fisherian and the good genes advantage of differential allocation of female parental effort can be separated only by determining whether the secondary sexual character is an arbitrary trait reflecting attractiveness or a reliable signal indicating overall quality and condition.

The good parent and the differential allocation models make different predictions about the extent of male parental care. The good parent model suggests that the extent of paternal care is positively related to the expression of the sex trait while the differential allocation model predicts a negative relationship. These predictions will be tested for the barn swallow in sections 10.4 and 10.5. However, the different kinds of male parental care will first be described in section 10.3.

10.3 Kinds of male parental care

Male barn swallows provide extensive paternal care, like the males of many other monogamous bird species. There are four different kinds of paternal care in barn swallows. Males of all subspecies of barn swallows contribute to the building of nests and feeding and defence of offspring. Males of the North American subspecies *erythrogaster* also contribute to incubation of eggs and brooding of nestlings. These different kinds of paternal care are briefly described.

10.3.1 *Nest building*

Barn swallows build nests out of straw mixed with mud or cow dung. The cup-shaped nest is subsequently lined with soft materials such as feathers, hair or straw. Farmers in many areas used to believe that adult barn swallows tied their nestlings to the nest with horse hair and in that way prevented nestlings from falling out of the nest. This belief apparently arose as a consequence of nestlings gripping firmly in the nest lining when humans attempted to remove them.

Barn swallows start nest-building shortly after pair formation and the duration of the nest building period depends on the weather and the availability of mud. Nest building usually takes place in the morning, and very little is done during rain or cold spells. Lining of the nest starts when the nest cup is finished and dry. This part of nest building usually lasts only a couple of days. Females may continue to bring feathers to the nest throughout the egg-laying and the early incubation period. Female barn swallows start to remove feathers from the nest during the early nestling period, and when the nestlings fledge there are hardly any feathers left in the nest (Møller 1987*d*). The presence of feathers is probably beneficial because of their insulative properties (Møller 1991*h*). However, the presence of many feathers in the nest may be detrimental during the late nestling period if they increase the probability of nestling hyperthermia or prevent parents from efficiently performing nest sanitation duties (Møller 1987*d*). The duration of the nest-building period, which lasts from the carrying of the first mud pellets until the nest is completed, in the Kraghede study area was on average 7.49 days (SE = 0.29, *n* = 168). A new nest is usually built for the second clutch if the first is infested with ectoparasites such as the tropical fowl mite *Ornithonyssus bursa* (Møller 1990*c*). The energy cost of nest building is relatively small (Turner 1980).

Nest-building swallows

The rate of nest building in the morning is highly variable, ranging from 5 to 42 collecting trips per hour, with an average of 29.9 trips per hour (SE = 1.9, *n* = 168). The male contribution to nest building is also highly variable (Fig. 10.1). Some males do most of the nest building whereas others

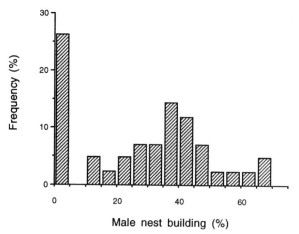

Fig. 10.1 The percentage of nest building by male barn swallows in different pairs. Sample size is 168 males.

do very little. The proportion of nest building performed by different males varied from 0.0 to 68.4%, with an average of 27.7% (SE = 0.8, n = 168).

10.3.2 *Incubation*

Only male barn swallows of the North American subspecies *erythrogaster* are known to incubate, and they usually incubate very little, on average 9.0% of all incubation (Smith and Montgomerie 1992). Unlike females, North American males do not develop a brood patch, but are nevertheless able to incubate efficiently and maintain an egg temperature sufficient for embryonic development (Ball 1983).

Incubating swallow

10.3.3 *Feeding of offspring*

Both parents provide offspring with food throughout the nestling period and the first couple of weeks of the fledgling period. Size and number of prey items per food bolus increase during the nestling period (Jones 1987). Parents usually provide whole boluses to nestlings, but a bolus is sometimes divided between two nestlings. Parent barn swallows work at a very high intensity during the nestling period, the metabolic rate reaching a maximum of 3.4 times basal metabolic rate (Turner 1980), which is relatively high compared with the maximum recorded in birds (Bryant and Tatner 1991).

Male barn swallows provide most of the nestlings' food during the first few days of their life while females brood. During the rest of the nestling period both male and female contribute large shares of nestling provisioning. Males undertake most of the feeding of first-brood fledglings while females immediately initiate a second clutch. The average feeding rate by male barn swallows during the nestling period varies considerably, with a mean of 6.9 feeds per hour for first broods (SE = 0.27, n = 189) and 10.2 feeds per hour

Food provisioning

for second broods (SE = 0.50, *n* = 89). Female barn swallows feed slightly more than males, on average 7.8 feeds per hour for first broods (SE = 0.28, *n* = 169) and 11.4 feeds per hour for second broods (SE = 0.43, *n* = 89). The percentage of all feeding visits contributed by the male varies considerably (Fig. 10.2), but the average is slightly below half of all feeding visits: 46.4% in first broods (SE = 0.8, *n* = 169 pairs) and 46.2% in second broods (SE = 0.7, *n* = 89).

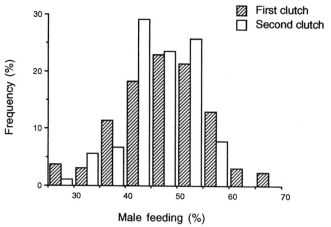

Fig. 10.2 The percentage of feeding of nestlings by male barn swallows in different pairs during rearing of the first and the second clutch. Sample sizes are 169 and 89 males, respectively.

10.3.4 *Defence of offspring*

Barn swallows regularly defend their offspring against potential nest predators such as feral cats *Felis catus* and sparrowhawks *Accipiter nisus*. Their
behaviour during nest defence ranges from emitting weak or strong alarm
calls to swoops towards or direct attacks at the predator. Studies of mobbing
behaviour in barn swallows have demonstrated that both male and female
parents mob potential predators (Smith and Graves 1978; Shields 1984). The
intensity of mobbing by female parents increases gradually throughout the
breeding cycle of the first and second clutch (Møller 1984*b*). Male barn
swallows mob at a low intensity during the first half of the breeding cycle,
when they are not investing in offspring, whereas there is a dramatic increase
in the intensity of mobbing during the nestling and fledgling periods (Møller
1984*b*). The generally higher intensity of mobbing by females and the
distribution of mobbing by the two sexes within the breeding cycle, suggest
that mobbing intensity reflects the level of parental effort by the respective
parent. Mobbing can be dangerous for mobbers as shown by the capture
of the nearest approaching individual barn swallow by sparrowhawks on
three occasions and by feral cats on two other occasions (Møller 1991*h*).

Swallows mobbing sparrowhawk

There is considerable variation in the intensity of mobbing by male barn
swallows when nestlings are 8–12 days old (Møller 1984*b*, 1991*h*). The
duration of time spent by males near a stuffed little owl *Athene noctua* model
during a five-minute trial ranged from hardly any to the entire period.
Similarly, the intensity of alarm calling by males ranged from none to
continual calling, and the proximity of approach to the model varied from
infrequent approaches to a closest distance of 5m to directly touching the
model. The three different scores of mobbing intensity were strongly
positively correlated with each other (Møller 1991*i*), suggesting that all three
represented willingness to defend offspring.

In conclusion, all four measures of paternal care demonstrated considerable variation between males. The relationship between these measures of

paternal care and the degree of male ornamentation is described in the next section.

10.4 Patterns of parental care and male ornamentation

There was considerable variation among male barn swallows with respect to the four kinds of paternal care. High-quality males with long tails might undertake a larger share of the parental duties if tail length is a reliable indicator of parenting ability.

The relative rate of female nest building (the percentage of all nest-building visits of a pair being performed by the female) was positively related to tail length of the mate (Fig. 10.3). Thus, female barn swallows mated to males with long tails contributed relatively more to nest building than did females mated to males with short tails. This relationship could potentially be caused by variables other than male quality reflected by male tail length, such as age or experience being positively correlated with tail length. However, male tail length, but not male age, accounted for a statistically significant proportion of the variation in female nest-building activity.[1] Female barn swallows mated to long-tailed males therefore appeared to invest relatively more in nest-building activity compared with females mated to short-tailed males.

The relative contribution of females to incubation among Canadian barn swallows of the subspecies *erythrogaster* was positively related to tail length of their mates (Smith and Montgomerie 1992). Male barn swallows with the

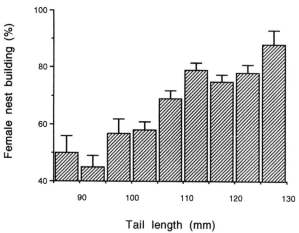

Fig. 10.3 Relationship between female nest building rate (% female nest-building visits) and tail length (mm) of their mates. Values are means (SE). Sample size is 168.

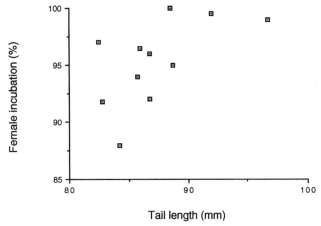

Female incubation (%)

Tail length (mm)

Fig. 10.4 Relationship between female incubation rate (% female incubation) and tail length (mm) of their mates. Adapted from Smith and Montgomerie (1992).

shortest tails contributed up to 14% of all incubation whereas long-tailed males hardly contributed at all (Fig. 10.4).

The percentage of all feeding visits to the nest by a pair being performed by a female was positively related to the tail length of its mate (Fig. 4.11). Female barn swallows mated to males with long tails contributed relatively more to feeding of nestlings than did females mated to males with short tails. Again, this relationship may have been caused by other variables such as age or experience being positively correlated with male tail length. However, male tail·length, but not male age, accounted for a statistically significant proportion of the variation in female feeding activity.[2] Female barn swallows mated to long-tailed males therefore appeared to invest relatively more in provisioning of offspring than did females mated to short-tailed males.

Female barn swallows could have invested in the defence of their offspring against a potential predator in direct proportion to the tail length of their mate. However, females mated to long-tailed males did not take higher risks in defence of their offspring compared with female barn swallows mated to short-tailed males.[3] Female barn swallows mated to long-tailed males therefore mobbed a predator model as much as females mated to short-tailed males.

In conclusion, several parenting activities of female barn swallows were positively related to tail length of their mates. This positive relationship can be due to either differential parental effort on the part of females mated to long-tailed males or females making up for the reduction of male parental care caused by the handicapping effects of its tail ornament. These alternatives are tested experimentally in the next section.

10.5 Differential parental effort

The only way to determine if investment by female barn swallows in their offspring might be causally related to the relative quality of their mates is to uncouple experimentally the relationship between female quality and male phenotype. I therefore did this together with F. de Lope in a field experiment (de Lope and Møller 1993). Male barn swallows were allowed to attract a mate before their tail length was experimentally altered (see Chapter 4). One group of males had their tails shortened by 20mm, another group's were elongated by 20mm, while a control group of males were left unmanipulated and thus maintained their original tail length. Reproductive effort in relation to the manipulated tail lengths was estimated at a number of different stages of the breeding season. The most dramatic effect was that females mated to male barn swallows with shortened tails suffered a lower nesting success in both their first and second clutches compared with control pairs and pairs with males with elongated tails. There was no effect of the experimental treatment on the body mass or tarsus length of surviving nestlings, suggesting that the offspring would probably enjoy similar probabilities of recruitment (review in Magrath 1991).

The tail-length manipulations of male swallows had a marked effect on the number of broods reared per season (de Lope and Møller 1993). Females mated to males with shortened tails reared on average 1.6 broods while females mated to males with elongated tails on average reared 2.1 broods per season. Female barn swallows mated to males with shortened tail reared on average 16% fewer offspring per season than did controls, while females mated to males with elongated tails reared on average 22% more offspring than did controls (Fig. 4.12). In conclusion, manipulation of male tail length after pair bonds had been formed resulted in marked differences in seasonal reproductive success. Reproductive success increased with increased tail length, and this could be due to improvement either in male or in female effort accompanying increased male tail length.

Male barn swallows participate in nest building and feeding of offspring, but provisioning of offspring is by far the most time-consuming and energetically expensive activity. Female barn swallows mated to males with elongated tails had significantly higher feeding rates than females mated to males with shortened tails (de Lope and Møller 1993). Males with elongated tails therefore contributed relatively less to the feeding of offspring than did tail-shortened males (Fig. 10.5). Male feeding efficiency was affected by the experimental manipulations both in terms of number and size of prey provided during each visit to the nest. Males with elongated tails provided more, but smaller prey items than males belonging to the control group, which provided fewer and smaller prey items than did males with shortened tails (de Lope and Møller 1993). The result of the present experiment is

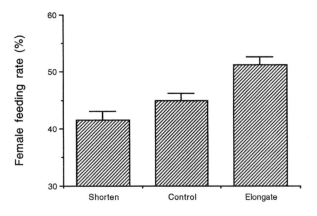

Fig. 10.5 Relative female feeding rate (% female feeding visits) in relation to tail treatment of their male after mate acquisition. Values are means (SE). Adapted from de Lope and Møller (1993).

remarkable because it was the barn swallow pairs consisting of males with an elongated tail that raised the largest number of offspring per season. The positive relationship between seasonal reproductive success and experimental tail length of male barn swallows therefore suggests that females were able to compensate — if not overcompensate — for the impaired foraging efficiency of their mates and still reproduce at a higher rate. This strongly suggests that the increased reproductive success with increasing experimental tail length was caused by differential female reproductive effort. A second tail-manipulation experiment on barn swallows also clearly revealed that females mated to the most attractive males invested differentially in their offspring (Møller 1993*b*).

10.6 Summary

Two different models relate male parental care to the expression of secondary sexual characters. The good parent model predicts that the extent of male parental care is positively related to the expression of secondary sexual characters. The differential allocation model predicts that the extent of male care is negatively related to the expression of secondary sex traits, because differential female parental effort is the cost females have to pay for acquiring males of relatively high quality.

The relative intensity of male nest building, incubation, and feeding of offspring was negatively related to the degree of male ornamentation. This may be due to a direct causal relationship or to some third variable accounting for the negative relation. There was a negative correlation between differential female allocation of parental effort and tail length of their mates in two field experiments. This suggests that females invest in reproduction as predicted by the differential allocation hypothesis, whereas the good parent hypothesis was not supported.

STATISTICS

1. Multiple regression: $F = 5.45$, $df = 2,165$, $P < 0.01$; β (tail length) = 0.33, $P < 0.01$, β (age) = 0.02, NS.
2. Multiple regression: $F = 4.38$, $df = 3,126$, $P = 0.021$; β (tail length) = 0.28, $P = 0.021$, β (age) = −0.003, NS, β (number of nestlings) = 0.0003, NS.
3. Linear regression: $F = 0.60$, $df = 1,51$, NS.

11

Sperm competition and sexual selection

11.1 Introduction

Contrary to the majority of other organisms, most bird species are socially monogamous. Males and females establish a pair bond, and individuals of both sexes contribute to the rearing of offspring. Monogamy may superficially mean mate fidelity. Charles Darwin (1859, 1871) wrongly assumed in his writings that monogamous females were faithful to their mates and therefore only copulated with their own mates. It is well known among farmers and poultry breeders that female domestic chickens and ewes regularly copulate with more than a single male, and this common knowledge must also have been available to Darwin through his extensive correspondence. The point of view that monogamous females copulate only with a single male may have been a consequence of the predominant Victorian moral attitudes of the 19th century. Recent research demonstrates that female copulations with multiple males are ubiquitous in animals (and plants), and that strictly monogamous relationships with complete mate fidelity are rare.

   Why should females copulate with more than a single male? One possibility is that females use copulations with multiple males as a means of increasing the level of sperm competition. Sperm competition is competition between sperm provided by different males for the fertilization of a single clutch of eggs of a female and all associated processes leading to and resulting from this competition for fertilizations. Sperm competition may result in post-copulatory sexual selection which may arise for two different reasons, as described in section 11.2. First, females may not have reproductive interests coinciding with those of their mates, and conflicts of interest between and within the sexes therefore arise. Copulations with multiple males and the resulting sperm competition can thus be viewed as a subtle form of post-copulatory male–male competition for access to unfertilized eggs, or as post-copulatory female choice of the most vigorous sperm

(Harvey and Bennett 1985). Second, females may be unable to acquire mates of a preferred phenotype because such males already are mated to other females. Sperm competition can therefore be viewed as a way in which females circumvent constraints on optimal mate choice set by the mating system, and mate choice thus may be continually adjusted during the reproductive cycle in an attempt on the part of females to obtain the most attractive males.

A relatively large proportion of female barn swallows engage in extra-pair copulations. Such copulations are apparently forced on females of some bird species, but those of female barn swallows are unforced since females copulate without behaving aggressively towards the extra-pair male. The behaviour during extra-pair copulations is described in section 11.3.

Extra-pair copulations can result in sexual selection only if males of different phenotypes vary in their ability to copulate successfully with female non-mates (section 11.4), and if extra-pair copulations result in fertilization of eggs (section 11.5). The expression of male secondary sexual characters may be one factor explaining variation in extra-pair copulation success. Successful males should have larger sex traits than the average male in the population, and females engaging in extra-pair copulations should be mated to males with small secondary sex traits, if female participation in extra-pair copulations is to be explained by the constraint hypothesis. The second requirement for sexual selection to result from extra-pair copulations was tested by determining the relationship between the frequency of extra-pair copulations and the frequency of extra-pair paternity by the use of DNA-fingerprinting.

Males do not passively watch their mates participate in extra-pair copulations. A number of different paternity guards have been described (Birkhead and Møller 1992), and these include male aggressive behaviour, mate guarding, frequent within-pair copulations, the use of deceptive alarm calls, and defence of a protective zone around the female (section 11.6). The use of paternity guards should vary with the probability of fertilization and thus the stage of the reproductive cycle if they are to be effective in preventing extra-pair copulations. Male barn swallows differing in tail length may be able to guard their mates with different intensities if tail length is a general measure of male quality.

Extra-pair copulations have a number of different costs and benefits to males and females as discussed in section 11.7. These are similar to the direct and indirect fitness benefits of mate choice, and the fitness benefits have to exceed the costs in order to explain the regular occurrence of extra-pair copulations. Females of many bird species engage in copulations with male non-mates, and such extra-pair copulations can either be apparently forced or unforced (Birkhead and Møller 1992). Extra-pair copulations do not have to result in a net fitness benefit to females if males forcibly copulate with females, but that has to be the case for unforced copulations.

11.2 Sperm competition and sexual selection

Superficially, sperm competition may not be related to sexual selection. However, the idea of male and female pair members constituting a working relationship for reproduction without conflicts has been long abandoned. Even if males and females establish a pair bond and share parental duties, this relationship is the outcome of sexual conflicts of interest rather than an agreement without conflicts. Males and females obviously differ in their reproductive interests, mainly because of differences in commitments to reproduction (Bateman 1948; Trivers 1972). Whereas females of many animals invest heavily in reproduction from the very beginning because of production of large energy-rich eggs, male contributions are usually more limited. This basic inequality in reproductive investment has important consequences for male and female participation in copulations with multiple mates.

Males and females can best be viewed as three parties at conflict with each other, male and female mates, and male non-mates (Parker 1984). The first evolutionary game arises because of the conflict between males and their mates. Females may want to copulate with males other than their mate if that confers any extra advantages such as offspring of higher quality or extra direct fitness benefits contributed by these males. On the other hand, males may attempt to prevent their mates from copulating with additional males because they thereby would avoid wasting their reproductive investment by providing resources for non-kin. The second evolutionary game consists of the conflict between females and male non-mates. Females may want to copulate only with particular extra-pair males, for example, if only some males are of high heritable quality or are free from a contagious disease, whereas all males may want to overcome the mating resistance of females because that would tend to increase male reproductive success. Third, there is a conflict of interest between male mates and male non-mates. Mates may intend to protect their certainty of paternity against the threats of cuckoldry from other males, and ever more efficient paternity guards will evolve. Male non-mates intend to overcome these paternity guards and copulate with mated females, partly with the help of females which are at conflict with their mates. These three inter-connected conflicts of interest will lead to escalating coevolutionary processes which result in ever higher levels of salesmanship and sales resistance among male non-mates and females and more sophisticated paternity guards and cuckoldry behaviour among males.

Sperm competition may result in sexual selection for at least two different reasons. First, sperm competition can be seen as nothing but post-copulatory male–male competition and female choice. Second, sperm competition may result as a consequence of constraints on the opportunity for females freely to choose a male of their preference. These possibilities are considered in the following two sections.

11.2.1 *Sperm competition as sexual selection*

Sperm competition can be viewed as nothing but sexual selection in the reproductive tract of females. Instead of subjecting themselves to the crude processes of pre-copulation sexual selection, females may simply copulate with a couple of the most preferred males and then resolve the male–male conflict and mate choice within their own body. Male–male competition would take place if sperm provided by different males competed with each other for access to eggs, while female choice would be the mechanism if females mechanically, chemically, or immunologically were able to choose among sperm for fertilizations (Birkhead and Møller 1993a). Male–male competition may occur either as (1) competition for access to and monopolization of a particular fertile female, or (2) male–male competition at the sperm level within the reproductive tract of a particular female. Sperm provided by different males contain different genetic material and are therefore at conflict over fertilizations. These conflicts may theoretically take several different forms. These include:

(1) a race between sperm of different males to reach the site of fertilization first;

(2) prevention of competitor sperm from reaching the site of fertilization; or

(3) more aggressive interactions between sperm such as displacement of sperm delivered by other males, or destruction of the fertilization ability of sperm.

The female reproductive tract of vertebrates is huge compared with the size of sperm, and sperm have to travel a long way in order to reach the site of fertilization. Sperm movements within the female tract are the result of interactions between sperm and the female, and competition between sperm to reach the site of fertilization first is strongly influenced by the female. However, the most important factor is probably the timing of insemination rather than competition between simultaneously inseminated sperm. In birds copulations usually do not take place immediately before fertilization because females do not provide males with information on the time of fertilization. Females ensure competition between sperm by internal sperm storage and by copulating with males for a protracted period before and during the time of fertilization.

Sperm may prevent competitor sperm from reaching the site of fertilization by blocking narrow passages of the female reproductive tract. The idea that sperm may have different functions such as blocking of the female reproductive tract has been put forward repeatedly (Silberglied *et al.* 1984; Baker and Bellis 1988). Certain sperm phenotypes apparently do not have a fertilizing function, and this is definitely the case for apyrene sperm of insects which lack nuclear DNA (Silberglied *et al.* 1984). However, there is no definitive evidence that so-called blocking sperm provided by one male prevent the sperm of another male from passing through the female reproductive tract and reaching the site of fertilization. It is not necessarily the case that sperm with different morphologies can be attributed different functions, because sperm with deviating morphologies simply may be the result of production errors, meiotic drive, or conflicts between cytoplasmic and nuclear DNA (Cohen 1967; Hurst 1990; Møller 1993*j*).

It has been suggested that some sperm morphs participate in aggressive interactions with sperm of other males and thereby reduce their probability of fertilization (Silberglied *et al.* 1984; Baker and Bellis 1988). This could take place as displacement of sperm delivered by other males, or as direct destruction of sperm and their fertilization ability. There is no direct evidence of such an aggressive function of sperm.

Sperm competition can also be viewed as female choice because the conflict, at least for internally fertilizing animals, takes place within the body of a female, and it is likely that females usually have control over processes within their own bodies. Females could theoretically allow particular sperm access to the site of fertilization at the expense of others. Chemical processes during sperm–egg fusion may also result in subtle mate choice much more intricate than the pre-copulatory process of mate choice. This view is supported in the most general sense by evidence about hostility of the female reproductive tract.

The single most important function of copulation is successful transfer of sperm and resultant fertilization. If the only function of copulations were to ensure fertilization of eggs we might expect females to facilitate this process. However, the female reproductive tract is a hostile environment to sperm, apparently making it very difficult for sperm to reach the site of fertilization, and this can most satisfactorily be explained by sexual selection theory (Birkhead *et al.* 1993). Female reproductive tracts often consist of extensive, convoluted tubes with narrow passages through which sperm have to pass. The chemical environment of the female reproductive tract is very different from the optimal environment for sperm storage in the male reproductive tract, and it can most easily be described as chemical warfare between the sexes. This analogy also holds in another respect. Shortly after copulation, the female tract of mammals is invaded by huge quantities of leucocytes whose primary objective is to engulf and destroy sperm (Birkhead *et al.* 1993). Finally, females of several mammals raise immune defences against sperm. In particular, the sperm of regular sexual partners, such as the mates of these females, may have difficulties in reaching the site of fertilization because of intense female immune responses to sperm, whereas the sperm of novel males may avoid this immune response geared to a particular male (Birkhead *et al.* 1993). The evolution of this apparently extremely hostile female reproductive tract cannot easily be explained by the fertilization hypothesis, but sexual selection through female choice provides a simple explanation for the phenomenon.

Females may originally have had a short, simple reproductive tract which facilitated fertilization. If there were genetic variation for this structure, females with slightly longer and more complicated reproductive tracts might have been fertilized by more vigorous sperm. Sperm vigour might have been a heritable trait that would allow sons to fertilize females with more complicated reproductive tracts. The quality of sperm may therefore have coevolved with the female reproductive tract to ever more extreme phenotypes if females continuously attempted to disfavour a fraction of males from fertilizing their eggs. Similar arguments can be made for all other anti-sperm properties of the female reproductive tract. However, females should ensure fertilization of their eggs, and this would be facilitated by copulation with multiple males before or during a single fertilization event. Thus, the sperm quality of a mate could easily be tested against that of other males, and the anti-sperm immune response of females could be viewed as a way in which females reduced the fertilization advantages of their regular mate.

The coevolutionary process between the quality of sperm and the hostility of the female reproductive tract could be the outcome of a runaway process, as suggested by Fisher for pre-copulatory sexual selection (Birkhead *et al.* 1993). However, female choice of sperm could just as well be explained by the good genes process of female choice, if sperm vigour genes were in linkage disequilibrium with viability genes (Birkhead *et al.* 1993). Alternatively, sperm vigour genes may have later pleiotropic effects on offspring viability.

In conclusion, there is considerable scope for sexual selection to take place at the behavioural level of sperm competition, and competition between sperm and interactions between sperm and the female reproductive tract may result in sexual selection. Direct evidence for these mechanisms is still virtually non-existent.

11.2.2 *Sexual selection and constraints on female choice*

Sperm competition may result as a consequence of constraints on female mate choice (Møller 1992*a*). Female choice is usually costly because it takes time and energy to make a mate preference match with a male phenotype. Princes on white horses do not wait outside every door! Females will usually, after a time, choose a mate, copulate and reproduce. However, this does not necessarily mean that they have found a mate of a preferred phenotype. Preferred males may not be available. For example, males of a monogamous animal species pair up consecutively and the first female which is ready to choose will ideally select the best male; this male is then no longer available for other females. The quality of unmated males will drop as more and more pairs become established. If the quality of males is normally distributed, the first females to choose will acquire males varying dramatically in quality, while modal females will obtain males that differ very little in quality. The last females to mate will again acquire males that differ considerably in quality.

The discrepancy between the most preferred male phenotype in the population and that with which the female mates, widens as more and more females become paired (Møller 1992*a*). This also means that the conflict of interest between male and female pair members is expected to increase with mating order. Late mated females should be able to acquire only males of relatively poor quality, and the only reason for accepting such a male is that it will provide some help with reproduction which may not be accomplished without the participation of two individuals. It would also pay the male to engage in such a relationship because it would be better to produce some offspring than none at all. The propensity for females to engage in copulations with other males should increase with the extent of the conflicts of interest between male and female pair members. Females that made their mate choice early and therefore were able to acquire a high-quality mate should be reluctant to engage in extra-pair copulations, while late-mating females should be particularly likely to participate in extra-pair copulations. In other words, sperm competition through extra-pair copulations can be viewed as adjustment of pre-copulation mate choice as constrained by the choice of other females.

Similar arguments can also be made for mating systems other than monogamy, because even males of lekking animals are unable to process all females. The first priority of a female attending a lek might be to ensure fertilization of all eggs. If the most preferred male has copulated with a number of females on a particular morning, late-arriving females might copulate with lower ranking males to ensure fertilization. However, it is likely that constraints on mate choice in this situation would result in repeated copulation by the female with a more preferred male. Constraints on female choice are likely to be ubiquitous although they may be particularly common among animals with monogamous mating systems (Møller 1992*a*).

There is considerable evidence in accordance with the constraints explanation for sperm competition (Møller 1992*a*) and, as we shall see later, constraints on female choice may also affect the propensity of female barn swallows to engage in extra-pair copulations.

11.3 Extra-pair copulation behaviour

Male barn swallows take a considerable interest in females other than their mates, particularly their nearest neighbours (Møller 1985). Males regularly chase female non-mates and the mates of these females retaliate by chasing away the intruder, often while giving alarm calls. Chases of a female reach a clear peak of frequency in its fertile period and drop to a very low level early during the incubation period (Møller 1985). Male barn swallows continue to chase neighbouring females after the fertile period of their own mate has ended, and their chasing activity first drops during the incubation period of their own clutch. Male mates are present during most chases by neighbouring males because of intense mate guarding (Møller 1985). However, the male is absent during 12% of all chases and more than half of such chases result in extra-pair copulation attempts. The chasing activity of male barn swallows is a relatively good predictor of their extra-pair copulation success because the frequency of extra-pair copulations is strongly positively related to the frequency of chases for different males[1] (Møller 1987*e*). Male barn swallows always interfere successfully with chases and extra-pair copulation attempts involving their own mate, and they therefore never witness their own mate engaged in extra-pair copulations.

Chasing swallows

However, males may use the frequency of chases involving their own mate as an indirect measure of the sexual interest that other males take in their female.

The behaviour of a male barn swallow during an extra-pair copulation attempt differs from that during a within-pair copulation attempt. The copulation call is given much more vigorously and at a higher rate than during pair copulations while the extra-pair male hovers over the female. Neighbouring males are sometimes attracted by these intense copulation calls and regularly approach a male and a female engaged in an extra-pair copulation attempt. The female barn swallow determines the outcome of the male approach. It will either give a threat display with exposed carpal joints and an open beak, or fly away if the male continues its attempt to copulate with the female, and this results in an unsuccessful attempt, or it will stay put and lean forward to allow the extra-pair male cloacal contact and thus complete the copulation. A male will sometimes attempt a second or a third extra-pair copulation if the female's mate has not appeared. Extra-pair copulations occur later in the fertile period than within-pair copulations, and 24% of extra-pair copulation attempts are successful which is almost the same as the success rate of within-pair copulations (23%) (Møller 1985). Two thirds of all extra-pair copulations are with the nearest neighbouring female, and I have never seen males copulate with females more than four territories away from their own territory (Møller 1985).

11.4 Patterns of extra-pair copulations

Prolonged sperm storage has been demonstrated in all bird species studied, and the minimum mean duration of sperm storage is 5 days (Birkhead and Møller 1992). I have thus assumed that barn swallows can store sperm for at least 5 days, and their fertile period (the period during which copulations may result in fertilizations) is assumed to last from 5 days before start of egg laying until the day when the penultimate egg is laid, which is the day when the last egg is fertilized. A minimum fertile period of 9 days for the modal clutch size of 5 eggs has thus been assumed. Of course, the possibility that sperm storage is longer than 5 days cannot be excluded.

Barn swallows start to copulate as soon as the female has chosen a mate and the pair bond has been established. The rate of within-pair copulations has a high, stable level from three weeks before start of egg laying until the first days of laying, and it drops to zero at the beginning of the incubation period (Fig. 11.1). It is common for many bird species to stop copulating before the end of the fertile period (Birkhead and Møller 1993*b*). Interspecific variation in the termination of within-pair copulations can be explained by

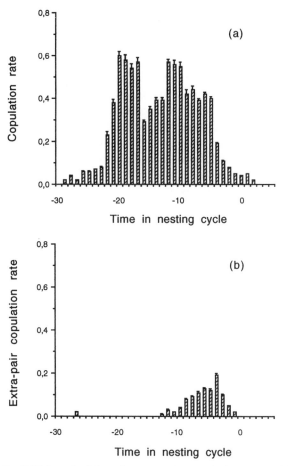

Fig. 11.1 Within-pair (a) and extra-pair copulation rates (b) of female barn swallows in relation to their breeding cycle. Day 0 is the first day of the incubation period. Values are mean rates per hour (SE).

a sexual conflict of interest over the optimal timing of copulations (Birkhead and Møller 1993*b*). Females generally attempt to copulate as early as possible before the start of laying in order to minimize the costs of copulations, whereas males attempt to copulate as late as possible in order to increase the probability of fertilizing all eggs. Early copulations also give females control of paternity while late copulations increases the level of control by males (Birkhead and Møller 1993*b*), because all birds studied had last-male sperm precedence, that is the last male to copulate fertilizes a disproportionate share of the eggs (Birkhead and Møller 1992). Within-pair copulations therefore cease late among bird species with intense sperm competition, but relatively early among species with weak sperm competition (Birkhead and Møller 1993*b*). In accordance with this hypothesis, barn swallows cease copulating relatively late in the fertile period.

Extra-pair copulations comprised 10.7% of all copulations recorded. The breeding cycle pattern of extra-pair copulations differed from that of within-pair copulations, as many extra-pair copulations occurred late, just before and during the start of egg laying (Fig. 11.1).

Only a minority of barn swallow males and females engaged in extra-pair copulations during my daily observations. Almost all males, mated as well as unmated, chased female non-mates and attempted extra-pair copulations at some time during their breeding cycle, but only a relatively small fraction of these resulted in apparent cloacal contact. Some male barn swallows were very successful in acquiring extra-pair copulations whereas others never succeeded. The frequency distribution of extra-pair mating success of males was therefore very skewed, with only a minority of males ever succeeding (Møller 1992*c*). The imbalance resembled that of mating success in many lekking animals where one or a couple of males account for the majority of all copulations.

Successful extra-pair males differed from unsuccessful males in respect of tail length (Fig. 4.5). Successful male barn swallows had longer tails than unsuccessful males, and females participating in extra-pair copulations had mates with shorter tails than females not participating (Fig. 4.5; Møller 1992*c*). Male and female engagement in extra-pair copulations may be directly related to male tail length because of a causal relationship, or because of a third variable being related to tail length and frequency of extra-pair copulations. These possibilities were tested in a tail-manipulation experiment in which the confounding effects of other variables were removed by random assignment of males to treatments. Male barn swallows with an elongated tail were much more successful in their extra-pair copulation attempts than males with shortened tails even though these males made a similar number of attempts (Fig. 11.2; Møller 1988*c*). Females engaged in extra-pair copulations were more often mated to males with shortened tails, whereas the mates of males with elongated tails only rarely participated in extra-pair copulations (Fig. 11.2). There was thus a direct causal relationship between male success in extra-pair copulations and their tail length, and between female engagement in extra-pair copulations and the tail length of their mates. These results are consistent with the hypothesis that females use extra-pair copulations as a way of adjusting their mate choice as constrained by the choice of previously mated females.

Even though male tail length accounted for some of the variance in extra-pair copulation success, this does not exclude the possibility that additional factors may affect the frequency of copulations with non-mates. These factors include:

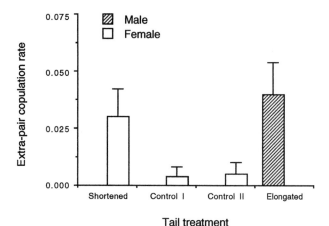

Fig. 11.2 Effects of tail treatment of male barn swallows on their extra-pair copulation behaviour and that of their mates. Values are mean rates per hour (SE). Adapted from Møller (1988c).

(1) male and female age;

(2) breeding synchrony;

(3) degree of coloniality; and

(4) intensity of mate guarding.

1. Age may be important if old experienced male barn swallows are more often able to attempt extra-pair copulations at the right stage of the female breeding cycle. Successful males predominate among early breeders whereas female barn swallows participating in extra-pair copulations tend especially to be late breeders, and early breeders tend to be older than late breeders (Møller 1985). However, only male tail length was able to explain variation in extra-pair copulations when both male age and male tail length were entered as variables in a multiple regression analysis. Similarly, female age did not explain a statistically significant amount of variance in the frequency of extra-pair copulations when the effect of the tail length of their mate was accounted for (Møller 1992c).

2. Breeding synchrony may affect the opportunity for males and females to engage in extra-pair copulations if mate guarding and extra-pair copulations are mutually exclusive activities for male barn swallows (Birkhead and Møller 1992). This appears to be the case, as male barn swallows tend to copulate with female non-mates after the end of the fertile period of their own mate (Møller 1985). Highly asynchronous

females thus may encounter more non-guarding males than highly synchronous females and therefore more often be subject to extra-pair copulation attempts. However, the frequency of male extra-pair copulations was unrelated to the deviation of the date for start of egg laying from the population mean when the effect of male tail length was controlled statistically.[2] The frequency of female extra-pair copulations was also unrelated to breeding synchrony as measured by the deviation of the date for start of egg laying from the population mean when the effect of mate tail length was controlled statistically.[3]

3. Coloniality must affect the opportunities for both male and female barn swallows to participate in extra-pair copulations (Møller 1991*k*). Solitarily breeding pairs have few opportunities to encounter extra-pair partners, while barn swallows living in a colony with more than 50 pairs encounter ample opportunities. None of the males and females from 35 solitarily breeding pairs of barn swallows ever engaged in extra-pair copulations, whereas 24.8% of 129 colonial males and 33.3% of 129 colonial females participated in at least one extra-pair copulation. This difference is statistically highly significant.[4] Once there was more than a single barn swallow pair breeding in a particular site there was a positive relationship between colony size and the probability of males or females engaging in at least one extra-pair copulation.[5] These results were not confounded by effects of male tail length because long-tailed males did not preferentially aggregate in colonies of a particular size.

4. The intensity of mate guarding may affect the opportunity for males as well as females to engage in extra-pair copulations. For example, the frequency of extra-pair copulations by female barn swallows was weakly negatively related to intensity of mate guarding when other factors were controlled statistically in a previous analysis (Møller 1987*f*). This result was not upheld in a larger data set from the years 1984–1992. Female engagement in extra-pair copulations was unrelated to guarding intensity by their mate when the effects of male tail length and colony size were controlled statistically.[6] Colony size had to be entered as a control variable because male barn swallows in large colonies generally guard their mates more intensely than other males (see section 11.6).

In conclusion, the extra-pair copulation success of male barn swallows was determined by male tail length and to some extent colony size, while female engagement in extra-pair copulations was determined by tail length of their mate and colony size. Male and female age, the degree of breeding synchrony, and the intensity of mate guarding were not important determinants of extra-pair copulations.

11.5 Extra-pair copulations and extra-pair paternity

In common with all bird species examined, female barn swallows have sperm storage tubules in their reproductive tract (T. R. Birkhead pers. comm.), and this allows the prolonged storage of sperm. Extra-pair copulations could therefore result in extra-pair paternity provided that sperm is transferred during an extra-pair copulation. I have tested whether extra-pair copulations lead to fertilizations in two different ways, by (1) sex-specific heritability of morphological characters, and (2) DNA-fingerprinting.

Offspring should resemble both their parents equally for sexually mono-morphic, quantitative characters because offspring inherit half their genes from the father and half from the mother. This fact can be used to estimate the frequency of extra-pair paternity since extra-pair copulations and the resulting extra-pair paternity should result in offspring resembling their mother more than their father, and the relative difference in heritability between the sexes therefore gives a direct estimate of the proportion of nestlings fathered by male non-mates (Alatalo *et al.* 1984). When applied to tarsus length of the barn swallow the annual variation in relative frequency of extra-pair offspring ranges from 15.7 to 44.4% with a mean estimate of 26.2% (Møller 1987*e*, 1989*d*). This differs significantly from the expected value of no difference in heritability between parents in the absence of extra-pair paternity.

The use of the sex-specific heritability method to estimate the relative frequency of extra-pair paternity has been criticized because it relies on a number of assumptions which have not always been tested directly (Alatalo *et al.* 1989; Lifjeld and Slagsvold 1989; Møller 1989*d*). For example, the method assumes that the mean and the variance of the size of a character are similar for the two sexes. A more direct test of the reliability of the method can be obtained by estimating the frequency of extra-pair paternity by the sex-specific heritability method and one of the biochemical methods. Such a comparison for nine different bird species revealed a strongly positive relationship indicating that the heritability method gave reliable estimates of extra-pair paternity (Møller and Birkhead 1992). The sex-specific herita-bility method can only be used to estimate the frequency of extra-pair paternity in a large sample and cannot be used to estimate it in particular broods or assign paternity to particular offspring. DNA-fingerprinting and other molecular techniques are innovative solutions for these kinds of problems.

DNA-fingerprinting relies on the fact that nuclear DNA contains tan-demly repeated sequences of DNA, so-called mini-satellites, which have a unique DNA sequence for each individual (Jeffreys *et al.* 1985). Avian red blood cells contain DNA, and a small blood sample gives enough DNA for several analyses. Nuclear DNA isolated from a blood sample can be cut into smaller pieces with the use of restriction enzymes, and these smaller sequences give unique bands after autoradiography and electrophoresis. These bands are usually inherited independently in a Mendelian fashion, and putative parents can be excluded by band matching between offspring and their supposed parents (Burke and Bruford 1987; Wetton *et al.* 1987). Similarly, extra-pair offspring can be assigned to a particular male by analysis of band sharing.

The DNA-fingerprinting method was used to compare the relative frequency of extra-pair offspring in barn swallow broods with the relative frequency of extra-pair copulations. The copulation behaviour of a number of barn swallow pairs was followed daily for one hour throughout the fertile period and the number of within-pair and extra-pair copulations was recorded. A small blood sample was taken from the adult barn swallows attending the nest and from the nestlings when they were 13 days old. The blood was subsequently used for DNA-fingerprinting. The prediction was that the percentage of extra-pair offspring in a brood should be positively related to the percentage of all copulations being extra-pair copulations. This prediction was fulfilled as the frequency of extra-pair offspring was strongly positively related to the relative frequency of extra-pair copulations (A.P. Møller and H. Tegelström unpublished data). A number of broods held extra-pair offspring even though the female had not been seen to engage in extra-pair copulations during my observations, and some broods consisted entirely of extra-pair offspring even though extra-pair copulations only accounted for a small number of all copulations. There should not necessarily be proportionality between extra-pair copulations and extra-pair offspring because last male sperm precedence may result in relatively more offspring being fathered by extra-pair males than suggested by the relative frequency of extra-pair copulations (A.P. Møller and H. Tegelström unpublished data). However, the analyses clearly indicate that there was a general agreement between extra-pair copulations and extra-pair paternity in the barn swallow. This result allows previous observations of extra-pair copulations to be interpreted as reflecting extra-pair paternity.

The results from DNA-fingerprinting of barn swallows can also be used to identify the determinants of extra-pair paternity. Male tail length has previously been identified as a determinant of extra-pair copulation success of males and the engagement of females in extra-pair copulations (see section 11.3). The proportion of extra-pair offspring in barn swallow broods was in fact negatively related to male tail length (A.P. Møller and H. Tegelström unpublished data). Long-tailed male barn swallows were therefore more certain of their paternity than short-tailed males. This result was not confounded by other factors such as male age, degree of breeding synchrony, colony size, or intensity of mate guarding which were relatively unimportant in determining the extent of extra-pair paternity (A.P. Møller and H. Tegelström unpublished data).

In conclusion, in my study extra-pair copulations translated into extra-pair paternity as determined from DNA-fingerprinting, and long-tailed male barn swallows were more certain of their paternity than other males.

11.6 Paternity guards

Barn swallow males do not passively accept that their mates engage in extra-pair copulations, and a number of male behaviours appear to prevent the female from engaging in extra-pair copulations or to increase the certainty of paternity if the female may already have engaged in an extra-pair copulation. These paternity guards include:

(1) aggressive behaviour directed towards potential extra-pair males;

(2) territorial behaviour and territory size;

(3) mate guarding;

(4) frequent within-pair copulations; and

(5) deceptive alarm calls.

The use of these five paternity guards and their relations with tail length of male barn swallows is briefly analysed.

1. *Aggressive behaviour*. The most obvious way in which males can defend their certainty of paternity is by behaving aggressively towards potential threats of cuckoldry. Aggressive behaviour involves not only aggressive displays and fighting, which are used at short range, but also singing and other long-distance signals of aggressive intent. Male barn swallows usually chase intruders showing an interest in their mate, and the behaviour of a male ranges from threat displays with singing and exposed carpal joints to direct fights which sometimes are fatal. Fighting is an effective paternity guard in the barn swallow because extra-pair males never succeed in performing an extra-pair copulation in the presence of the male mate (Møller 1985).

Males of many bird species demonstrate a peak in their singing activity during the fertile period of their mate, and it may appear counter-intuitive that males should thereby directly announce the fertility status of their mate to neighbouring males (Møller 1991*j*). However, males of most bird species closely guard their mates against extra-pair males during the fertile period of the female, and neighbouring males therefore probably are already aware of the fertility status of the female. Song often competes with other essential activities such as foraging, and males which spend long periods singing must be able to acquire their resources during a short time span. Male song activity therefore can be viewed as an advertisement of male quality. Males may intrude on neighbouring territories during the fertile period of the female relative to the amount of song produced by the male territory owner, and this will force all males to sing as much as possible in order to reduce their share of all intrusions (Møller 1991*j*). Song activity therefore can be regarded as a paternity guard used by other males to assess the quality of neighbours and the probability of successfully achieving an extra-pair copulation (Møller 1991*j*). Barn swallow males also have a peak in their song activity during the fertile period of their mate, and this peak occurs during both first and second clutches (Møller 1991*j*). Song activity, measured as the number of times males were singing during scan observations every second minute for daily one-hour observation periods throughout the fertile period, varied from 0 to 53 with a mean of 11.2 songs (SE = 1.0, $n = 163$) during the first clutch and from 0 to 21 with a mean of 4.5 songs (SE = 0.5, $n = 95$) during the second clutch. Song activity was only weakly positively related to male tail length,[7] and these correlations were not confounded by colony size because male tail length did not vary consistently with size of the colony where they settled.

2. *Territorial behaviour and territory size*. Breeding territories are defended by the males of most bird species even though the resources of the territory vary from a mere nest and perching site as in the barn swallow to a nest site and some food resources, or an all-purpose territory with all resources needed for successful reproduction (Hinde 1956). Breeding territories have usually been assumed to provide females with essential resources. However, most intruders on avian breeding territories are neighbouring males, which intrude during the fertile period of the female (reviewed in Møller 1987*c*). Therefore, the most important function of the territory could be male defence of the mate against potential extra-pair copulation partners. If this were the case one should expect males to invest disproportionately in territory defence during the fertile period, and territorial behaviour and territory size might be expected to peak at that time of the breeding cycle. This was in fact the case. Male barn swallows defended larger breeding territories during the fertile period than during incubation and the early nestling period (Møller 1990*b*). Territory size increased again late in the nestling period of the first clutch just before a second clutch is usually initiated, but not in the late nestling period of the second clutch when the female is not fertile (Møller 1990*b*). Territory size in the barn swallows and a number of

other bird species with different types of breeding territories peaked during the fertile period of the resident female, and that is in accordance with the anti-sperm competition explanation for the evolution of breeding territoriality (Møller 1990b).

It remains to be determined whether certainty of paternity is related to territory size in the barn swallow, and whether long-tailed males defend larger territories than short-tailed male barn swallows.

3. *Mate guarding.* If males guard their fertile females against other males we should expect:

(a) males to follow their mates whenever they moved while the opposite should not be the case;

(b) the distance between a male and his mate should reach a minimum during the fertile period; and

(c) the male should attempt to maximize the amount of time spent within a safe distance of its mate (Birkhead *et al.* 1987; Birkhead and Møller 1992).

I collected quantitative information on the first and the last of these measures of mate guarding in barn swallows.

Male barn swallows followed their mates during most moves initiated by the female from three weeks before start of egg laying until the end of the laying period (Fig. 11.3). Female barn swallows only rarely followed their mates whenever they initiated a move, and there was a clear decrease in the frequency of following during the breeding cycle (Fig. 11.3). Male following of their mates was clearly an efficient way of preventing extra-pair copulations as male barn swallows were always able to thwart extra-pair copulation attempts if within a distance of 5m from their mate (Møller 1985).

Mate-guarding swallows

The percentage of time male barn swallows spent within a distance of 5m from their mates was high throughout the pre-laying period and dropped dramatically during the egg-laying period to a minimum during

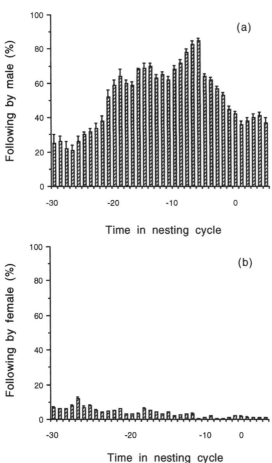

Fig. 11.3 Flight initiations by paired barn swallows leading to following by the other pair member (within 30s, when both pair members were within 5m of each other) in relation to time in the nesting cycle. (a) Male following of the female. (b) Female following of the male. Day 0 is the first day of the incubation period. Values are means (SE).

incubation (Fig. 11.4). Male presence near the mate was an effective paternity guard because short-term detention of males resulted in significantly increased levels of chases of the temporarily widowed female by neighbouring males and elevated levels of extra-pair copulation attempts and extra-pair copulations (Møller 1987g). Male barn swallows guarded their mates more intensely in larger colonies and if the operational sex ratio in the colony was male biased (Møller 1987a, 1987f). Termination of mate guarding, defined as the day when the male spent less than 50% of its time within a distance of 5m from its mate,

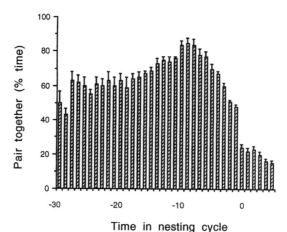

Fig. 11.4 Percentage of time spent within a distance of 5m by barn swallow pairs in relation to time in the nesting cycle. Day 0 is the first day of the incubation period. Values are means (SE).

was related to male body mass, because heavy males ceased guarding later than others (Møller 1987*f*). The operational sex ratio also affected termination of mate guarding because males experiencing male-biased sex ratios ceased guarding later than others (Møller 1987*f*).

Male barn swallows also adjusted their mate-guarding intensity to short-term changes in the perceived risk of cuckoldry. Temporary detention of some male barn swallows during the fertile period of their mates resulted in an increased sexual interest taken by other males in the single female. This resulted in an increase in the intensity of mate guarding among other males having a fertile female, but not among males with a non-fertile mate (Møller 1987*g*).

Mate-guarding intensity, estimated as the percentage of time males spent within a distance of 5m from their mates during daily one hour observation periods throughout the fertile period, varied considerably during the first clutch from 32.3 to 100% with a mean value of 75.7% (SE = 1.3, *n* = 162) and during the second clutch from 36.8 to 100% with a mean value of 75.9% (SE = 1.5, *n* = 93). The intensity of mate guarding during the fertile period, measured as the percentage of time spent within a distance of 5m from the female, was not significantly related to male tail length.[8] This result was not confounded by effects of colony size because the tail length of a male was not consistently related to the size of the colony where it settled.

4. *Frequent copulations.* Frequent within-pair copulations may be an efficient paternity guard because the probability of fertilization is

proportional to the relative number of sperm provided by each male, and because the probability of being the last male to copulate with a female is proportional to the number of copulations (Birkhead *et al.* 1987; Birkhead and Møller 1992). It is important to be the last male to copulate with a female before fertilization because the probability of fertilization is higher for the last male to copulate among all the bird species studied (last male sperm precedence) (Birkhead and Møller 1992).

Colonial barn swallows copulated more frequently than solitarily breeding barn swallows, and this may be an attempt on the part of colonial male barn swallows to increase their certainty of paternity (Møller 1985). Male barn swallows apparently adjusted their copulation rate to the number of chases by other males directed towards their mate, and this was particularly the case during the fertile period (Møller 1987*e*). If female barn swallows had been chased by another male, the time interval to the next within-pair copulation was shorter than expected by chance. This response may increase the certainty of paternity because extra-pair copulations were more likely to have occurred if a female frequently had been chased by other males (Møller 1987*e*).

The copulation rate, estimated as the number of copulations per hour during daily one-hour observation periods throughout the fertile period, varied considerably during the first clutch from 0 to 20 with a mean value of 1.5 (SE = 0.2, $n = 162$) and during the second clutch from 0 to 13 with a mean value of 0.8 (SE = 0.2, $n = 94$). The frequency of within-pair copulations during the fertile period was positively related to male tail length.[9] This result was not confounded by effects of colony size because the tail length of a male was not consistently related to the size of the colony where it settled.

Long-tailed male barn swallows may also be able to increase their probability of both within-pair and extra-pair fertilization because they had considerably larger testes than short-tailed males (Fig. 11.5). Testes volume was positively related to male tail length and male body mass,[10] but tail length was more important than body mass.[11] Male barn swallows with the longest tails had a combined testes volume more than 40% larger than that of males with the shortest tails. It is well known from both intra- and inter-specific studies that larger testes produce larger amounts of sperm (Møller 1988*d*, 1988*e*, 1989*f*). If long-tailed males deliver more sperm per ejaculate, this should result in an increased probability of fertilization because of the direct relationship between the probability of fertilization by a male and its relative contribution to the sperm pool within a female. Alternatively, as a consequence of their larger testes size and their presumably larger sperm production, long-

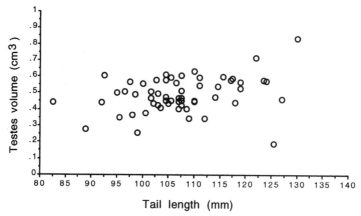

Fig. 11.5 Testes size (cm³) of male barn swallows in relation to their tail length (mm). Data from Chernobyl, Ukraine.

tailed male barn swallows may be able to copulate at a higher rate (as they do) without suffering from sperm depletion.

5. *Deceptive alarm calls.* Male barn swallows cannot simultaneously maximize mate-guarding activities and extra-pair copulations since they are mutually exclusive activities. There is thus a trade-off between the benefits of increased certainty of paternity and increased probability of fertilizing the eggs of female non-mates. Female barn swallows spend an increasing amount of time on their nest each day as soon as egg laying starts, and they are therefore less exposed to extra-pair copulations. Male barn swallows also run lower risks of losing paternity because an increasing proportion of the clutch is fertilized as egg laying progresses. The female is guarded by its mate a decreasing amount of time as egg laying advances (Fig. 11.5). The male regularly returns to its nest during the egg-laying period while giving contact calls, but such visits are much less common during the early incubation period (Møller 1990*f*). The male disappears immediately afterwards if the female is present on the nest. However, when the female was absent during the egg-laying period, the male almost always gave intense alarm calls while flying around in the colony. Male barn swallows only rarely gave alarm calls when the female was absent during the incubation period. These alarm calls were given apparently without any predators being present, and many of the barn swallows responded to the alarm calls by flying up and outside the breeding site (Møller 1990*f*). Female barn swallows regularly engaged in extra-pair copulations during their late fertile period (Møller 1985), and alarm calls given by the mate apparently

thwarted extra-pair copulation attempts on five occasions (Møller 1990*f*). Male barn swallows thus appeared to use alarm calls deceptively in an attempt to prevent extra-pair copulations if they were unaware of the whereabouts of their mate (Møller 1990*f*).

Barn swallows also use alarm calls when potential predators are present, and it is therefore difficult to know whether males gave alarm calls because the female was absent or because of the presence of a predator undetected by the observer. This possibility was excluded by an experiment in which the female barn swallow was chased away from its nest either during nest building, egg laying, or incubation (Møller 1990*f*). Colonially breeding male barn swallows gave alarm calls much more often during the egg-laying period than during the two other periods, while solitarily breeding males only rarely gave alarm calls. This result was expected because extra-pair copulations were never observed among solitarily breeding barn swallows, and a solitary male unaware of the whereabouts of its mate thus would not run high risks of losing paternity (Møller 1990*f*).

Male barn swallows were in fact able to adjust their alarm calling behaviour to current threats of extra-pair copulations. Presentation of a male barn swallow model near the nest of a solitarily breeding pair in the absence of the female resulted in alarm calls by the male mate more often than when the model was absent (Møller 1990*f*). Solitarily breeding males only responded to the presence of the model during the nest-building and particularly the egg-laying periods, when paternity was at stake, but not during incubation after the end of the fertile period (Møller 1990*f*). Deceptive alarm calls were therefore used by male barn swallows whenever there was an increased threat of losing paternity.

Almost all colonial male barn swallows gave alarm calls in the absence of their mate during the egg-laying period whereas solitarily breeding males did not. There was therefore no statistically significant relationship between male behaviour and male tail length, as male tail length was unrelated to the size of the colony in which it settled.

In conclusion, male barn swallows adopted a number of different paternity guards which appeared to be effective as determined from the negative relationship between the intensity of the paternity guard and the engagement of the female mate in extra-pair copulations. However, the intensity of only one of these paternity guards, frequent within-pair copulations, was clearly positively related to tail length of male barn swallows.

11.7 Costs and benefits of extra-pair copulations

Females may experience a number of potential costs and benefits of copulations with multiple males (reviewed in Birkhead and Møller 1992). These are briefly summarized in Table 11.1. The fitness benefits of copulations with multiple males can most readily be classified as those gained directly by the female itself and those gained indirectly, through its reproducing offspring. Direct fitness benefits of copulations with multiple males include:

(1) quantity or quality of sperm;

(2) resource acquisition;

(3) male parental care;

(4) avoidance of infanticide; and

(5) avoidance of rejection costs.

Indirect fitness benefits include:

(6) attractiveness genes; and

(7) good genes.

The fitness costs of copulations with multiple males include the direct fitness costs of:

(1) copulations;

(2) contagious disease;

(3) mate desertion;

Table 11.1. Benefits and costs of extra-pair copulations to female barn swallows.

Benefits	Costs
Direct fitness benefits	*Direct fitness costs*
Quantity or quality of sperm	Copulation costs
Resource acquisition	Contagious disease
Male parental care	Mate desertion
Avoidance of infanticide	Reduced paternal care
Avoidance of rejection costs	Sib competition
Indirect fitness benefits	*Indirect fitness costs*
Attractiveness genes	Unattractive genes
Good genes	Bad genes

(4) reduced paternal care, and

(5) increased sibling competition,

and the indirect fitness costs of:

(6) reduced attractiveness of offspring, and

(7) reduced viability of offspring because of 'bad' genes.

11.7.1 *Benefits of extra-pair copulations*

Direct fitness benefits to females of copulations with multiple males include several different possibilities.

1. *Quantity or quality of sperm* (Walker 1980; Gibson and Jewell 1982; Drummond 1984; McKinney *et al.* 1984). Males may differ in their quality or quantity of sperm. Females should be able to replenish their sperm reserves just by copulating repeatedly with the same male. The average number of within-pair copulations per clutch in the barn swallow may exceed fifty, and it is therefore unlikely that females run out of sperm. Females may benefit directly from copulating with more than one male if males differ in their quality of sperm. There is some evidence of total or partial infertility in barn swallows (see section 5.4.1), but this infertility is unrelated to tail length of males which is one of the main determinants of extra-pair copulation success in the barn swallow.

2. *Resource acquisition* (Thornhill and Alcock 1983; Birkhead and Møller 1992). Females may acquire resources by copulating with more than a single male. For example, males may feed the female or give it access to a food resource following copulations. This is not the case in the barn swallow because males never provide courtship food and they do not defend food resources. Alternatively, ejaculates may contain nutritious substances which can be incorporated by the female into production of additional offspring. There is no evidence for nutrient transfer in ejaculates of vertebrates, including birds (Birkhead and Møller 1992).

3. *Male parental care* (Birkhead and Møller 1992). Females may gain help with provisioning of their offspring by copulating with more than a single male. This is for example the case in several polyandrous birds where the number of males feeding offspring and the amount of food provided by different males is directly related to their copulation activity (Birkhead and Møller 1992; Davies 1992). This explanation is not appropriate for the barn swallow, because males only feed offspring in their own nest.

4. *Avoidance of infanticide* (Hrdy 1977). Females may benefit from copulating with potential infanticidal perpetrators if copulation access provides males with some paternity and thereby prevents infanticidal behaviour. Unmated male barn swallows regularly commit infanticide, and they also attempt extra-pair copulations. However, there is no evidence of unmated males ever succeeding in their extra-pair copulation activities (section 8.5; Møller 1988a).

5. *Avoidance of rejection costs* (Birkhead and Møller 1992). If the duration of a copulation is relatively short, it might be less costly for a female to accept an extra-pair copulation than to waste time rejecting persistent male non-mates. However, this direct fitness benefit will only apply if males are of a uniform quality, and if overall population density is low. Most barn swallows breed in colonies, and extra-pair males are sufficiently common not to make acceptance of extra-pair copulation attempts a feasible alternative.

Indirect fitness benefits.

6. *Attractiveness genes* (Møller 1988c; Birkhead and Møller 1992). Females may benefit from copulating with extra-pair males if heritable features of these males are particularly attractive. The offspring of females engaging in extra-pair copulations thus may inherit the female mate preference genes or the genes for the expression of the attractive male secondary sexual character. The female mate preference and the male secondary sex trait therefore could coevolve to ever more extreme expressions as a consequence of the runaway process (Fisher 1930). This hypothesis has not been evaluated in any species, although it potentially could result in the evolution of strong mate preferences working outside the context of normal pair formation.

7. *Good genes* (Møller 1988c; Birkhead and Møller 1992). Females may benefit indirectly from copulating with extra-pair males if male secondary sexual characters used in extra-pair mate choice reflect good genes. Females copulating with preferred non-mates with the most extravagant secondary sexual characters would then raise offspring with higher viability. It is known that long-tailed male barn swallows are more successful in acquiring extra-pair copulations and fertilizations, long-tailed males have higher survival probability, and they raise more offspring with higher survival probability (Smith *et al.* 1991; Møller 1991e, 1993n; A.P. Møller and H. Tegelström unpublished data). A study of extra-pair paternity in the blue tit *Parus caeruleus* revealed that the most preferred male mates also acquired more extra-pair offspring, but suffered fewer cases of extra-pair paternity in their own broods. Preferred male blue tits also had higher survival probability and

recruited more offspring than less preferred males (Kempenaers *et al.* 1992). These results are consistent with the good genes hypothesis.

In conclusion, female barn swallows do not benefit from copulating with multiple males because of sperm quality or quantity, resource acquisition, male parental care, or avoidance of infanticide, while females may benefit from raising offspring with attractiveness genes or good genes.

11.7.1 *Costs of extra-pair copulations*

Direct fitness costs.

1. *Copulation costs* (Birkhead and Møller 1992). It is costly for females in terms of time and energy to copulate repeatedly, but the increased time and energy use due to extra-pair copulations is probably always very limited. Copulation costs may also include risks of injury by the extra-pair male during forced extra-pair copulations (Birkhead and Møller 1992). For example, females of several duck species may drown as a result of forced extra-pair copulations (Birkhead and Møller 1992). Extra-pair copulations in the barn swallow do not appear to be forced on the female because females are able to terminate any attempt by their aggressive behaviour or by simply flying away.

2. *Contagious parasites* (Birkhead and Møller 1992). Extra-pair copulations may be costly because extra-pair males may suffer from infections with contagious parasites and diseases, and these may not be readily observable during the short extra-pair copulation display. There is little evidence of transmission of contagious parasites or venereal diseases in bird populations, but this may be due to a lack of studies (Birkhead and Møller 1992). However, venereal diseases in birds are known sometimes to cause reduced fertility or even mortality (Sheldon 1993). Female barn swallows potentially could suffer from this cost of extra-pair copulations, and it may explain why some males are very successful while others never obtain extra-pair copulations.

3. *Mate desertion*. Males have been suggested to desert their mates as a response to involvement of their mates in extra-pair copulations (Trivers 1972; Xia 1992). This option may only be feasible if the deserting male has a high probability of finding another mate, and if the survival prospects of the deserted clutch is not greatly reduced. There is no evidence of mate desertion in birds in connection with extra-pair copulations (Birkhead and Møller 1992).

4. *Reduced paternal care*. Males may engage less in parental care as a result of engagement by their mates in extra-pair copulations (Dawkins 1976;

Whittingham *et al.* 1992). If males provide parental care, they obviously incur a cost in terms of time and energy use, and this investment would be wasted if some or all offspring were sired by other males. Parents therefore usually only provide parental care for offspring in their own nest and not for the offspring of neighbours. It is potentially costly for males to reduce the amount of care or not to provide paternal care at all to offspring in their own nest, if the reduction is not fully compensated by the female mate. The reason for this is that their own offspring as well as offspring of the extra-pair male will suffer from the reduction. The condition for a reduction in paternal care depends on the reduction in residual reproductive value of the male and the probability of increasing certainty of paternity in the future.

The relative feeding rate of male barn swallows is:

(a) positively related to the number of pair copulations;

(b) negatively related to the number of extra-pair copulations by their mates; and

(c) negatively related to the relative number of extra-pair copulations by their mates (Møller 1988*b*).

These results could suggest that males either adjust paternal care to certainty of paternity, or that males that suffer from cuckoldry because of their poor quality are also unable to provide extensive parental care. These possibilities were tested experimentally by short-term removal of male barn swallows during the fertile period of their mates, and by subsequently recording paternal care. Temporary male detention resulted in increased levels of extra-pair copulations by the temporarily widowed females for colonially, but not for solitarily breeding barn swallows (Møller 1988*b*). Experimental male barn swallows breeding in colonies were able to assess the involvement of their mates in extra-pair copulations after their release from the number of chases by extra-pair males directed towards females. Male but also female barn swallows subsequently reduced the absolute amount of parental care in the colonially breeding experimental group and this reduction in parental care was associated with brood reduction (Møller 1988*b*). The relative provisioning rate of males was reduced among colonial experimental males, but not among solitarily breeding experimental males, or among the control groups (Møller 1988*b*). The reduction in absolute female provisioning was unexpected, as was the brood reduction, and it is not clear whether the relatively larger reduction in male provisioning of offspring among colonially breeding experimental pairs resulted as a consequence of the treatment or the brood reduction (Wright 1992).

This explanation also applies to defence of offspring by male barn swallows against a potential predator (Møller 1991*i*).

5. *Sibling competition.* Extra-pair copulations often result in extra-pair paternity with the consequence that average relatedness between offspring in a brood is reduced. This should result in conflicts between nest mates of different genetic origin because of their lack of common interests (Hamilton 1964). For example, the intensity of sibling competition may be higher in broods with unrelated nest mates, and the level of begging for food may be elevated among broods with a low degree of relatedness compared with broods of full siblings. This cost has not been studied among birds, but could be a potentially direct fitness cost of extra-pair paternity.

Females may also suffer indirect fitness costs from engaging in extra-pair copulations.

6. *Unattractive genes* (Birkhead and Møller 1992). Offspring may not appear attractive if females copulate with unattractive extra-pair males.

7. *'Bad' genes.* Similarly, if the main fitness benefit of extra-pair copulations for females is acquisition of viability genes for their offspring, females may incur a cost in terms of 'bad' genes by copulating with low-quality extra-pair males (Birkhead and Møller 1992).

In conclusion, female barn swallows do not suffer from extra-pair copulations because of direct costs of copulations or mate desertion, but the costs of contagious disease, reduced paternal care, increased sibling competition, and unattractive and bad genes for offspring may all be potentially important. The indirect fitness benefits of extra-pair copulations to females must be of a considerable magnitude in order to overcome these direct and indirect fitness costs.

11.8 Summary

Female barn swallows regularly copulate with males other than their mates and this results in post-copulatory sexual selection. Long-tailed males are very successful in their extra-pair copulation attempts, and female participants tend to be the mates of short-tailed male barn swallows. Sex differences in the heritability of tarsus length suggested that extra-pair offspring were common. DNA-fingerprinting revealed that extra-pair paternity in barn swallow broods was positively related to the relative frequency of extra-pair copulations. Therefore, long-tailed male barn swallows were more certain of the paternity of the brood in their nest than short-tailed males.

A number of paternity guards appear to increase the certainty of paternity. These include male aggressive behaviour directed towards male non-mates approaching the female mate, increased intensity of male territorial behaviour during the fertile period of the female mate, mate guarding, frequent within-pair copulations, and deceptive alarm calls of the male used to disrupt extra-pair copulation attempts. Long-tailed male barn swallows had higher within-pair copulation rates than short-tailed males, and this may have resulted in a higher certainty of paternity because of a relatively larger contribution of sperm to the mate, and because of a higher probability of being the last male to copulate. Long-tailed males also had larger testes, a factor that may further increase their certainty of paternity.

A number of potential fitness costs and benefits accrue to females engaged in extra-pair copulations. Female barn swallows only appear to obtain indirect fitness benefits in terms of attractiveness genes or good genes to their offspring, whereas the fitness costs of extra-pair copulations are acquisition of contagious parasites, reduced paternal care, increased sibling competition and acquisition of unattractive genes or bad genes to offspring.

STATISTICS

1. Spearman rank correlation: $r = 0.53$, $n = 56$, $P < 0.001$.

2. Kendall partial rank-order correlation: tau $= 0.06$, $n = 161$, $z = 1.04$, NS; second clutch: tau $= -0.06$, $n = 92$, $z = 0.86$, NS.

3. Kendall partial rank-order correlation: tau $= -0.02$, $n = 161$, $z = 0.28$, NS; second clutch: tau $= 0.04$, $n = 92$, $z = 0.55$, NS.

4. Log-likelihood ratio tests: males: $G^2 = 32.67$, $df = 1$, $P < 0.001$; females: $G^2 = 23.31$, $df = 1$, $P < 0.0001$.

5. Spearman rank correlation: males: $r = 0.22$, $n = 129$, $P < 0.01$; females: $r = 0.24$, $n = 129$, $P < 0.01$.

6. Kendall partial rank-order correlation: tau $= -0.07$, $n = 161$, $z = 1.22$, NS; second clutch: tau $= 0.06$, $n = 92$, $z = 0.82$, NS.

7. Spearman rank correlation: first clutch: $r = 0.14$, $n = 161$, $P = 0.08$; second clutch: $r = 0.15$, $n = 95$, $P = 0.14$.

8. Pearson product-moment correlation: first clutch: $r = -0.07$, $n = 160$, NS; second clutch: $r = 0.07$, $n = 93$, NS.

9. Spearman rank correlation: first clutch: $r = 0.28$, $n = 160$, $P = 0.0004$; second clutch: $r = 0.25$, $n = 93$, $P = 0.016$.

10. Multiple regression: $F = 12.94$, $df = 2,45$, $P < 0.001$.

11. Standardized partial regression coefficients: tail length: $\beta = 0.44$, $P = 0.0008$; body mass: $\beta = 0.33$, $P = 0.011$.

12

Sexual size dimorphism and female ornaments

12.1 Introduction

In many animal species males differ in size from females. For example, females are considerably larger than males in many invertebrates, fish, amphibians, and reptiles, whereas males are larger than females in most birds and mammals (Darwin 1871). Why is this the case? Size differences between the two sexes are most readily explained by selection processes affecting individuals differently (section 12.2). Sexual size dimorphism may result from natural or sexual selection, as already emphasized by Charles Darwin (1871). Natural selection will affect the morphology of males and females both during reproduction and the non-reproductive period. For example, males and females may have slightly different foraging behaviour and habitat choice, and the different natural selection pressures under two kinds of environmental conditions may result in sexual differences in morphology. Intersexual competition for food resources may thus lead to sexual size dimorphism due to natural selection acting on heritable variation in morphology. The most prominent example is the now extinct Hawaiian huia *Heteralocha acutirostris* in which males and females had completely different bill shapes and equally different foraging ecologies (Darwin 1871).

Sexual size dimorphism may also result from the process of sexual selection affecting males and females differently. Sexual selection arises as a consequence of variation in some aspect of mating success owing to competition with individuals of the same sex being associated with behavioural or morphological characters. Mating success will usually not be similarly related to morphology in the two sexes because of sexual differences in activities such as foraging or reproduction. Variation in mating success in the chosen sex, usually males, will result in a micro-evolutionary change in characters associated with variance in mating success. Morphological characters in the two sexes are usually controlled by identical genes, and there is often a strong genetic correlation between the characters in the two sexes. Sexual selection on a male phenotype will imprison the female phenotype which can be dragged along evolutionarily by the genetic correlation. The female trait therefore evolves as a response to selection on males, and not primarily because of selection on females. Eventually the female character may become so detrimental to viability that genes controlling the character become sex-limited which results in a partially or completely sex-limited expression.

Alternatively, the vestigial expression of a male secondary sexual character in females may be attributed to direct sexual selection on females. Individual females with the most extreme expression of a secondary sexual character may gain a mating advantage for the same reasons as males with extravagant ornaments achieve a mating advantage. The expression of a secondary sexual character in females is likely to be less extreme than in males because of the usually larger role played by females in reproductive activities. These natural and sexual selection explanations for the evolution of sexual dimorphism are discussed further in section 12.2.

Secondary sexual characters are often expressed to some degree in the choosy sex (usually females), but this fact has received much less attention than the 'real' ornaments of the chosen sex (usually males). Female barn swallows have relatively long and variable tails compared with many close relatives and compared with juvenile barn swallows (section 12.3). The evolution and the maintenance of variation in female tail length therefore begs an explanation. There are three hypotheses for the evolution of sexual size dimorphism:

1. Males and females may differ in morphology as a result of different intensities of natural selection such as differences in foraging behaviour (Darwin 1871).

2. Males and females may differ in morphology because of sexual selection in one sex, usually males, being more intense than in the other, and the male secondary sexual character being expressed in females because of a correlated response to selection on the male trait (Fisher 1930; Lande 1980).

3. Males and females may differ in morphology because of intense sexual selection on one sex, and the secondary sexual character being expressed in females has a similar sexual selection function as the male trait (Trivers 1972).

These three hypotheses for the evolution of sexual size dimorphism are tested on female barn swallows in sections 12.4 and 12.5.

12.2 Evolution of sexual dimorphism

Morphological characters are sometimes expressed similarly in both sexes, but different selection pressures on the two sexes may subsequently lead to accumulation of genes with different effects in males and females (Fisher 1930). Sexual selection may, for example, lead to directional selection on tail length in males. Tail length may be subject to directional sexual selection in males, but not in females, because of the viability costs of a long tail, and the tail trait may subsequently become more exaggerated in males than in females because of accumulation of genes with a sex-limited expression. Many different morphological characters demonstrate sexual size dimorphism, and variation in such characters is usually polygenic (reviewed in Lande 1980). Selection on a character such as tail length in one sex causes a response in the selected sex as well as a correlated response in the other sex. Such correlated responses to selection on a character in one sex are usually attributed to the effects of pleiotropy of genes affecting the homologous characters in both sexes. The reason for this is that the genetic correlation between the characters in males and females is usually high (Falconer 1981). Sexual selection is a very strong evolutionary force, and male characters as well as homologous female characters may change as a response to sexual selection. The change in a male character such as tail length will only much later be followed by a slow change in the degree of sexual size dimorphism because of strong genetic correlations (Lande 1980, 1987). The reason why this second phase of the evolution of sexual dimorphism is slow is that most mutations are not sex-limited in their expression. If a sex-limited mutation arises, natural selection on females will favour its spread. The intensity of sexual selection is balanced by oppositely directed natural selection, and the spread of a sex-limited mutation among females leads to a decrease in the intensity of natural selection. This will result in an increase in non-sex-limited genes for increased size of the male trait, which will retard the evolution of dimorphism. Therefore, a morphological trait particularly expressed in one sex, such as tail length in males, can be expressed for a large number of generations in the other sex because of the genetic correlation between the sexes. This will be the case even though the trait may not be advantageous or perhaps even detrimental in the sex demonstrating the vestigial expression of the trait. An evolutionary change in sexual size dimorphism will therefore be an extremely slow process compared with, for example, a change in a morphological character in one sex (Lande 1980, 1987).

The genetic correlation between the sexes will be broken up only very slowly by natural selection working on females because evolution of the character in males as well as females primarily is determined by strong directional sexual selection on males (Lande 1980, 1987). Eventually, the tail

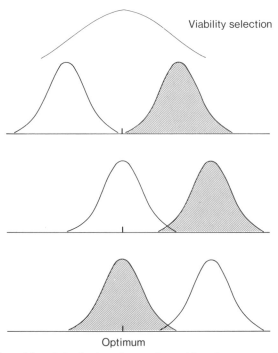

Fig. 12.1 Male and female body size when males and females compete for resources (top), when males are under stronger sexual selection (centre), and when females are under stronger sexual selection (bottom). The fitness curve under viability selection is indicated at the top. Male phenotypes are shown as hatched frequency distributions.
The optimum phenotype under viability selection is indicated at the bottom.

trait may become so exaggerated in males that natural sexual on females results in a complete sex-limitation of the character (Fisher 1930).

Sexual size dimorphism usually arises as a result of one or both sexes being displaced from the optimum phenotype under natural selection (Fig. 12.1). Females of most animal species perform the majority of reproductive duties and they should thus be under stronger natural selection than males. Sexual size dimorphism may therefore be caused by males being displaced from the natural selection optimum by sexual selection, whereas female morphology is maintained close to the natural selection optimum (Fig. 12.1). The other alternative is that male and female mates compete for limited food or other resources within a territory. This results in the intensity of competition for food being reduced by the use of a slightly different foraging niche, and males and females may subsequently diverge in morphology as a result of adaptation to their respective niches. Both males and females are therefore displaced from the optimum morphology under natural selection (Fig. 12.1). Finally, males contribute disproportionately to reproduction in a few species with sex role reversal. This is the case in, for example, pipefishes and seahorses, and in jacanas. Males of such species are under much stronger natural selection than females because of their reproductive burden, and sexual size dimorphism may result from females being displaced from the natural selection optimum body size by sexual selection (Fig. 12.1).

Sexual size dimorphism can be explained in at least three different ways:

1. Males and females differ in morphology as a result of different intensities of natural selection arising from differences in foraging behaviour.

2. Males and females differ in morphology because of sexual selection in one sex, usually males, being more intense than in the other, and the male secondary sexual character is expressed in females because of a correlated response to selection on the male trait. Therefore, the female trait has no current function in terms of natural or sexual selection.

3. Males and females may differ in morphology because of intense sexual selection on one sex, and the secondary sexual character being expressed in females has a sexual selection function as the male trait.

The first hypothesis suggests that natural selection may operate differently on males and females during reproduction or competition for resources, and this may result in sexual size dimorphism. It is well known that the foraging niche of many resident specialists such as several species of woodpeckers and the Hawaiian huia, overlap very little (Darwin 1871; Selander 1966, 1972; Ligon 1968; Downhower 1976; Slatkin 1984). Niche segregation is usually the outcome of competition for food between the sexes, and the intensity of competition may be reduced by behavioural niche segregation. Behavioural differences will form the basis for evolution of morphological differentiation and sexual size dimorphism, because different tools and thus different morphologies may be adaptations to different kinds of habitats. Sex differences in foraging behaviour and habitat use are usually already present among juveniles, and sexual size dimorphism caused by competition for food or other resources is thus expected to be expressed before the start of reproduction.

The second hypothesis for sexual size dimorphism posits that the morphology of one sex has been exaggerated as a result of sexual selection. This process leads to a change in morphology in the sex subject to sexual selection, but also a correlated response to selection in the other sex because of a high genetic correlation for the morphological trait between the sexes. Intense directional selection caused by female mate preferences can lead to a rapid exaggeration of male morphology by the Fisher process (Lande 1980; Lande and Arnold 1985). Alternatively, the male trait may become exaggerated as a result of a mate preference for direct fitness benefits or by the good genes process. The female morphological trait is dragged along by sexual selection on the male trait, and the subsequent decrease in the size of the character in females, which leads to sexual size dimorphism, is a very slow process governed by natural selection (Lande 1980; Kirkpatrick 1982; Lande and Arnold 1985). The female phenotype will eventually equilibrate at its natural selection optimum while the male phenotype is displaced from its natural selection optimum. The expression of the male secondary sexual character in the female will be merely a correlated response to selection in males, and the trait in females has no functional importance in terms of sexual selection.

The third hypothesis for sexual size dimorphism suggests that both the male and the female trait have been subject to directional sexual selection as a result of mate choice in both sexes (Trivers 1972). Both males and females should perform mate choice when there is parental investment by individuals of both sexes, and the intensity of sexual selection should be directly related to relative parental investment (Trivers 1972). The exaggeration process of the secondary sexual character thus results from sexual selection in males as well as in females. The male secondary sexual character may partly be expressed in females because of a correlated response to selection since sexual selection is likely to be more intense in one sex than the other. A genetic correlation between the expression of the trait in males and females is therefore also predicted. This hypothesis thus suggests that the secondary sex trait is advantageous in both sexes, and that choice of partners with the most extreme expression of the trait confers a selective advantage (Trivers 1972). The mating advantage of individuals of the least ornamented sex could be due to direct or indirect fitness benefits, just as suggested for male secondary sexual characters.

These three alternative explanations for the evolution of sexual size dimorphism in barn swallows are evaluated in the following paragraphs.

12.3 Sexual size dimorphism in barn swallows

12.3.1 *Size dimorphism*

Measurements of a large sample of adult male and female barn swallows allowed determination of the degree of sexual size dimorphism in a number of different morphological characters. Dimorphism is most readily calculated as the difference between the log-transformed mean values for males and females, which gives male body size as the relative deviation from female body size. Barn swallows demonstrate relatively little size dimorphism in most morphological characters (Table 12.1). Nine out of 11 different characters revealed small, but nonetheless statistically significant deviations of less than 2% from monomorphism. The size of the throat patch is an areal measurement, and the estimate of size dimorphism thus has to be square-root transformed in order to be comparable to the other measurements. Body mass is a volume and therefore has to be cubic-root transformed in order to be comparable. The only deviating measurements are keel length and tail length (Table 12.1). Keel length was 3.4% larger in males than in females. Keel length is a skeletal character like the three beak measurements and tarsus length, but differs in degree of size dimorphism. All five skeletal characters are fixed for life in juveniles. It is obvious from these measurements that it can be dangerous to use a single skeletal character as a representation of overall body size, because the degree of sexual size dimorphism may vary by as much as a factor of five. The only character with an extreme degree of sexual size dimorphism was tail length, which was on average almost one fifth longer in males than in females.

12.3.2 *Dimorphism in tail length*

Sexual size dimorphism in tail length clearly deviated from other morphological characters measured (Table 12.1). Male tails were on average 19.6% longer than those of females. The dimorphism was a property of only the outermost tail feathers, as the short, central tail feathers were actually slightly longer in females than in males (Table 12.1).

Table 12.1. Size of morphological traits in males and sexual size dimorphism of these morphological traits of the barn swallow in the Kraghede study area. Size dimorphism is $\log_{10}$(mean male size) minus $\log_{10}$(mean female size). Significance levels denote statistically significant size differences between males and females. Values are means (SE).

Morphological trait	Male size	Size dimorphism
Wing length (mm)	126.83 (0.11) ***	0.007
Wing span (mm)	329.87 (0.03) ***	0.008
Tail length (mm)	107.66 (0.32) ***	0.078
Short tail length (mm)	44.28 (0.09) **	−0.004
Throat patch area (mm²)	206.59 (0.03) *	−0.015
Beak length (mm)	7.66 (0.02) *	−0.003
Beak width (mm)	11.99 (0.02) ***	0.004
Beak depth (mm)	2.94 (0.01)	0.003
Tarsus length (mm)	11.42 (0.02) *	−0.003
Keel length (mm)	21.70 (0.04) ***	0.015
Body mass (g)	19.20 (0.04) ***	−0.026
n	1593	

t-tests, *$P < 0.05$, **$P < 0.01$, ***$P < 0.001$

Tail length in female barn swallows demonstrates considerable variability compared with other morphological characters in the same individuals. The coefficient of variation (CV) for female tail length is 6.6% which is significantly smaller than the value for male tail length (Møller 1991*a*). However, the CV for female tail length is considerably higher than for all other morphological characters in female barn swallows.

Some of the variation in female tail length can apparently be ascribed to additive genetic variance, because tail length of females resembles that of their mothers. The heritability estimate of 0.81 (SE = 0.23) is statistically significant (Møller 1993*k*). This estimate is biased because of assortative mating regarding tail length, sex differences in phenotypic variance, and maternal effects. However, the first two of these biases do not affect the heritability estimate for females to a large degree (Møller 1993*k*).

The genetic correlation between the sexes for tail length can be estimated by correlating the tail length of sons to that of their mother and the tail length of daughters to that of their father (Falconer 1981). The two estimates of the genetic correlation are 0.49 and 0.59, respectively, and both are statistically significant (Møller 1993*k*). A number of different biases can affect these estimates. For example, maternal and common environment effects may result in biased estimates of genetic correlations (Falconer 1981). However, the estimates are not easily adjusted for such biases.

12.3.3 *Evaluation of hypotheses on sexual size dimorphism*

There are three hypotheses which can explain sexual size dimorphism:

(1) the niche segregation hypothesis which is based on natural selection;

(2) the correlated response hypothesis which assumes that the size of the trait in females is a result of sexual selection on males; and

(3) the ornament hypothesis which posits that the female character also has a sexual selection function.

The first and the third hypotheses are adaptive explanations for male and female size differences, while the second hypothesis is a non-adaptive explanation for female size.

The niche segregation hypothesis predicts that:

(1) sexual size dimorphism should be present before sexual maturity if it allows competition for food or other resources to be reduced;

(2) sexual size dimorphism should be particularly prominent in structures used in foraging; and

(3) the sexes should exploit different kinds of food or habitats.

Sexual size dimorphism in the most dimorphic character was present in the barn swallow only after completion of the first moult in early spring (section 3.3.4). This suggests that dimorphism plays a role during reproduction rather than during foraging.

The second prediction that dimorphism should be particularly marked in structures used during foraging was not fulfilled (Table 12.1). The beak and to some extent feather characters are used by the aerially insectivorous barn swallow during foraging. The beak is used for capturing and manipulating insect prey, but the three beak measurements were some of the least dimorphic characters. Wings and tails affect the flight behaviour of barn swallows and thus indirectly their behaviour during foraging. The two wing measurements and the short-tail measurement were all only slightly dimorphic (Table 12.1). The length of the outermost tail feathers may affect flight behaviour and thus foraging in both sexes. The delayed dimorphism of tail length until early spring, just before start of the breeding season, and the direct use of tails in sexual display, suggest that sexual size dimorphism in tail length can primarily be attributed to sexual selection. It is therefore unlikely that sexual size dimorphism has arisen as a result of niche segregation among the sexes during foraging.

The third prediction that the food items taken by males and females should differ was tested by sampling insects brought by males and females

Foraging swallows

to their nestlings aged 8–12 days by fitting nestlings with neck collars. Food was sampled between 09:00 and 12:00 hours from first-brood nestlings in 1988. Barn swallows mainly select prey according to size (Turner 1980), and the prediction was tested by determining size differences in prey items simultaneously brought by males and females to their broods. The size of insect prey was very similar for male and female barn swallows,[1] and the data therefore did not suggest that males and females exploited differently sized prey items.

The correlated response hypothesis posits that (1) there is a genetic correlation between tail length of males and females, and (2) variation in tail length among females is selectively neutral or even slightly detrimental. The first assumption has been confirmed for the barn swallow as already stated in section 12.3.2. This assumption is not exclusive for the correlated response hypothesis because the ornament hypothesis also may assume a genetic correlation between the secondary sex trait in the two sexes. The prediction that variation in tail length among female barn swallows is selectively neutral or detrimental will be discussed later, when the ornament hypothesis is tested. Only one other study has tested for the different explanations of vestigial expression of secondary sexual characters in females. The size of the epaulets of female red-winged blackbirds *Agelaius phoeniceus* was apparently not currently associated with female quality or reproductive potential, and their presence was thus concluded to be the result of a correlated response to selection on males where the epaulets have a function in male–male competition (Muma and Weatherhead 1989).

The ornament hypothesis suggests that:

(1) there is a genetic correlation between tail length of males and females;

(2) females with a long tail have a mating advantage; and

(3) female tail length reflects female parenting ability or overall viability if tail length is a reliable indicator of quality.

The first assumption has already been tested and confirmed for the barn swallow (see section 12.3.2). The second prediction can most readily be tested if a fraction of females remain unmated. This is not the case since female barn swallows in my Kraghede study area have never been found to remain unmated for the breeding season. A second way to test whether particular females are preferred as mates is to determine the relationship between the duration of the pre-mating period and female tail length. If long-tailed female barn swallows are preferred mates, it should on average take less time for such females to become mated. In other words, the duration of the pre-mating period should be inversely related to their tail length. This prediction was fulfilled (Fig. 12.2). Whereas it took female barn swallows with the shortest tails an average of almost four days to become mated, females with

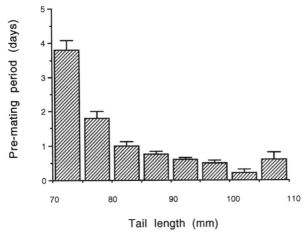

Fig. 12.2 The duration of the pre-mating period of female barn swallows in relation to their tail length (mm). Values are means (SE).

the longest tails spent an average of less than a day at the breeding sites before acquiring a mate. The duration of pre-mating periods in female barn swallows was much shorter than that of males[2] (see section 4.3.2), and this may reflect that males arrive on average 4.8 days before females and that females are more choosy than males. It should at least in theory be possible to test experimentally whether tail length of female barn swallows is subject to a male mating preference by manipulation of tail length before mate acquisition. The relationship between female tail length and duration of their pre-mating period is consistent with a sexual selection explanation of an exaggerated tail length of females being longer than that of juveniles.

Female barn swallows with long tails acquired a mate more rapidly than short-tailed females, and they should therefore be able to start reproduction relatively early during the breeding season. The pairing date of female barn swallows was in fact negatively related to female tail length (Fig. 12.3). Long-tailed females bred considerably earlier than short-tailed females. This relationship was independent of female age (Møller 1993*k*). This advancement of reproduction among long-tailed females was partially caused by their shorter pre-mating period and their earlier arrival date (see section 12.4). This observation is therefore also consistent with a sexual selection explanation of female tail ornaments.

If female barn swallows prefer long-tailed males, and if males prefer long-tailed females, we should expect assortative mating with respect to tail length. As predicted, the tail length of pair members was positively correlated, and long-tailed females tended to mate with long-tailed males more often than expected by chance (Fig. 12.4; Møller 1993*k*). This

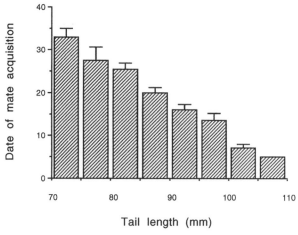

Fig. 12.3 Date of mate acquisition by female barn swallows in relation to their tail length (mm). 1 = 1 May. Values are means (SE). Adapted from Møller (1993*k*).

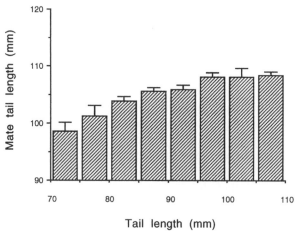

Fig. 12.4 Tail length (mm) of female barn swallows in relation to that of their mates. Values are means (SE). Adapted from Møller (1993*k*).

relationship was independent of female age (Møller 1993*k*). A process of assortative mating is consistent with a sexual selection explanation for male and female tail length in the barn swallow.

In conclusion, there is little evidence for a niche segregation explanation of sexual size dimorphism in tail length in the barn swallow. The correlated response hypothesis for sexual size dimorphism in tail length was not supported because female tail length was associated with a mating advantage and, as we shall see in the next sections, female tail length reliably reflects some properties of female quality and reproductive potential.

12.4 Female ornaments as signals of female quality

If female ornaments in the barn swallow are reliable signals of female quality we should expect:

(1) female timing of moult to be positively correlated with female tail length;

(2) female arrival date to be positively related to female tail length; and

(3) survival prospects of females to be positively related to female tail length.

If long-tailed female barn swallows are of high quality, they should moult relatively earlier than other females because that would allow them to arrive early during spring at the breeding grounds. A sample of female barn swallows from the winter quarters in Namibia did show a negative relationship between timing of overall moult or tail moult and female tail length, even when controlling for age and other potentially confounding variables (Møller *et al.* 1993). Therefore, long-tailed females do not moult earlier than others.

The long-distance migration of the barn swallow passes the hostile Sahara, and the trip of many thousand kilometres is a major feat for a bird weighing less than 20g. Speed of migration and timing of arrival at the breeding grounds is thus a severe test of quality. Female barn swallows with long tails arrived earlier at the breeding sites than short-tailed females (Fig. 12.5). This result was not confounded by differences in the timing of

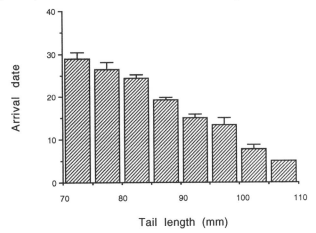

Fig. 12.5 Arrival date of female barn swallows in relation to their tail length (mm). Arrival date 1 = 1st May. Values are means (SE). Adapted from Møller (1993*h*).

moult in relation to tail length. Long-tailed females were therefore able to migrate faster than short-tailed females, or refuel their energy reserves more rapidly during stop-overs on migration.

A final test of female quality and female viability is the ability of female barn swallows to survive. Good genes may provide females with an ability to evade dangers such as predators and detrimental parasites and to overcome the dangers of stressful periods of reproduction, migration and moult. Female survival ability was not associated with tail length as female survivors had an almost identical tail length to non-survivors (Møller 1991*k*). This was also the case when the confounding effect of female age was controlled statistically. Tail length of female barn swallows thus was not a reliable indicator of female viability.

In conclusion, there was rather little evidence that tail length of female barn swallows was a reliable indicator of female viability. Long-tailed females arrived earlier than short-tailed ones, but this could simply reflect the higher reproductive potential of female barn swallows with long tails.

12.5 Female ornaments and reproduction

If female ornament size reflected reproductive potential we should expect:

(1) female tail length to be positively related to reproduction and ability to lay multiple clutches;

(2) seasonal reproductive success to be positively related to female tail length; and

(3) female provisioning of offspring to be positively correlated with female tail length.

If tail length of female barn swallows reflected their reproductive potential, long-tailed females should lay larger and perhaps also more clutches than short-tailed females. The reproductive ability of females measured in terms of clutch size or brood size of the first breeding attempt was unrelated to their tail length (Møller 1993*k*). However, females with long tails laid a second clutch more often than short-tailed females (Fig. 12.6). The frequency of second clutches increased from 56% among short-tailed female barn swallows to 100% among females with the longest tails. This relationship between double-clutching ability and female tail length remained after controlling for confounding variables like year effects and female age (Møller 1993*k*).

A more critical test of the prediction that female reproductive potential is related to tail length is to consider the relationship between changes in tail length and changes in reproductive parameters between years. On

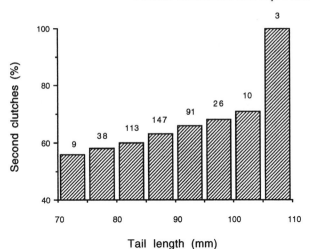

Fig. 12.6 Percentage of female barn swallows laying a second clutch in relation to their tail length (mm). Numbers are number of females. Adapted from Møller (1993*k*).

average, female tail length increases slightly from one year to the next although this increase is only statistically significant from the first to the second reproductive season (Møller 1991*a*). An increase in female tail length from one year to another was associated with an increase in the size of both the first and the second clutch and brood (Møller 1993*k*). An increase in female tail length was also associated with an increase in the frequency of second clutches. There was thus clear evidence of an association between female reproductive potential and tail length of female barn swallows.

The reproductive success of long-tailed female barn swallows should be higher than that of short-tailed females if tail length reflected female reproductive ability. This was actually the case because seasonal reproductive success for females increased from slightly more than four fledglings among short-tailed females to more than seven fledglings among long-tailed females (Fig. 12.7). This difference in seasonal reproductive success was not confounded by age or year differences in reproduction (Møller 1993*k*). Long-tailed female barn swallows therefore had larger lifetime reproductive success than short-tailed females because female survival prospects were unrelated to their tail length.

Finally, female tail length may reflect provisioning ability if tail length is an overall reflection of metabolic efficiency. Provisioning by female barn swallows was measured both as absolute and relative feeding rates (percentage of food provided by the female) throughout the nestling period. Female tail length did not account for a statistically significant amount of the variation in female feeding rates during the first clutch (Møller 1993*k*). This was also the case when potentially confounding variables such as annual

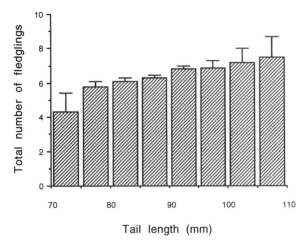

Fig. 12.7 Seasonal reproductive success in relation to tail length (mm) of female barn swallows. Values are means (SE). Adapted from Møller (1993*k*).

variation, female age, and brood size were taken into consideration. However, among females that laid a second clutch, long-tailed females fed their nestlings more often than short-tailed females, even after controlling for potentially confounding variables (Møller 1993*k*). Nestling body mass and tarsus length were not related to the tail length of their mothers, and thus, the offspring of long-tailed females were not of a better quality than the average offspring (Møller 1993*k*). This suggests that female tail length was not a reliable predictor of female provisioning of offspring.

In conclusion, there is little direct evidence for the good genes explanation of tail-length variation in female barn swallows since female survival prospects were unrelated to tail length. Female tail length could be an exaggerated character because long tails provided their possessors with a mating advantage as suggested for the Fisher process. There was clear evidence for the good parent process because male barn swallows, which mated with long-tailed females, acquired a reproductive advantage due to the higher double-clutching ability of long-tailed females.

12.6 Summary

Sexual size dimorphism in morphology may arise because of natural selection or sexual selection. Barn swallows only demonstrate slight sexual size dimorphism in morphology with the exception of the longest, outermost tail feathers, which are considerably longer in males than in females. There was little evidence for the hypothesis that natural selection could account for the maintenance of sexual size dimorphism in barn swallow morphology.

Sexual selection may account for sexual size dimorphism in two different ways. First, the non-adaptive explanation posits that dimorphism is a result of the female morphological character developing as a correlated response to selection on males because of a strong genetic correlation between the sexes for the trait. This hypothesis was not supported for the barn swallow because variation in the size of female tail length was clearly positively related to various measures of female quality and reproductive potential.

The second sexual selection explanation suggests that female tail length is an ornament currently under sexual selection because of female–female competition and male choice. It took less time for long-tailed females to become mated and their pairing date was therefore earlier than that of short-tailed females. Long-tailed female barn swallows also tended to mate with long-tailed males and there was a clear indication of assortative mating with respect to tail length.

Long-tailed females arrived earlier at the breeding grounds, more often laid second clutches, and consequently had a higher seasonal reproductive success than short-tailed females; this supports the good parent hypothesis. Male barn swallows mated to long-tailed females experienced a direct fitness advantage because of the reproductive potential of such females. Female barn swallows have longer tails than juveniles because of sexual selection, and the good parent process can fully account for the mating advantage of long-tailed females.

STATISTICS

1. Mean size of prey items for each brood: Males: 12.2mg (SE = 1.0), females: 11.3mg (SE = 0.8), $n = 22$, paired t-test, $t = 0.91$, $df = 21$, NS.
2. Females: 0.80 days (SE = 0.22), $n = 122$; males: 5.61 days (SE = 0.31), $n = 198$.

13

Geographic variation in ornament size

13.1 Introduction

Most field studies are restricted to a single locality, and field workers therefore often have the misconception that the typical behaviour or morphology of a species is that observed at their own study site. All individuals of a species rarely look similar across the distributional range because environmental conditions and the intensities of various selection pressures change geographically. Animals usually demonstrate geographic variation with sometimes characteristic patterns such as those described by the zoogeographic rules (Mayr 1942). For example, Bergmann's rule states that individuals of a species living in more northern parts of the distributional range are larger whereas Allen's rule posits that individuals in more northern latitudes exhibit smaller extremities. This kind of clinal variation in morphology is what is expected when natural selection is at work. Individuals with a smaller surface to volume ratio and smaller extremities are at an energetic advantage at more northern latitudes.

Clinal variation in morphology, behaviour or any other character could lead to speciation in the absence of geographic variation (Bush 1975; Endler 1977; Coyne 1992). This was first proposed by Fisher (1930), who suggested that geographical or ecological races, which were adapted to local environmental conditions, may hybridize at a reduced frequency if hybrids are at a selective disadvantage. Genetic variation in mating preferences could form the mechanism for the evolution of reproductive isolation. Another mechanism of sexual isolation and character divergence along a cline is Fisher's runaway process, which may occur across wide geographic ranges as a result of coevolution of a female mate preference and a male secondary sex trait to ever more extreme expressions (Fisher 1915, 1930; Lande 1982). Handicap traits may demonstrate geographic patterns reflecting variation in condition-dependence or costs of a female mate preference, and geographic variation in the size of a male character can be seen as a result of the optimization process to maximize the net benefits of the character. The importance of sexual selection as a process in character divergence and speciation is described in section 13.2.

Barn swallows have a wide geographical distribution and occur throughout most parts of the Holarctic region. The species is divided into six subspecies which replace each other in different parts of the range. Several morphological characters of the nominate subspecies demonstrate clinal variation, with latitudinal increases in wing and particularly in tail length. Clinal variation in the size of the secondary sexual character may be due to the effects of the Fisher or the handicap process. Clinal variation in tail ornaments and other morphological characters of the barn swallow is described in section 13.3.

The most likely explanation for clinal variation in morphological characters is that geographic variation in environmental conditions determines the costs of production or maintenance of a secondary sexual character. The hypotheses for clinal variation in secondary sexual traits suggest that the costs and benefits of ornaments vary geographically. This idea has received very limited empirical attention, probably because of the difficulty of experimentally manipulating the size of morphological characters. Geographic variation in the costs and benefits of long tails in male barn swallows is described in section 13.4.

A total of six different subspecies have been described for the barn swallow. The characteristics of these subspecies and the implications of their morphological variation for theories of sexual selection in clines are evaluated in section 13.5.

13.2 Clinal variation in ornaments

Clinal variation in morphology, behaviour or any other character could lead to speciation even when geographical barriers are absent (Bush 1975; Endler 1977; Coyne 1992), as first proposed by Fisher (1930). Three different processes could lead to speciation in a cline:

(1) evolution of sexual isolation after secondary contact;

(2) the runaway process and the resulting rapid evolution of a secondary sexual character and a female mate preference; and

(3) the good genes process.

13.2.1 *Speciation after secondary contact*

Geographical or ecological races may form as a result of clinal variation and adaptation to local environmental conditions. Hybrids between such races might be at a selective disadvantage if they are less well adapted to local conditions than the pure bred races (Bush 1975; Endler 1977; Coyne 1992). Genetic variation in female mate preferences could form the basis for the evolution of sexual reproductive isolation as only females with the 'right' preference would avoid or less frequently engage in hybridization. A narrow hybrid zone may then become established between two races when hybrids are at a substantial selective disadvantage (Endler 1977). This may eventually result in the formation of reproductive character displacement and a reversed cline in the contact zone. Genetic differences in mate preference or other pre-mating isolation mechanisms may subsequently diffuse by migration to other parts of the range of the two races. Hybrids in the contact zone usually have low fitness compared with nearby pure bred individuals, and extensive gene flow from vast breeding areas outside the hybrid zone may swamp the evolution of sexual isolation or limit it to the zone of secondary contact. Temporally and spatially stable hybrid zones are known from several bird species, such as the carrion and hooded crow (*Corvus corone, C. cornix*) and the pied and collared flycatcher (*Ficedula hypoleuca, F. albicollis*), without any widespread sexual isolation or reproductive character displacement.

The local origin and slow rate of spread of reproductive isolation occurring as a result of hybridization differs markedly from the potentially widespread and rapid sexual isolation arising as a result of the runaway process.

13.2.2 *The runaway process in a cline*

Sexual isolation and character divergence along a cline may arise as a result of Fisher's runaway process, which may occur across wide geographical ranges. Genetic variation in a mating preference and a secondary sexual character may result in coevolution of these two characters to ever more extreme expressions in a Fisher process (Fisher 1915, 1930). The female mate preference and the male sex trait become coupled in linkage disequilibrium if females with more extreme preferences mate with males with the most extreme sex traits. An unstable equilibrium between the mate preference and the male sex trait may become perturbed, for example by natural selection on the male secondary sexual character. This may result in a self-reinforcing runaway process because females are selecting males with the most extreme sex traits, and also because the genetic correlation between the mate preference and the male sex trait selects for more extreme mating preferences. The runaway process may rapidly lead to character divergence in secondary sexual characters and thus reproductive isolation over large areas even in the absence of geographical barriers (Lande 1982).

Speciation in a cline by the runaway process of sexual selection has been modelled by Lande (1982) and a few others (Nei *et al.* 1983; Wu 1985). The quantitative genetic model of Lande is based on a male trait and a female mating preference. Female choice is not costly because polygynous males are readily available, and there is thus no selection directly on the female mating preference. This assumption may not be particularly likely (Parker 1983*b*; Pomiankowski 1987*b*), but the model could be modified to incorporate costly mate choice by introducing mutation bias on the male character (Pomiankowski *et al.* 1991). The initial condition is spatially homogeneous, mean male and female phenotypes following rapid colonization or after a dramatic change in a uniform environment. In the case of the barn swallow, this could have taken place when human settlements were initially colonized and the species was released from severe breeding habitat limitation. Three different mating preferences were analysed:

(1) a psychophysical mating preference in which the perceived intensity of a stimulus is proportional to a power of the actual intensity;

(2) an absolute preference for a certain male phenotype; and

(3) a relative preference for the upper fraction of the frequency distribution of males.

The male secondary sexual character and the female mate preference coevolve from the initial state to ever more extreme expressions because of increasing levels of linkage disequilibrium. The female mating preference changes the amount of geographic variation and the characteristic length, which is the minimum distance over which the equilibrium cline responds to spatial variation in selective forces. If there is no sexual selection, clinal variation in male phenotype simply equals the spatial average of the optimal phenotype under natural selection. Genetic variation in the mate preferences affects clinal variation in the male secondary sexual character. Genetic variance in the psychophysical and the relative mate preference enhances geographic variation in the male character beyond that produced by natural selection (Fig. 13.1). The picture for genetic variance in an absolute mate preference is more complicated because geographic variation is diminished when the additive genetic covariance between the mate preference and the male character is smaller than the breeding value for the male character. Geographic variation is enhanced when the additive genetic covariance between the mate preference and the male character is larger than the breeding value for the male character. The rate at which the male secondary sexual character approaches the equilibrium cline is retarded by the female mate preference. The evolution of female mating preferences may, depending

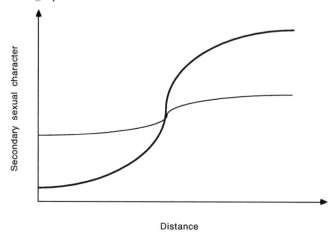

Distance

Fig. 13.1 Equilibrium cline in a male secondary sexual character (heavy line) which has evolved in a runaway process from an initial state of no geographic variation. The cline in a male character may be greatly exaggerated in comparison with the optimal cline under natural selection (thin line). Adapted from Lande (1982).

on the type of preference, greatly amplify clinal variation in a male secondary sexual character beyond that expected under natural selection. The rate of equilibration is proportional to the ratio of the dispersal distance to the characteristic length, which is the minimum distance over which the equilibrium cline responds to spatial variation in selective forces.

The runaway process can be started by spatial variation in natural selection on the male trait because of a genetically unstable female mate preference (Lande 1982). Alternatively, secondary contact between sexually isolated races may initiate a runaway process of reproductive character displacement in the hybrid zone. The condition for genetic instability is that the ratio of the genetic covariance between the female mate preference and the male character to the breeding value for the male character must exceed a threshold value that depends both on stabilizing natural selection on the male trait and the form of the mate preference (Lande 1981). This ratio gives the correlated response in a female mate preference relative to the direct response to selection on males.

The polygenic cline model has, like any model, a number of assumptions, such as a relatively shallow cline and persistence of genetic variation in the mate preference and the male sex trait. These assumptions have been relaxed in a single-locus genetic model of a female mate preference and a male character in a cline (Tomlinson 1989). Examples of such clines include the colour morphs of the two-spot ladybird *Adalia bipunctata* (Tomlinson 1989) and clinal variation in predation in the guppy (Breden and Stoner 1987). If the female mate preference is present above some threshold level throughout the cline, it will be sufficiently strong to overcome natural selection against the male trait. The male character and the mate preference will coevolve in a runaway fashion with the secondary sexual character eventually reaching fixation everywhere, as described in the polygenic model by Lande. If the runaway process does not occur, the coevolution of the male trait and the female preference will take place in two stages. First, there is a relatively rapid increase in the mate preference and the male sex trait in areas where the sex trait is advantageous (or only slightly disadvantageous). The mating preference reaches a maximum frequency before the male secondary sexual character, and the clines in the two traits therefore do not necessarily mirror each other. Second, when the preference genes and the male character genes reach their maximum frequency, the preference is gradually lost from the cline and the cline in the male sex trait eventually returns to that seen under random mating. Tomlinson (1989) used a number of different initial conditions for the mate preference, and the results suggest that female choice often will have little effect on the frequency of the male character in a cline. The only exception is if the mate preference enters the population at a high frequency, as in Lande's model, because the male character then may be driven by the mate preference to higher frequencies where it is selectively disadvantageous. Therefore, mating preferences are usually unlikely to result in sympatric speciation under the single-locus model of sexual selection analysed by Tomlinson (1989).

13.2.3 *The handicap process in a cline*

All previous theoretical analyses of clinal variation in secondary sexual characters have focused on the Fisher process. Handicap traits that reflect viability differences or differences in parenting ability among males may also demonstrate geographic variation. The expression of handicapping secondary sexual characters depends on the quality of a male. Only high-quality individuals will be able to develop the most extreme secondary sexual characters because the cost of a sex trait is relatively higher for low- than for high-quality individuals (Heywood 1989; Grafen 1990*a*; Iwasa *et al.* 1991; Price *et al.* 1993). Clinal variation in handicapping traits may occur as a result of geographic variation in the cost of the female mate preference or in the expression of the male sex trait.

Clinal variation in a female mating preference may result in a cline in the male sex trait because the male trait evolves to more extreme expressions in the presence of a more intense female mate preference (Grafen 1990*a*; Iwasa *et al.* 1991). The costs of a female mate preference and the intensity of female mate choice may vary geographically for a number of reasons, and this may result in clinal variation in a male secondary sexual character. Potential reasons for geographic variation in a mate preference include geographic variation in:

(1) the tertiary sex ratio;

(2) the operational sex ratio; and

(3) social organization.

First, the sex ratio may vary geographically if the cost of reproduction increases with increasing latitude and females therefore tend to die earlier than males. Second, the social organization of a species may vary geographically. For example, colonially breeding species may aggregate in large colonies especially in areas where the unpredictability of food is high, and this may more often be the case at the edge of the distributional range of a species. Third, the sexes of migratory species may differ geographically in their time of arrival because of sex-specific cost functions of early arrival (see Chapter 7). Early arrival may be relatively more costly for females at high latitudes. A larger proportion of males may therefore have arrived before females at more northern latitudes, and female choice would be less costly where the operational sex ratio is more male biased.

Clinal variation in the expression of a male sex trait may occur for two main different reasons. These are (1) geographic variation in the heritability of the male sex trait, and (2) geographic variation in condition-dependence of the sex trait. First, the heritability of a male secondary sexual character may increase with latitude if the environmental variance component of the sex trait decreases under more extreme environmental conditions such as in warmer or dryer climates (Falconer 1981). This would result in a stronger response to selection at higher latitudes even in the presence of a geographically constant intensity of the female mate preference. The mating preference may in fact increase with increasing latitude, as described above. Second, condition-dependence of the male secondary sexual character may vary geographically. This can either be as a result of geographic variation in male condition or in developmental control of a male sex trait. Male condition may vary geographically either because of variation in environmental conditions or because of temporally more severe constraints on the development of a secondary sex trait. For example, the duration of the growing season decreases with increasing latitude and this may also affect the period available for development of a secondary sexual character. The extent to which a male secondary sexual character reflects male condition may also depend on the extent to which development of the male sex trait is genetically controlled. Secondary sexual characters are usually characterized by a low degree of developmental control compared with ordinary morphological characters, and they therefore generally exhibit large degrees of phenotypic variance (Møller and Pomiankowski 1993b).

Environmental conditions are usually more extreme at the edge of the distributional range of a species, the phenotypic variance is therefore higher and the effects of developmental control are more pronounced in marginal populations (Møller 1993*l*). This should result in higher phenotypic variability and in higher degrees of fluctuating asymmetry in male secondary sexual characters in marginal populations (Møller 1993*l*). Alternatively, the genetic control of development may itself be subject to selection pressures if the conditions for growth of the sex trait vary geographically. Extreme environmental conditions may select for a higher degree of developmental control, which would lead to a smaller phenotypic variance and a lower degree of condition-dependence. Clinal variation in a male sex trait therefore may reflect geographic variation in heritability of the sex trait or in its genetic control of development.

If there is geographic variation in the expression of a male secondary sexual character or a female mate preference, this would result in clinal variation in a male secondary sexual character which would deviate from the cline developing entirely under natural selection.

Geographic variation in the tail ornaments of barn swallows and the extent to which this variation can be explained by current models of sexual selection in a cline are described in the next section.

13.3 Geographic variation in barn swallow morphology

Barn swallows occur throughout most parts of the Holarctic, with the exception of Arctic regions. Morphometric data are available from a large number of barn swallow populations from the Western Palaearctic (Fig. 13.2). Clinal variation in morphology and sexual size dimorphism is analysed in the next two sections.

13.3.1 *Clinal variation in morphology*

Geographic variation in morphological traits of the barn swallow is latitudinally clinal, with larger body size at the northernmost latitudes. For example, wing length increases with increasing latitude in both males and females (Fig. 13.3). Geographic variation in wing length tends to be larger in females than in males. The same applies to the short central tail feathers, which demonstrate very little sexual size dimorphism (Fig. 13.4). The length of the short central tail feathers increases with increasing latitude in both male and female barn swallows, but the increase is larger in females. Tail length demonstrates a clearly significant increase with increasing latitude in both males and females (Fig. 13.5; Møller 1993*m*). Mean male ornament size increases by 23mm, from slightly more than 90mm at the southern limit

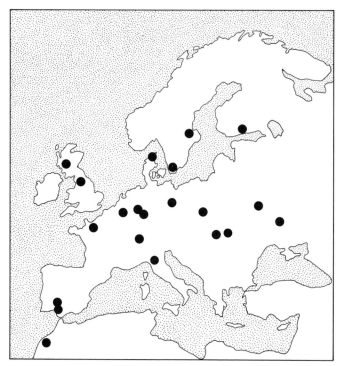

Fig. 13.2 Sampling localities in the Western Palaearctic for morphometric data on barn swallows. One locality in Afghanistan is not shown.

of the distribution to almost 115mm at the northern limit. The increase in female tail length is slightly less; only 16mm, from 79mm at the southern range to 95mm at the northern range (Fig. 13.5). Sex differences in the latitudinal rate of increase in morphological characters of the barn swallow suggest that geographical patterns of sexual size dimorphism vary for different morphological characters.

Longitudinal variation in tail length, other morphological characters and in sexual size dimorphism is slight and statistically non-significant (Møller 1993*m*). Longitudinal effects are therefore not considered further here.

13.3.2 *Clinal variation in sexual size dimorphism*

Barn swallows show considerable sexual size dimorphism in tail length in my Kraghede study area (section 12.3). Geographic variation in ornament size dimorphism demonstrates a clear latitudinal pattern (Fig. 13.6). There is a dramatic latitudinal increase across the distributional range from only 6% longer male tails at the southern limit of the range to almost 25% longer

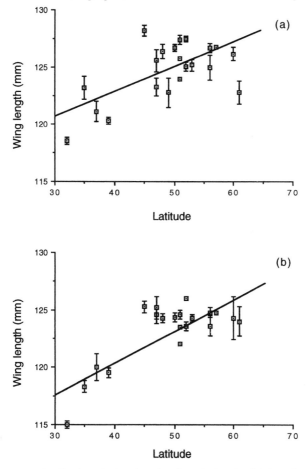

Fig. 13.3 Latitudinal variation in wing length (mm) of (a) male and (b) female barn swallows in the Western Palaearctic. Localities are given in Fig. 13.2. Model I regression lines are given. Values are means (SE). Adapted from Møller (1993*m*).

male tails at the northern limit of the range (Møller 1993*m*). Dimorphism in tail length therefore increases in the same manner as the size of morphological characters increases with latitude.

The pattern of geographic variation in sexual size dimorphism of other morphological characters, however, is strikingly different. The largest degree of dimorphism occurs at the southern distributional limit of the range (Møller 1993*m*). This is the case for wing length, tarsus length, and even for the length of the short central tail feathers (Fig. 13.7). These differences in geographic patterns of sexual size dimorphism are related to the hypotheses on sexual selection in clines, as we shall see in the next section.

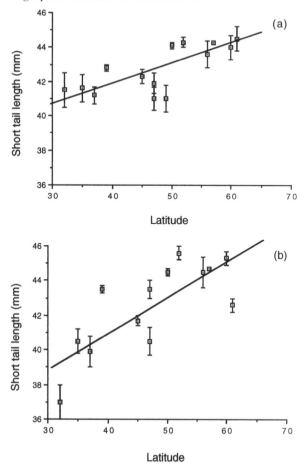

Fig. 13.4 Latitudinal variation in the length of the short tail feathers (mm) of (a) male and (b) female barn swallows in the Western Palaearctic. Localities are given in Fig. 13.2. Model I regression lines are given. Values are means (SE). Adapted from Møller (1993*m*).

13.3.3 *Evaluation of hypotheses on sexual selection in clines*

Barn swallows are aerial insectivorous migratory birds, which rely on a temporally and spatially unpredictable food source. Predictability of insect food may vary geographically and therefore provide an intense selection pressure on barn swallow morphology, which is primarily adapted to aerial insectivory. Morphological characters of the barn swallow generally increase with increasing latitude, which is in accordance with Bergmann's rule. A generally larger body size could be feasible at higher latitudes because of the lower surface to volume ratio. This may not be a particularly likely

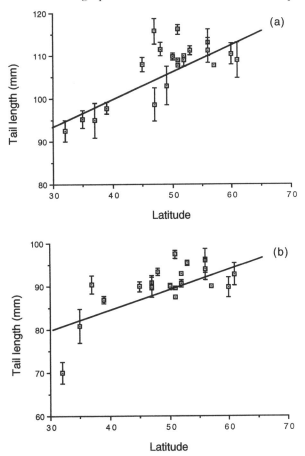

Fig. 13.5 Latitudinal variation in tail length (mm) of (a) male and (b) female barn swallows in the Western Palaearctic. Localities are given in Fig. 13.2. Model I regression lines are given. Values are means (SE). Adapted from Møller (1993*m*).

explanation for a transequatorial migrant that spends only the summer in the northern hemisphere. Alternatively, larger body size may be an adaptation to increasing distances of migration with increasing latitude of the breeding range. The geographical patterns of dimorphism in ordinary morphological characters of the barn swallow are probably primarily governed by natural selection arising from aerial insectivory. Male barn swallows are more similar in size to females or become smaller than females at the northernmost latitudes. Here the male phenotype is therefore probably closer to the female phenotype, which is expected to be near the optimum under natural selection (Price 1984).

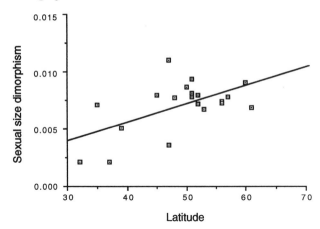

Fig. 13.6 Latitudinal variation in sexual size dimorphism in tail length of barn swallows in the Western Palaearctic. Dimorphism is calculated as $\log_{10}$(mean male tail length) minus $\log_{10}$(mean female tail length). Localities are given in Fig. 13.2. The model I regression line is given. Adapted from Møller (1993*m*).

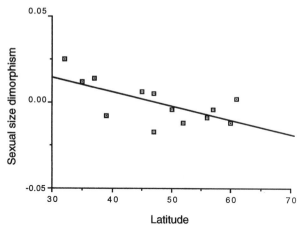

Fig. 13.7 Latitudinal variation in sexual size dimorphism in the length of the short tail feathers of barn swallows in the Western Palaearctic. Dimorphism is calculated as $\log_{10}$(mean male short tail length) minus $\log_{10}$(mean female short tail length). Localities are given in Fig. 13.2. The model I regression line is given. Adapted from Møller (1993*m*).

The reverse cline in tail dimorphism in the barn swallow (Fig. 13.6) cannot be explained as a result of a positive genetic correlation between tail length and other morphological characters and resulting morphological covariance. Increasing sexual size dimorphism in tail length with increasing latitude has to be attributed to a process different from that affecting other morphologi-cal characters, and that is most likely sexual selection. Geographical variation in tail-size dimorphism is clearly the reverse of that demonstrated by other morphological characters. Clinal variation in sexual size dimor-phism in tail length cannot be explained by the model of speciation after secondary contact because there is no evidence of formation of subspecies in the Western Palaearctic and the patterns of geographic variation are gradual rather than a steep cline around a hybrid zone of secondary contact. The pattern of sexual size dimorphism cannot be explained by sexual selection by a female mating preference greatly amplifying large-scale geographic variation in the male secondary sexual character as suggested by Lande's polygenic model of incipient speciation in a cline. The pattern of geographic variation in tail length of male barn swallows is not the greatly amplified clinal variation in tail length of females.

The third alternative is the good genes model of clinal variation which potentially may explain the reversed patterns of sexual size dimorphism in tail length as opposed to other morphological characters in the barn swallow. The good genes model of clinal variation in secondary sexual characters posits that condition-dependence or the costs of a female mate preference vary geographically.

Geographic variation in condition-dependence of the male secondary sex trait may depend on (1) the heritability of the character, and (2) the genetic control of development. First, there is only one heritability estimate for tail length in male barn swallows (section 6.2.2), and the prospects of obtaining estimates from other parts of the distributional range in the near future are small. Second, the genetic control of development of the male secondary sexual character can be estimated indirectly from (1) the degree of fluctuating asymmetry in male tail length, and (2) the coefficient of variation in tail length. Developmentally well-controlled characters are expected to demon-strate small degrees of fluctuating asymmetry and phenotypic variation (Møller and Pomiankowski 1993*b*). The degree of fluctuating asymmetry in tail length increased significantly with increasing latitude in both male and female barn swallows (Fig. 13.8; Møller 1993*m*). This suggests that either (1) the degree of developmental control decreased, or (2) the environmental variance increased with increasing latitude. The first of these alternatives is likely if the costs of the tail handicap decrease with increasing latitude. The second alternative is also likely if weather conditions for an aerial insectivore like the barn swallow may deteriorate farther north.

The phenotypic variance measured as the coefficient of variation in tail length increased with increasing latitude in male barn swallows, but decreased in females (Fig. 13.9). A similar relationship was found for wing length (Møller 1993*m*). These results suggest that the environmental variance increased with increasing latitude, but that the two sexes responded differ-

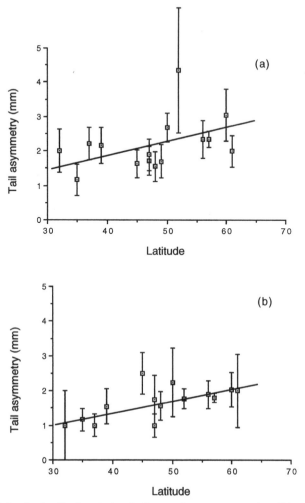

Fig. 13.8 Latitudinal variation in fluctuating asymmetry in tail length (mm) of (a) male and (b) female barn swallows in the Western Palaearctic. Localities are given in Fig. 13.2. Model I regression lines are given. Values are means (SE). Adapted from Møller (1993*m*).

ently to this increase. Whereas the phenotypic variance increased latitudinally among males, there was a clear decrease among females. Female barn swallows may suffer less from poor environmental conditions than males if female tail and wing length are closer to the optimum determined by natural selection. The lower rate of latitudinal increase in tail length among females compared with males may suggest that the female phenotype is primarily determined by natural selection while the male phenotype to a large extent is determined by sexual selection. The phenotypic variance in the male secondary sexual character thus increased with increasing latitude, and the extent of condition-dependent expression of the secondary sexual character may have demonstrated a parallel increase as determined by fluctuating asymmetry and variability in the tail trait. Female tail length demonstrated less condition-dependence with increasing latitude as determined by the coefficient of variation in tail length.

Geographic variation in the cost of the female mating preference also may result in clinal variation in a male secondary sexual character. These costs of the mate preference may depend on the tertiary sex ratio, the operational sex ratio, and social organization. There is little quantitative information on these three costs of the female mating preference. First, information on the tertiary sex ratio was only available for my Kraghede study area. Second, the operational sex ratio can be determined from detailed information on sex specific arrival and mate acquisition, but such data are only available from my Kraghede study area. Third, barn swallows breed in colonies of variable size, and the cost of the female mating preference is probably larger in areas with smaller colony size. There is some quantitative information on geographic variation in colonies. The frequency of solitary breeding did not change with latitude in the Western Palaearctic,[1] neither did the proportion of barn swallows breeding in colonies with more than 10 pairs.[2] This does not suggest that the cost of the female mate preference changes with latitude.

In conclusion, clinal variation in sexual size dimorphism in the tail ornament of the barn swallow demonstrated a pattern opposite to that of other morphological characters. There is thus little evidence for an amplification of large-scale geographic variation caused by a female mating preference, as suggested by the runaway model of speciation in a cline. The geographical pattern of sexual size dimorphism is consistent with the good genes model of clinal variation in secondary sexual characters if the degree of condition-dependence of the male sex trait increases with latitude, or if the costs of the female mate preference decrease with latitude. There is some empirical evidence for a latitudinal increase in condition-dependence of the expression of the tail ornament in male barn swallows.

13.4 Geographic variation in costs and benefits of ornaments

There was a clear latitudinal increase in the phenotypic variance of tail length of male barn swallows, whereas that was not the case in females. This may

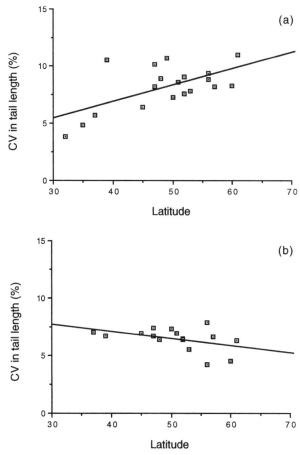

Fig. 13.9 Latitudinal variation in the coefficient of variation (CV, %) in tail length of (a) male and (b) female barn swallows in the Western Palaearctic. Localities are given in Fig. 13.2. Model I regression lines are given. Adapted from Møller (1993*m*).

indicate increasing latitudinal condition-dependence of the expression of the tail ornament, as suggested in the previous section. How can this latitudinal variation in condition-dependence of the tail ornament of barn swallows be explained? Aerial insectivores with a spatially and temporally unpredictable food source obviously are very susceptible to variation in environmental conditions. Wind, precipitation and ambient temperature are important determinants of the abundance of flying insects eaten by barn swallows (Turner 1980). Only temperature demonstrates a clear clinal variation from Northern Africa and Central Asia to Northern Europe. Insect physiology is very temperature dependent and the heterothermic temperature regulation of insects renders their activity very dependent on ambient temperatures.

Insect flight and the ability of flying insects to escape avian predators also depend on ambient temperatures. The success of aerial insectivory depends on the size of prey items and ambient temperatures. Barn swallows are specialists on large, actively flying Diptera (Bryant and Turner 1982), which obviously will be able to move faster at higher ambient temperatures. There are two pieces of information on foraging barn swallows that support this notion. First, barn swallows in my Kraghede study site fed their nestlings at a decreasing rate with decreasing cloud cover (A. P. Møller unpublished data), and the ambient temperature at micro-sites where insects are living will increase with decreasing cloud cover. Second, the diurnal pattern of nestling provisioning differs dramatically between Southern Spain and Denmark. In Southern Spain feeding rates are high in the early morning and the late evening, but drop dramatically in the early afternoon. In Denmark feeding rates are low in the early morning and to some extent in the evening, but remain high from late morning and throughout the rest of the day (A. P. Møller and F. de Lope unpublished data). These observations suggest that feeding rates are depressed at high ambient temperatures, and this may potentially result in a lower prey capture rate. The cost for a male barn swallow of maintaining its extravagant tail ornament may be less at higher latitudes where foraging is easier due to environmental conditions. A relatively lower cost of the secondary sexual character will select for less developmental control of the expression of the tail ornament of male barn swallows (Møller and Pomiankowski 1993*b*). Thus, the degree of fluctuating asymmetry and the phenotypic variance in male tail length will increase with increasing latitude because of reduced developmental control. The geographic pattern of female phenotypic variance was the opposite of that in males. The latitudinal decrease in phenotypic variance among female barn swallows may be causally related to geographic variation in reproductive parameters (Møller 1984*a*). Clutch size increases considerably with increasing latitude (Møller 1984*a*), and parental investment by female barn swallows may also increase latitudinally. If females invest relatively more in reproductive activities at higher latitudes, this should result in more intense natural selection on the female phenotype, an increase in developmental control, and thus a decrease in variance around the mean phenotype.

The costs of barn swallow tail ornaments should vary geographically if the scenario on geographic variation in the effects of ambient temperature on insect physiology is a likely explanation. Tail-manipulation experiments have been performed in both Southern Spain and Denmark and the effects of tail manipulation on the size of captured prey have simultaneously been assessed (Møller and de Lope 1993). Elongation of male barn swallow tails resulted in a reduction in the size of captured prey whereas shortening of tails resulted in an increase in prey size (Fig. 6.14). The effect of the tail-manipulation experiment on the size of captured prey items was in fact larger in Southern Spain than in Denmark. Tail length was manipulated to a similar absolute magnitude in the two sites although mean tail length of males in the Danish population is 108mm whereas mean tail length in the Spanish population is 98mm. The relative effects of tail manipulation on the size of insect prey could still be assessed by regressing mean prey size on the proportional change in tail length at the two study sites. The result of this exercise is that a similar proportional change in tail length results in a larger change in insect prey size in the Spanish population (A. P. Møller and F. de Lope unpublished data). Therefore, male tail length is shorter in the area where a long tail is a larger handicap in terms of foraging ability. This is consistent with the view that the costs of a long tail ornament in the barn swallow decrease with increasing latitude. The higher phenotypic variance in male tail length at higher latitudes is probably not a direct consequence of the higher foraging costs for a reproducing male with a long tail ornament because an equally large variance was found among yearlings, which had not yet reproduced.

13.5 Barn swallow subspecies

The barn swallow demonstrates discontinuous geographic variation, which has resulted in six different subspecies occurring in different parts of the distributional range (Fig. 3.1). The subspecies differ slightly in the colour of their ventral side and the extent of the breast band. These characteristics are qualitative differences. For example, the North American subspecies *erythrogaster* has chestnut or cinnamon-rufous ventral sides, but approximately 4% of the barn swallows in my Kraghede study area also have this colour. The subspecies *transitiva* from the Middle East has dark rufous-buff underparts, the subspecies *savignii* from the Nile deep rufous-chestnut underparts and a complete breast-band, the subspecies *tytleri* from Siberia and Mongolia cinnamon-rufous to chestnut underparts and an incomplete breast-band, and the subspecies *gutturalis* from Asia has whitish underparts and an incomplete breast-band, while *erythrogaster* has chestnut or cinnamon-rufous underparts and an incomplete breast-band (Turner and Rose 1989). The subspecies intergrade with each other and populations from areas of overlap demonstrate intermediate characteristics. The subspecies *tytleri* and *gutturalis* have until recently been separated geographically in Eastern Siberia, but now interbreed because of recent human settlement (Smirensky and Mishchenko 1981).

There is also a decline in body size from north-west (*rustica*) and north-east (*tytleri*) to the south and east (*transitiva, savignii* and *gutturalis*) (Table 13.1). The body size of the two subspecies from the Eastern Mediterranean region (*transitiva* and *savignii*) can be compared with that

Table 13.1. Morphometric data on subspecies of the barn swallow. Values are means (SE). All values in mm.

Subspecies	Sex	Tail length	Short tail length	Wing length	Tarsus length	Beak length	Beak depth	Beak width	*n*
transitiva	male	99.10	43.60	126.10	11.37	7.79	2.97	11.92	16
		(2.00)	(0.67)	(0.91)	(0.21)	(0.12)	(0.06)	(0.14)	
	female	86.11	44.00	120.89	11.24	7.64	2.99	11.97	12
		(1.68)	(0.55)	(0.72)	(0.16)	(0.12)	(0.14)	(0.18)	
savignii	male	92.00	42.89	118.77	11.13	7.54	2.89	11.56	35
		(0.88)	(0.33)	(0.25)	(0.09)	(0.08)	(0.06)	(0.08)	
	female	85.64	44.79	118.79	11.55	7.85	2.76	11.53	19
		(1.99)	(0.46)	(0.76)	(0.16)	(0.13)	(0.07)	(0.16)	
gutturalis	male	88.23	41.37	116.29	10.57	7.55	2.66	11.88	35
		(2.32)	(0.44)	(0.90)	(0.11)	(0.11)	(0.05)	(0.11)	
	female	76.14	40.77	114.71	10.53	7.75	2.58	11.57	17
		(3.52)	(0.85)	(1.15)	(0.12)	(0.15)	(0.05)	(0.16)	
tytleri	male	107.67	41.56	119.33	11.16	7.57	2.76	11.23	9
		(3.93)	(0.80)	(1.54)	(0.17)	(0.05)	(0.04)	(0.09)	
	female	89.50	42.50	117.50	11.03	7.16	2.43	11.49	7
		(5.10)	(0.75)	(0.23)	(0.04)	(0.08)	(0.05)	(0.10)	
erythrogaster	male	88.42	42.89	120.89	11.45	7.88	2.92	12.14	26
		(1.47)	(0.63)	(0.53)	(0.09)	(0.13)	(0.07)	(0.13)	
	female	75.86	41.93	119.79	11.08	7.57	2.88	12.20	15
		(1.30)	(0.75)	(0.64)	(0.20)	(0.15)	(0.11)	(0.14)	

predicted from clinal variation in the nominate subspecies. In particular, females of the subspecies *transitiva* from the Middle East have longer tails than predicted from Fig. 13.3 (males: 99mm versus the predicted 97mm, females: 86mm versus the predicted 81mm). Wings are also longer than predicted from Fig. 13.4 (males: 126mm versus the predicted 122mm, females: 121mm versus the predicted 119mm). Females of the subspecies *savignii* also have longer tails, though not wings, than predicted (tail length: males: 92mm versus the predicted 91mm, females: 85mm versus the predicted 77mm; wing length: males: 119mm versus the predicted 120mm, females: 119mm versus the predicted 117mm). These calculations suggest that females in particular of the subspecies *transitiva* and *savignii* have attained a larger body size since separation from the nominate *rustica*. The Egyptian *savignii* demonstrates little sexual size dimorphism in tail length compared with the other subspecies (*savignii*: 7% versus 15% to 20% in the other five subspecies). Sexual size dimorphism in wing length is similar to that in the other subspecies (*savignii*: 0% versus 1% to 4% in the other five subspecies). The smaller degree of sexual size dimorphism can probably be explained by the extreme temperature regime experienced by Egyptian males. A large handicap therefore will be more costly in the Egyptian *savignii* than in the other subspecies.

It is perhaps striking that the different subspecies of the barn swallow have diverged in coloration of their ventral sides and in the shape of the breast band rather than in the morphology of the much more conspicuous tail ornament. The runaway model of female mate choice suggests that a female mate preference may result in exaggeration of a secondary sexual character and divergence in this character even in the absence of any obvious geographical barriers (Lande 1981, 1982). There is no such clear divergence in the barn swallow. Tail length of the Egyptian subspecies of the barn swallow and its sexual size dimorphism can be explained by an extreme cost of the secondary sexual character. This is consistent with the handicap idea that geographic variation in tail length results from geographic variation in condition-dependence of the secondary sexual character or the cost of the female mating preference.

13.6 Summary

The nominate subspecies of the barn swallow demonstrates clear patterns of clinal variation in morphology, with increasing size in both sexes at higher latitudes. Sexual size dimorphism in the length of the tail ornament increases with increasing latitude whereas size dimorphism in other morphological characters decreases latitudinally. Phenotypic variance in tail length increases with latitude in males, but not in females, as demonstrated by the coefficient of variation in morphology. The genetic control of development decreases with increasing latitude as demonstrated by an increasing latitudinal cline in fluctuating asymmetry. These geographic patterns of morphological variation suggest that the tail character has diverged geographically as a result of the good genes process. Two tail-manipulation experiments in Southern Spain and Denmark suggest that the tail ornament of male barn swallows is relatively more costly in terms of foraging ability at southern latitudes. The main reason for this is probably that large, actively flying insects are more difficult to catch at the higher ambient temperatures in Southern Europe. The six subspecies of barn swallows have mainly diverged in coloration of their ventral sides and in shape of the breast band whereas divergence in the size of the tail ornament is slight.

STATISTICS

1. Linear regression: $F = 0.004$, $df = 1,9$, $P = 0.95$.
2. Linear regression: $F = 0.005$, $df = 1,10$, $P = 0.95$.

14

Synthesis

14.1 Introduction

The extensive, long-term study of sexual selection in the barn swallow has several general implications which are the main theme of this final chapter. The first of these general implications concerns the role of sexual selection in the life of animals. Sexual selection traditionally attempts to explain the evolution and the maintenance of extravagant secondary sexual characters which are supposed to have arisen as a consequence of the advantages that certain individuals have over conspecifics of the same sex in relation to reproduction. However, the process of sexual selection and the presence of secondary sexual characters affect in one way or another almost every aspect of the life of animals. This point is addressed in section 14.2.

Sexual selection does not arise only as a consequence of differential mating success: a large number of different selection components may give rise to sexual selection. Almost all aspects of the reproduction of organisms may result in sexual selection, and the process will entirely overlay the reproductive cycle. The different episodes of sexual selection occurring throughout the reproductive cycle can be viewed as steps in a continuous process during which females, with varying success, approach their goal of obtaining a preferred mate. The evolutionary causes and consequences of this process of sexual selection are described in section 14.3.

Female mating preferences are often clear and persistent, although the benefits of such stringent preferences may be less obvious. The fitness benefits of female mate preferences have traditionally been divided into direct and indirect benefits, which in turn have been subdivided into a number of categories. These benefits and their complex nature are described in section 14.4.

Most studies of sexual selection have focused on the functional aspects of the process. For example, a long tail in male barn swallows may function as a means of attracting mates. However, the way in which mates are attracted and the role of the secondary sex trait during this process of mate attraction may provide additional evidence for why the tail trait and its expression are of importance. Similarly, tail length of male barn swallows is highly variable, despite the fact that there is a large premium on males being able to produce the most extravagant tails. A study of the mechanism of the optimization process determining tail length will help us understand the maintenance of reliable signalling and the function of long tails. The importance of functional and mechanistic studies of sexual selection is discussed in section 14.5.

Extravagant feather ornaments are relatively rare among the 9000 species of extant birds, and the number of independent evolutionary events may be as small as a couple of dozen cases. Why is it that some taxa have evolved exaggerated feather sex traits while others have not? One explanation is that constraints may have affected the evolution of secondary sexual characters. The importance of constraints is treated in section 14.6.

Secondary sexual characters usually become more exaggerated in some mating systems than in others. Males in highly polygynous mating systems, such as those using leks, potentially can mate with a very large number of females and thus have much larger variance in mating success than males in monogamous mating systems. This may provide one of the bases for a larger degree of exaggeration in highly polygynous bird species. However, ornaments are likely to be much more costly in monogamous mating systems and thus more often reliably indicate the quality of their possessors. This will tend to increase the intensity of sexual selection arising from female mating preferences. This may explain why the males of some monogamous bird species have secondary sexual characters similar in size to those of lekking birds. The relationship between degree of ornamentation and mating system is analysed in section 14.7.

Life-history theory attempts to explain intraspecific and interspecific variation in life-history characters and the trade-off structure between these characters. Sexual selection relates to reproduction and thus one of the two major components of life histories, fecundity and survival. Sexual selection has at least two important consequences for the evolution of life histories. First, the different components of sexual selection and their relation to the expression of different kinds of secondary sexual characters (direct fitness traits, handicaps, arbitrary traits) may strongly affect the allocation of reproductive effort by individuals of the two sexes to mating effort and parental effort (Møller 1993c). Second, life-history traits evolve as a result of phenotypic selection on metric traits such as a secondary sexual character, and conflicting selection pressures on metric traits are common. For example, the expression of a male secondary sexual character may be positively related to mating success, but negatively to survival prospects. Such conflicting selection pressures may also result in trade-offs between different life-history traits (Schluter *et al.* 1991). These two relationships between life-history theory and sexual selection theory are treated in section 14.8.

14.2 Sexual selection in the life of barn swallows

Sexual selection was defined by Darwin (1871) as resulting from 'the advantage which certain individuals have over other individuals of the same sex and species, in exclusive relation to reproduction'. Empirical studies of sexual selection have concentrated on mate acquisition as the main aspect of the selection process, perhaps because other components were considered to be relatively unimportant in highly polygynous animals such as many lekking species. Sexual selection is, however, best viewed as a sequence of selection events which are naturally related to the different stages of the reproductive cycle (see section 14.3).

Darwin clearly stressed the link between sexual selection and reproduction. However, sexual selection also has considerable influence on events outside the breeding season because of the costs of mate choice and activities of reproductive competition (Table 1.2). Males may experience a number of different costs during the non-reproductive season. They include, among others:

(1) costs of production of secondary sexual characters;

(2) costs of maintenance of secondary sexual characters if these are carried during the non-reproductive season;

(3) costs of reduced body condition during courtship;

(4) costs of reproductive stress during courtship;

(5) costs of elevated hormone levels that suppress the immune system during reproduction;

(6) costs of early emergence or arrival; and

(7) costs of establishment of a display ground or a breeding territory.

Females are usually less adorned by secondary sexual characters than males, but they also have to pay a price for sexual selection. For example, female costs include:

(1) costs of the mate preference;

(2) costs of the correlated response to the male secondary sex trait;

(3) costs of reduced body condition during courtship;

(4) costs of reproductive stress during courtship;

(5) costs of elevated hormone levels that suppress the immune system during reproduction; and

(6) costs of early emergence or arrival.

The entire life of organisms thus may be influenced by sexual selection in one way or another, as demonstrated by the barn swallow.

Barn swallows are influenced by sexual selection in most aspects of their lives. First, the annual cycle of the two sexes and the timing of the main events such as migration, moult and reproduction are influenced by sexual selection. Migration and time of arrival at the breeding grounds affect the probability of mating and other aspects of reproduction. The timing of moult determines the probability of completing growth of the feather ornaments before the start of breeding. Second, the secondary sexual characters are costly to produce and maintain. Male barn swallows with long tails suffer a foraging cost and a mortality cost outside the reproductive season as a direct consequence of their sexual ornaments. Third, the size of wings in the barn swallow partly ameliorates the flight cost of the tail ornament, and the production of large wings as a supporting, cost-reducing structure is itself costly. Fourth, early-arriving male barn swallows have a sexual selection advantage, but it is also costly to arrive early at the breeding grounds because of the increased risks of mortality. Fifth, the production of tail ornaments may require elevated levels of circulating androgens during the non-reproductive season, and this could potentially result in a depressed immune defence (Folstad and Karter 1992). Sexual selection is therefore a factor permeating many aspects of the life of the barn swallow. This situation may apply to many other species, and it clearly warns against a restricted view of the sexual selection process.

14.3 Sexual selection as a continuously reinforcing process?

Sexual selection does not result just from the single event of mate acquisition. Many other reproductive activities may give rise to sexual selection. These include, among others, the timing of mating activity, mortality during the reproductive season, sperm competition, acquisition of mates differing in quality, selective fertilization, selective embryo development, differential reproductive effort, and differential parental care (Table 1.1).

Sexual selection may therefore best be described as a chain of processes, which may either reinforce or counteract each other. In the barn swallow there is clear evidence of continuous reinforcement of sexual selection with respect to the size of the male tail ornament (Chapter 4). Long-tailed male barn swallows more often acquire a mate and do so more rapidly than short-tailed males. Differential mating success of males with long tails is also demonstrated in frequency of copulations outside the socially monogamous pair bond, and long-tailed males are more successful in committing infanticide than short-tailed males. Long-tailed male barn swallows also acquire mates of higher quality, and these mates provide more parental care and have a higher reproductive effort than the mates of short-tailed males. Differential success of males with long tails after mate acquisition is therefore reinforced at a number of different stages during the breeding season, and the fitness difference between male barn swallows with long and short tails increases continuously during reproduction.

Female mating preferences are assumed to lead to the acquisition of a preferred mate by several theoretical models. The reality of the living world is often quite different. Mate preferences are often costly because females use time or energy or lose mating opportunities while searching for a mate matching the preferred phenotype (Parker 1983*b*; Pomiankowski 1987*b*). Preferred males are sometimes not even available if they have already been chosen. Such constraints on mate choice are probably particularly common in socially monogamous mating systems where males continually drop out of the mating pool, but even in lek mating systems the highest quality male is unable to copulate with all females (Møller 1992*a*). Constraints on mate choice will often result in a mismatch between the preferred male phenotype and that actually acquired. Females are partially able to circumvent constraints on mate choice by copulating with multiple males (Møller 1992*a*). The magnitude of the mismatch between the preferred and the acquired mate will determine the propensity of females to copulate with multiple males. Constraints on female mate choice therefore can be readily quantified from:

(1) the difference in phenotype between social mates and copulation partners;

(2) the frequency of copulations with multiple males; and

(3) the frequency of extra-pair paternity (Møller 1992*a*).

The different episodes of sexual selection occur at different stages of the annual reproductive cycle, and total sexual selection can therefore best be viewed as a chain of selection processes, which ultimately results in total fitness. The mating system of an organism has traditionally been defined in terms of the way in which mates are acquired, mating success, characteristics of the pair bond or the relationship between mates, and patterns of parental care (Davies 1991, 1992). A mating system can more readily be grasped as the sequence of the various components of sexual selection (Fig. 4.1). This view places the study of mating systems firmly within the theoretical framework of sexual selection. This sexual selection definition of mating systems is advantageous because of the natural chain of selection events which each feed onto subsequent events. The selection episodes of a mating system can be quantified and analysed by the standard techniques of selection analyses (Manly 1985; Endler 1986). Different mating systems can also be readily described and quantified from the relative importance of different sexual selection components. The road to quantitative comparative studies of the evolution of mating systems as determined by their selection episodes is therefore open to new advances.

14.4 Which are the benefits of female choice?

Most recent empirical and theoretical studies of sexual selection have concentrated on the benefits of mate choice. The main dichotomies are between (1) direct and indirect fitness benefits, and for the indirect benefits between (2) good genes and attractiveness benefits. The dichotomous view of the sexual selection world is most clearly expressed in the theoretical literature where some authors, for example, have claimed that the handicap process or the sexy son process do not work. Such claims are unfortunate because in reality the world is not 'black or white'. Female mating preferences may evolve and be maintained for a number of different reasons, and it is perhaps unlikely that one particular kind of benefits can account for all or even most mating preferences. Attempts to uncover the relative importance of the forces affecting the evolution of mate preferences are therefore much needed.

Females of a number of different bird species consistently demonstrate mate preferences for particular males, and such clear preferences beg for an explanation. Females may achieve a number of direct fitness benefits from their mate choice (Table 5.1). The only major direct benefit of the mating preference in barn swallows was avoidance of males infested with a contagious haematophagous mite. Female barn swallows did not achieve a benefit from their mate choice in terms of paternal care, as is usually expected in species with biparental care. On the contrary, females had to pay a direct fitness cost in terms of parental care by mating with the most preferred males, because long-tailed male barn swallows provided less care than short-tailed males.

In addition to direct fitness benefits, females may obtain indirect benefits from their mate preference. Indirect benefits include attractiveness genes and good genes. Indirect benefits of a mate preference are the most difficult to evaluate for the empirically oriented biologist because complicating factors make quantitative assessment difficult or virtually impossible. For example, indirect fitness benefits of a female mate preference due to enhanced attractiveness of sons can only be demonstrated if:

(1) the male sex trait and the female preference have a genetic basis;

(2) preferred males are consistently attractive when tested under standardized conditions; and

(3) the expression of the male sex trait is unrelated to male condition or viability.

Another example concerns the effects of good genes on viability of offspring which are likely to be confounded by effects of differential reproductive investment by females. Female barn swallows appeared to obtain good genes for resistance to parasites and general viability genes from their mate preference. It is possible that females also experienced an indirect advantage because of the enhanced attractiveness of the sons of preferred male barn swallows. The Fisher process is usually assumed to be less important under monogamy because of the limited variance in male mating success. However, extra-pair copulations may considerably increase the variance in mating success. There were thus clear indications of indirect fitness benefits of the female mating preference.

In conclusion, female barn swallows achieve both direct and indirect fitness benefits from their mate choice, and among the indirect fitness benefits females probably gained both an attractiveness and a good genes mating advantage.

14.5 Mechanisms versus functions in sexual selection

Behavioural ecology is a branch of evolutionary biology with a strong emphasis on functional aspects of behaviour. During the last few decades, the functional approach has resulted in an immense increase in our knowledge of the adaptive bases of behaviour, but it has also led to a limitation in our understanding of how specific fitness differences arise. The study of sexual selection is not just about functional aspects of competition between conspecifics of the same sex in relation to reproduction. The exact mechanism of how variance in mating success arises as a result of competition is also important for a complete understanding. The apparent function is not always clear-cut even when an experimental approach is used. This can easily be illustrated with an example. The long tail of male barn swallows was hypothesized to result from sexual selection. This functional hypothesis was subsequently tested experimentally by manipulating tail length of male barn swallows before they acquired a mate (Møller 1988c). The result of this experiment was clear: males with elongated tails took less time to acquire a mate than males with shortened tails, while there was no effect of the treatment on the frequency of male–male competition (Møller 1988c). This experiment demonstrated only that females apparently used some aspect of long tails as a cue in their mate choice, and the mechanism of mate choice supposedly was female assessment of an aspect of the tail ornament display. The simplest explanation was that tail length *per se* was an important cue in mate choice. This idea subsequently had to be modified when individual fluctuating asymmetry in tail length was shown to be strongly negatively correlated with tail length (Møller 1990g). The results of the tail-length experiment were not useful for distinguishing between the tail-length and the tail-asymmetry explanation for mate choice because the relative degree of asymmetry was changed as well when tail length was manipulated. Tail elongation resulted in a reduction in relative asymmetry and tail shortening resulted in an increase in relative asymmetry. When both tail length and tail asymmetry were manipulated independently, there was a clear effect of both on mate choice while the frequency of male combat was again unaffected (Møller 1992d). This suggests that female barn swallows use both ornament size and asymmetry in their assessment of potential mates. If it had been possible to determine directly the aspect of the secondary sexual character which was assessed by females, it could have been concluded much earlier that asymmetry as well as size was important.

Behaviour is always based on morphological characters in one way or another, and selection arising from variation in behaviour will also result in selection on the morphological basis of the behavioural pattern. Sexual selection is usually characterized by a strong directional component, and this results in a reduction of the developmental control of the trait under selection (Møller and Pomiankowski 1993b). Secondary sexual characters therefore demonstrate elevated levels of phenotypic variance and elevated levels of fluctuating asymmetry compared with ordinary morphological traits. Many secondary sexual characters also demonstrate characteristic patterns of fluctuating asymmetry because individuals with the largest sex traits usually have the smallest degree of asymmetry (Møller and Pomiankowski 1993b). This probably results from the handicap process since individuals with the highest level of advertising show the smallest signs of cost of production of the sex trait (Møller and Pomiankowski 1993b). Many secondary sexual characters may demonstrate these patterns of phenotypic variance and fluctuating asymmetry irrespective of the sensory modality to which they are tuned because of the ubiquitous morphological basis of displays. This also provides a mechanism for maintenance of genetic variance in secondary sex traits since biased mutation will invariably result in continuous increases in the level of fluctuating asymmetry of secondary sexual characters. Females thus should continuously be able to assess the quality of males from the asymmetry of their displays.

The direct mechanism of mate choice in the barn swallow has been studied by observing female behaviour during mate assessment (Møller 1988*c*, 1990*a*). Females are attracted by male sexual displays of their tail ornaments and more rapidly so if the tail is long and symmetric. This relationship arises because female barn swallows more often leave a displaying male if it has a short rather than a long tail. Short-tailed males are therefore visited by a larger number of potential mates which subsequently desert.

The function of long tails in male barn swallows is obviously the rapid attraction of high-quality females. Long tails therefore should be more efficient signals than short tails, but many male barn swallows still have short tails. The mechanism of how tail length is optimized among male barn swallows was studied experimentally. Elongation of tails resulted in an impaired foraging ability as determined from a negative relationship between experimental tail length and the size of prey items captured by males (Møller 1989*a*; Møller and de Lope 1993). The foraging ability of male barn swallows subsequently affected their condition and tail growth during the next moulting season. Whereas males with elongated tails reduced their tail length, those with shortened tails increased their tail length in the following year (Møller 1989*a* and unpublished). The response of males to experimental manipulation of their tail length clearly depended on their original tail length. Male barn swallows with tails above the median were better able to cope with the extra costs of an elongated tail while males with naturally short tails were less able to cope with the handicap (Møller 1989*a*; Møller and de Lope 1993). The mechanism behind the optimization of tail length in male barn swallows is therefore clearly that it is costly for males to have a long tail, but it is less costly for long- than for short-tailed males. These are the exact conditions for reliable signalling of quality according to the handicap mechanism (Zahavi 1975, 1977). The long tail apparently is a reliable signal of male quality, while the mechanism controlling reliability in signalling is the differential cost of long tail ornaments to males differing in phenotypic quality. The combined study of function and mechanism thus provide a clear picture of the process of sexual selection.

14.6 Constraints on the evolution of secondary sexual characters

Feather ornaments of birds are usually large and conspicuous morphological characters. Just imagine the long tail of the barn swallow and compare it with the short tail of the sexually monomorphic red-rumped swallow *Hirundo daurica*. The mere size of feather ornaments suggests that they are costly to produce and maintain. Maintenance of a large morphological secondary sexual character may also be costly for animals that spend a considerable amount of time in flight, which is the energetically most expensive form of locomotion. The feathers of secondary sexual characters are larger than needed for efficient locomotion, and this affects the amount of drag produced by the ornament and its moment of inertia (Evans and Thomas 1992).

Extravagant feather ornaments are not particularly widespread among the 9000 different bird species. In fact, extravagant feather ornaments have evolved only a couple of dozen times among extant birds. The distribution of feather ornaments across avian taxa appears to be non-random. Many families do not have a single evolutionary event while others may have two. Why is this the case? It is likely that different kinds of constraints have affected the evolution of ornaments. Bird taxa in which extravagant feather displays are very costly to produce or maintain are unlikely ever to evolve such characters for locomotory, ecological or reproductive reasons. For example, bird species that spend a considerable amount of time on energetically very expensive locomotion such as flight or diving are unlikely to evolve extravagant feather ornaments because they will interfere with efficient locomotion during foraging. It is perhaps no coincidence that feather ornaments have never evolved among birds of prey and owls, which capture relatively large moving prey.

Ecological constraints occur when factors in the environment affect the likelihood of evolution of extravagant sexual characters. For example, the kind of food items exploited by an organism and their predictability in space and time will put upper limits to the evolution of elaborate morphology. This can easily be illustrated with some examples. First, frugivory is particularly common among birds in the tropics where fruits occur throughout the year. Frugivorous birds have evolved feather ornaments only in the tropics where the food source is temporally predictable, while among temperate or subtropical frugivores such ornaments have never evolved. Among tropical birds a change from an invertebrate to a frugivorous diet is usually associated with acquisition of feather ornaments. This suggests that feather ornaments may arise when a particular taxon is released from the ecological constraint set by the predictability of food. Second, large, rapidly flying insects are more predictable in time and space than small insects with weak flying abilities. The evolution of feather ornaments therefore should not be as constrained among bird taxa that feed on relatively large insects.

Parenting is usually energetically costly and parental duties therefore may constrain the evolution of extravagant secondary sexual characters. The cost of parental care would be much higher in the presence of a large handicapping feather ornament than in its absence. It is well known that males of highly polygynous bird species often have remarkable and beautiful feather displays and that such males usually play only a minor role in parental duties. Extravagant feather ornaments are clearly much more common among polygynous avian taxa than predicted by chance. The higher degree of ornamentation of polygynous taxa could be due to an absence of a parenting constraint or to more intense sexual selection under polygyny. However, even within mating systems there is a clear relationship betwen male parental roles and ornamentation. Non-ornamented males provide more parental care than males of ornamented taxa. Male birds that play a large role in parental care therefore appear more rarely to have evolved extravagant feather ornaments than males with small or no roles in parental care. For example, the extremely ornamented standard-winged nightjar *Macrodipteryx longipennis* has no paternal care, whereas males of nightjar species without ornaments provide parental care. Similarly, males of the scissor-tailed flycatcher *Tyrannus forficata*, which have elongated tail feathers, only feed nestlings, whereas males of other tyrant flycatchers such as the eastern kingbird *Tyrannus tyrannus* participate in nest building, incubation, and feeding of offspring. The nominate subspecies of the barn swallow with long tail feathers as compared with the North American subspecies with shorter tails provides a third example.

There are exceptions to the rule that males which spend a considerable amount of time on costly locomotory activities, which capture prey that occur unpredictably in time and space, and which provide extensive paternal care never evolve elaborate feather displays. One of these exceptions has been described in this book, the barn swallow. When ornaments occur despite a number of different kinds of constraints, the secondary sex traits are bound to be very costly to produce and maintain. High costs of a character are also prerequisites for such characters being revealing of individual quality. Only males of the highest quality will be able to fly around with a very costly adornment while performing a number of costly activities.

14.7 Mating systems, ornaments and sexual selection

The intensity of sexual selection is determined by the magnitude of non-random variance in aspects of reproduction caused by competition for mates among males and females. Sexual selection is usually assumed to be more important in polygynous than in monogamous species. We should therefore expect more extravagant and elaborate ornaments in highly polygynous species: this generally appears to be the case (Darwin 1871). The male secondary sexual character of a monogamous species may not reach extreme expressions far from the survival optimum because of the limited variation in female fecundity, which in some models is the driving force behind the evolution of male sex traits (Darwin 1871; Fisher 1930;

Kirkpatrick *et al.* 1990). The larger the variance in the nutritional effect on female fecundity, the larger the limits on the expression of the male sex trait. This limitation is less important among polygynous species because of the larger number of mates acquired by the most preferred males.

Direct selection on the female mating preference may also pose a limit on the intensity of sexual selection in monogamous organisms. Males of most monogamous taxa contribute parental care and other resources to the female mate. Darwin (1871) suggested that females should choose mates on the basis of their vigour and the expression of their secondary sexual characters. Male vigour may potentially affect the quality of male parental care, and if females mated to males providing the highest quality parental care, they would leave more offspring. This should result in direct selection on the female mate preference. The evolutionary result of this scenario is, in the case of condition-independent male sex traits, an equilibrium at which the mean female preference maximizes female fitness, while the male trait equilibrates away from the survival optimum (Kirkpatrick *et al.* 1990). When the expression of the male sex trait is condition-dependent, the female mate preference will not maximize female fitness if the expression of the male trait has a direct deleterious effect on female fecundity (Grafen 1990*a*; Price *et al.* 1993). The reason for this is that the negative effects of the male trait on female fecundity balance the positive effects of a high body condition at the equilibrium. Other kinds of direct selection on the female mate preference may occur, and these include a failure to mate among females with the most extreme mate preference.

The most extravagantly ornamented males may not necessarily provide the largest contribution to parental care, if males with extreme traits are less efficient parents, or if males attract predators to the nest or themselves (Baker and Parker 1979). If paternal care favours female mate preferences for less ornamented males, this will limit exaggeration of male secondary sexual characters. Females of several monogamous bird species such as barn swallows, blue tits *Parus caeruleus*, zebra finches, and house sparrows *Passer domesticus* (this study; Burley 1986; R. Johnston pers. comm.; H.-U. Reyer pers. comm.) contribute relatively more to the raising of offspring if paired with highly preferred males. This strongly suggests that good genes or other indirect fitness benefits may enter and affect the evolution of secondary sexual characters in monogamous birds. The extent to which this effect of differential reproductive effort is able to overcome the effects of direct selection on a female mate preference needs theoretical attention.

Extravagant secondary sexual characters are not equally costly to males of all organisms (Møller and Pomiankowski 1993*b*). For example, males of highly polygynous taxa rarely or never provide any parental care and therefore allocate all their reproductive effort to mating activities. They can therefore afford production and maintenance of an extravagant secondary sex trait. Only greatly exaggerated ornaments are therefore likely really to compromise male survival. The potential number of mates and thus the benefits of display are very large, and so are the Fisherian benefits. The costs of female choice are usually reduced because males of polygynous taxa often aggregate on leks or mating arenas, and hence many sex traits are likely to evolve in highly polygynous taxa. Among monogamous taxa ornaments will often demonstrate signs of condition dependence for two different reasons. First, males often provide extensive parental care and it will be particularly difficult to do so while carrying an exaggerated sex trait. Even slight exaggeration thus will be very costly and most ornaments are likely to retain or attain an association with condition. Second, the aftermaths of reproductive activity usually carry over to the non-reproductive season. Males that have provided extensive parental care will have worse body condition, which may delay the start of their autumn migration or their moult. This will have consequences for the development of the ornament. The Fisher process is usually much weaker in monogamous species because of restrictions on multiple mating (Fisher 1930; Kirkpatrick *et al.* 1990) and because of constraints on female mate choice due to males dropping out of the mating pool once paired (Møller 1992*a*). Males of monogamous species, such as the barn swallow, are likely to evolve one or a few sex traits which are likely to have a good genes signalling function (Møller and Pomiankowski 1993*b*). This is supported by the higher frequency of multiple ornaments among avian taxa with polygynous mating systems, and the higher frequency of condition-dependence expressed by ornaments of taxa with single as compared with multiple ornaments (Møller and Pomiankowski 1993*b*).

The rule that the expression of secondary sexual characters is more limited in monogamous taxa is not without its exceptions. For example, the quetzal *Pharomachrus mocinno* and the paradise flycatcher *Terpsiphone viridis* are monogamous, but still have some of the largest feather ornaments recorded in any bird species. Similarly, extravagant feather ornaments or conspicuous coloration in birds are not clearly related to mating systems (Møller 1986; Höglund 1989). Variance in female fecundity caused by nutritional state has traditionally been assumed to be the most important or even the only component causing exaggeration of a male sex trait under monogamy (Kirkpatrick *et al.* 1990). This is obviously not the case in the barn swallow.

14.8 Life-history consequences of sexual selection

Life-history theory attempts to explain intraspecific and interspecific variation in life-history traits such as age of first reproduction, egg size and clutch size (Stearns 1976, 1992; Charlesworth 1980; Partridge and Harvey 1988).

The resources available for maintenance and reproduction are limited and allocation of resources thus has to be optimized. Individuals that allocate resources appropriately to fecundity and survival within the constraints set by the environment will leave the largest number of descendants and thus be favoured by natural selection. The optimization process will result in specific suites of life-history traits which in combination constitute the life history of an organism. Life-history variables may constrain the extent of sexual selection and the kind of sexually selected traits in at least three different ways (Partridge and Endler 1987):

(1) life-history traits may influence the opportunity for sexual selection by affecting the operational sex ratio;

(2) the kinds of traits affected by sexual selection are likely to be partially determined by patterns of mortality and development; and

(3) the life history and the various costs of sexually selected traits will eventually halt the sexual selection process.

Life-history variables may therefore act as constraints on sexual selection, but sexual selection is also able to influence directly the evolution of life histories. Life-history theory is directly related to sexual selection in two different ways:

(1) sexual selection on metric characters results in direct effects on life-history traits; and

(2) life-history traits evolve as a result of sexual selection.

The first way in which sexual selection may affect life-history evolution concerns the effects of selection (Schluter *et al.* 1991). Phenotypic selection acts on metric characters related to specific components of fecundity and survival. The number of mates acquired by a male is often positively related to the expression of its secondary sexual character, but a high mating success is either acquired at the expense of survival costs as in the Fisher process, or as a result of overall high quality as in the handicap process. Trade-offs between life-history traits arise as a consequence of the inability of individuals simultaneously to maximize all components. No individual is able to reproduce at a very high rate and survive indefinitely. Trade-offs between life-history traits are the outcome of optimization of conflicting demands for reproduction or survival in order to maximize lifetime reproductive success. Conflicting selection pressures on male secondary sexual characters arise if the most extravagantly ornamented males acquire a mating advantage but have to pay for this in terms of reduced survival prospects. The expression of the male secondary sexual character may also result from the optimization of the handicap process. Conflicting selection pressures appear to be absent because males with the most extravagant ornaments both acquire the largest number of mates and survive better than males with small sexual displays (Møller 1991*e*, 1992*c*). This apparent absence of trade-offs in life-history traits is typically the result of effects of condition masking the true trade-off structure of the life history (van Noordwijk and de Jong 1986; Price and Schluter 1991; Schluter *et al.* 1991). Supposedly contradictory results presented by Borgia (1993) are consistent with this point of view. Perturbation experiments such as the tail-manipulation experiments on the barn swallow clearly reveal that males pay a viability cost by producing and carrying a large secondary sexual character (Møller 1989*a*; Møller and de Lope 1993). The relatively larger cost of the tail ornament for males with the smallest secondary sex traits generate the positive phenotypic correlation between life-history characters.

Second, life-history traits may evolve as a direct result of competition between conspecifics of one sex in relation to reproduction. Sexual selection consists of a number of sequential selection processes which constitute the mating system of an organism. The relative magnitude of these components is partially determined by the kind of male quality that is signalled by a secondary sexual character. This may be either direct fitness benefits such as paternal care, male genetic quality, or male attractiveness. If a male secondary sexual character reliably predicts male parenting ability, this should result in the expression of the male secondary sexual character being related strongly to the ability of males to provide courtship food to their mates and the ability of males to provide parental care. These direct fitness benefits of a mate preference should allow females to alter several life-history characters. For example, females may be able to start reproducing earlier, lay larger and more eggs, and reproduce more times per season than females which do not obtain any direct fitness benefits from their mates.

If the male secondary sexual character reliably predicts male genetic quality, this may also result in males providing more food for their mate and offspring, but females may also be willing to invest relatively more in reproduction as a result of having achieved a high-quality mate (Burley 1986). Differential parental effort will be able to enhance the effects of sexual selection on life-history traits predicted for male sex traits signalling direct fitness benefits. Females will only be able to invest differentially in reproductive effort without paying any costs if they are able to acquire additional resources (Tuomi *et al.* 1983). Thus, differential allocation of reproductive resources should result in a reduction in longevity of females, which will have important consequences for the trade-off structure of the life history.

If the male secondary sexual character reflects only male heritable attractiveness, males with the most extravagant ornaments should allocate all their reproductive effort to mating effort and none to parental effort (Maynard Smith 1977; Low 1978). The most ornamented males thus should not provide courtship food or parental care, which would result in a reduction in egg size, clutch size, and number of reproductive events per season, and a delayed reproduction. Differential reproductive effort by females mated to the most extravagantly ornamented males may partly alleviate this effect of the Fisher process on life-history evolution.

These examples clearly suggest that sexual selection theory may provide important insights into life-history theory.

14.9 Summary

The study of sexual selection will progress by the use of a number of different approaches. Theoretical and empirical studies have far too long been separated by a wide sound although individuals on both shores obviously benefit from interactions. Future progress also clearly relies on a comparative approach to the study of coevolution between female mate preferences and male secondary sexual characters when more detailed case studies of sexual selection become available. It is my hope that this book will inspire others to contribute to this task by increasing our knowledge.

References

Aarestrup, W.C. and Møller, A.P. (1980). Landsvaler *Hirundo rustica* på en overnatningsplads i Nordjylland. *Dansk ornithologisk Forenings Tidsskrift* **74**, 149–52.

Alatalo, R.V., Gustafsson, L., and Lundberg, A. (1984). High frequency of cuckoldry in pied and collared flycatchers. *Oikos* **42**, 41–7.

Alatalo, R.V., Gustafsson, L., and Lundberg, A. (1989). Extra-pair paternity and heritability estimates of tarsus length in pied and collared flycatchers. *Oikos* **56**, 54–8.

Alatalo, R.V., Gustafsson, L., and Lundberg, A. (1990). Phenotypic selection on heritable size traits: Environmental variance and genetic response. *American Naturalist* **135**, 464–71.

Alatalo, R.V., Höglund, J., and Lundberg, A. (1988). Sexual selection models and patterns of variation in tail ornaments in birds. *Biological Journal of the Linnean Society* **34**, 363–74.

Alatalo, R.V., Höglund, J., and Lundberg, A. (1991). Lekking in the black grouse — a test of male viability. *Nature* **352**, 155–6.

Alatalo, R.V., Lundberg, A., and Glynn, C. (1986). Female pied flycatchers choose territory quality and not male characteristics. *Nature* **323**, 152–3.

Altmann, J. (1974). Observational study of behaviour: Sampling methods. *Behaviour* **49**, 227–67.

Altmann, J., Altmann, S.A., and Hausfater, G. (1978). Primate infant's effects on mother's future reproduction. *Science* **201**, 1028–30.

Anderson, R.M. and May, R.M. (ed.). (1982). *Population biology of infectious diseases.* Springer, Berlin.

Andersson, M. (1982a). Sexual selection, natural selection and quality advertisement. *Biological Journal of the Linnean Society* **17**, 375–93.

Andersson, M. (1982b). Female choice selects for extreme tail length in a widowbird. *Nature* **299**, 818–20.

Andersson, M. (1986a). Evolution of condition-dependent sex ornaments and mating preferences: Sexual selection based on viability differences. *Evolution* **40**, 804–20.

Andersson, M. (1986b). Sexual selection and the importance of viability differences: A reply. *Journal of theoretical Biology* **120**, 251–4.

Arnold, S.J. (1983). Sexual selection: The interface of theory and empiricism. In *Mate choice* (ed. P. Bateson), pp. 67–107, Cambridge University Press, Cambridge.

Asbirk, S. (1971). Danske svalers trækforhold. *Flora og Fauna* **77**, 119–21.

Baker, R.R. and Bellis, M.A. (1988). 'Kamikaze' sperm in mammals? *Animal Behaviour* **36**, 936–9.

Baker, R.R. and Parker, G.A. (1979). The evolution of bird coloration. *Philosophical Transactions of the Royal Society of London B* **287**, 63–130.

Bakker, T.C.M. (1993). Positive genetic correlation between female preference and preferred male ornament in sticklebacks. *Nature* **363**, 255–7.

Ball, G.F. (1983). Functional incubation in male barn swallows. *Auk* **100**, 998–1000.

Balmford, A. and Thomas, A. (1992). Swallowing ornamental asymmetry. *Nature* **359**, 487.

Bańbura, J. (1992). Mate choice by females of the barn swallow *Hirundo rustica*: Is it repeatable? *Journal für Ornithologie* **133**, 125–32.

Barclay, R.M. (1988). Variation in the costs, benefits, and frequency of nest reuse by Barn Swallows (*Hirundo rustica*). *Auk* **105**, 53–60.

Barnard, P. (1991). Ornament and body size variation and their measurement in natural populations. *Biological Journal of the Linnean Society* **42**, 379–88.

Barnett, S.A. (1958). *A century of Darwin.* Harvard University Press, Cambridge.

Baron, R.W. and Weintraub, J. (1987). Immunological responses to parasitic arthropods. *Parasitology Today* **3**, 77–82.

Barton, N.H. and Turelli, M. (1991). Natural and sexual selection on many loci. *Genetics* **127**, 229–55.

Basolo, A. (1990). Female preference predates the evolution of the sword in swordtails. *Science* **230**, 808–10.

Bateman, A.J. (1948). Intrasexual selection in *Drosophila*. *Heredity* **2**, 349–68.

Bateson, P. (ed.) (1983). *Mate choice.* Cambridge University Press, Cambridge.

Bell, G. (1978). The "handicap" principle of sexual selection. *Evolution* **32**, 872–85.

Belt, T. (1874). *A naturalist in Nicaragua.* John Murray, London.

Bennett, G.F. (1987). Hematozoa. In *Companion bird medicine* (ed. E. W. Burr), pp. 120–8, Iowa State University Press, Ames.

Bensch, S. and Hasselquist, D. (1992). Evidence for active female choice in a polygynous warbler. *Animal Behaviour* **44**, 301–11.

Birkhead, T.R., Atkin, L., and Møller, A.P. (1987). Copulation behaviour in birds. *Behaviour* **101**, 101–38.

Birkhead, T.R. and Møller, A.P. (1992). *Sperm competition in birds: Evolutionary causes and consequences.* Academic Press, London.

Birkhead, T.R. and Møller, A.P. (1993a). Female control of paternity. *Trends in Ecology and Evolution* **8**, 100–4.

Birkhead, T.R. and Møller, A.P. (1993b). Why do male birds stop copulating while their partners are still fertile? *Animal Behaviour* **45**, 105–18.

Birkhead, T.R., Møller, A.P., and Sutherland, W.J. (1993). Why do females make it so difficult for males to fertilize their eggs? *Journal of theoretical Biology* **161**, 31–60.

Boake, C.R.B. (1985). Genetic consequences of mate choice: A quantitative genetic method for testing sexual selection theory. *Science* **227**, 1061–3.

Boake, C.R.B. (1989). Repeatability: Its role in evolutionary studies of mating behavior. *Evolutionary Ecology* **3**, 173–82.

de Bont, A.F. (1962). Composition des bandes d'hirondelles de cheminée dans son quartier d'hiver. *Gerfaut* **52**, 298–343.

Borgia, G. (1979). Sexual selection and the evolution of mating systems. In *Sexual selection and reproductive competition in insects* (ed. M.S. Blum and N.A. Blum), pp. 19–80, Academic Press, New York.

Borgia, G. (1993). The cost of display in the non-resource-based mating system of the satin bowerbird. *American Naturalist* **141**, 729–43.

Bradbury, J.W. and Andersson, M.B. (ed.). (1987). *Sexual selection: Testing the alternatives.* Wiley, Chichester.

Bradbury, J.W. and Davies, N.B. (1987). Relative roles of intra- and intersexual selection. In *Sexual selection: Testing the alternatives* (ed. J.W. Bradbury and M.B. Andersson), pp. 143–63, Wiley, Chichester.

Breden, F. and Stoner, G. (1987). Male predation determines female preference in the Trinidad guppy. *Nature* **329**, 831–3.

Bremermann, H.J. (1980). Sex and polymorphism and strategies of host-pathogen interactions. *Journal of theoretical Biology* **87**, 641–702.

Brombach, H. (1977). *Rauchschwalben.* DBV Ortsgruppen Leverkusen e. V. und Köln e. V., Leverkusen.

Brooke, R.K. (1972). Generic limits in Old World Apodidae and Hirundinidae. *Bulletin of the British Ornithologists' Club* **92**, 53–7.

Brooke, R.K. (1974). Nomenclatural notes on and the type localities of some taxa in the Apodidae and Hirundinidae (Aves). *Durban Museum Novitatis* **10**, 127–37.

Brooks, D.R. and McLennan, D.A. (1991). *Phylogeny, ecology and behavior.* University of Chicago Press, Chicago and London.

Bryant, D.M. and Tatner, P. (1991). Intraspecific variation in avian energy expenditure: Correlates and constraints. *Ibis* **133**, 236–45.

Bryant, D.M. and Turner, A.K. (1982). Central place foraging by swallows: The question of load size. *Animal Behaviour* **30**, 845–56.

Bubenik, A.B. (1982). Physiology. In *Elk of North America* (ed. J.W. Thomas and D.E. Toweill), pp. 125–79, Stackpole Books, Harrisburg.

Bulmer, M.G. (1983a). Models for the evolution of protandry in insects. *Theoretical Population Biology* **23**, 314–22.

Bulmer, M.G. (1983b). The significance of protandry in social hymenoptera. *American Naturalist* **121**, 540–51.

Bulmer, M.G. (1989a). The structural instability of sexual selection models. *Theoretical Population Biology* **36**, 195–206.

Bulmer, M.G. (1989b). Maintenance of genetic variability by mutation-selection balance: A child's guide through the jungle. *Genome* **31**, 761–7.

Burke, T. and Bruford, M.W. (1987). DNA fingerprinting in birds. *Nature* **327**, 149–52.

Burley, N. (1977). Parental investment, mate choice, and mate quality. *Proceedings of the National Academy of Science of USA* **74**, 3476–9.

Burley, N. (1981a). Sex-ratio manipulation and selection for attractiveness. *Science* **211**, 721–2.

Burley, N. (1981b). Mate choice by multiple criteria in a monogamous species. *American Naturalist* **117**, 515–28.

Burley, N. (1986). Sexual selection for aesthetic traits in species with biparental care. *American Naturalist* **127**, 415–45.

Burley, N. (1988). The differential-allocation hypothesis: An experimental test. *American Naturalist* **132**, 611–28.

Bush, G.L. (1975). Modes of animal speciation. *Annual Review of Ecology and Systematics* **6**, 339–64.

Cade, W.H. (1981). Alternative mating strategies: Genetic differences in crickets. *Science* **212**, 563–4.

Charlesworth, B. (1980). *Evolution in age-structured populations.* Cambridge University Press, Cambridge.

Charlesworth, B. (1984). The cost of phenotypic evolution. *Paleobiology* **10**, 319–27.

Charlesworth, B. (1987). The heritability of fitness. In *Sexual selection: Testing the alternatives* (ed. J. W. Bradbury and M.B. Andersson), pp. 21–40, Wiley, Chichester.

Christensen, P.V. (1981). Genfund af danske landsvaler i Danmark. *Dansk ornithologisk Forenings Tidsskrift* **75**, 47–50.

Clarke, B. (1976). The ecological genetics of host-parasite relationships. In *Genetic aspects of host-parasite relationships* (ed. A.E.R. Taylor and R. Muller), pp. 87–103, Blackwell, Oxford.

Clarke, B.C. (1979). The evolution of genetic diversity. *Proceedings of the Royal Society of London B* **205**, 453–74.

Clayton, D. (1991). The influence of parasites on host sexual selection. *Parasitology Today* **7**, 329–34.

Clayton, G.A. and Robertson, A. (1957). An experimental check on quantitative genetical theory. II. Long term effects on selection. *Journal of Genetics* **55**, 152–80.

Cohen, J. (1967). Correlation between sperm 'redundancy' and chiasmata frequency. *Nature* **215**, 862–3.

Cooke, F., Taylor, P.D., Francis, C.M., and Rockwell, R.F. (1991). Directional selection and clutch size in birds. *American Naturalist* **136**, 261–7.

Cott, H.B. (1940). *Adaptive coloration in animals.* Methuen, London.

Coyne, J.L. (1992). Genetics and speciation. *Nature* **355**, 511–5.

Cramp, S. (ed.) (1985). *Handbook of the birds of Europe, the Middle East and North Africa: The birds of the Western Palearctic.* Vol. 5. Oxford University Press, Oxford.

Crook, J.R. and Shields, W.M. (1985). Sexually selected infanticide by adult male barn swallows. *Animal Behaviour* **33**, 754–61.

Crossley, S.A. (1974). Changes in mating behavior produced by selection for ethological isolation between ebony and vestigial mutants of *Drosophila melanogaster*. *Evolution* **28**, 631–47.

Curtsinger, J.W. and Heisler, I.L. (1988). A diploid "sexy son" model. *American Naturalist* **125**, 788–810.

Curtsinger, J.W. and Heisler, I.L. (1989). On the consistency of sexy son models. *American Naturalist* **134**, 978–81.

Dale, S., Amundsen, T., Lifjeld, J.T., and Slagsvold, T. (1990). Mate sampling behaviour of female pied flycatchers: Evidence for active mate choice. *Behavioural Ecology and Sociobiology* **27**, 87–91.

Darwin, C. (1859). *On the origin of species by means of natural selection.* John Murray, London.

Darwin, C. (1871). *The descent of man, and selection in relation to sex.* John Murray, London.

Davies, N.B. (1991). Mating systems. In *Behavioural ecology: An evolutionary approach.* 3rd ed. (ed. J.R. Krebs and N.B. Davies), pp. 263–94, Blackwell, Oxford.

Davies, N.B. (1992). *Dunnock behaviour and social evolution.* Oxford University Press, Oxford.

Davis, J.W.F. and O'Donald, P. (1976). Sexual selection for a handicap: a critical analysis of Zahavi's model. *Journal of theoretical Biology* **57**, 345–54.

Davis, P. (1965). Recoveries of swallows ringed in Britain and Ireland. *Bird Study* **12**, 151–69.

Dawkins, R. (1976). *The selfish gene.* Oxford University Press, Oxford.

Dobzhansky, T. (1937). *Genetics and the origin of species.* Columbia University Press, New York.

Dominey, W.J. (1983). Sexual selection, additive genetic variance, and the "phenotypic handicap". *Journal of theoretical Biology* **101**, 495–502.

Downhower, J.F. (1976). Darwin's finches and the evolution of sexual dimorphism in body size. *Nature* **263**, 558–63.

Drummond, B.A. (1984). Multiple mating and sperm competition in the Lepidoptera. In *Sperm competition and the evolution of animal mating systems* (ed. R.L. Smith), pp. 291–370, Academic Press, Orlando.

Elton, C.S. (1927). *Animal ecology.* Sidgwick and Jackson, London.

Emlen, J.M. (1973). *Ecology: An evolutionary approach.* Addison-Wesley, Reading.

Endler, J.A. (1977). *Geographic variation, speciation, and clines.* Princeton University Press, Princeton.

Endler, J.A. (1986). *Natural selection in the wild.* Princeton University Press, Princeton.

Erskine, A.J. (1979). Man's influence on potential nesting sites and populations of swallows (Hirundinidae) in Canada. *Canadian Field-Naturalist* **93**, 371–7.

Eshel, I. (1978). On the handicap principle: A critical defence. *Journal of theoretical Biology* **70**, 245–50.

Eshel, I. and Hamilton, W.D. (1984). Parent-offspring correlations in fitness under fluctuating selection. *Proceedings of the Royal Society of London B* **122**, 1–14.

Evans, M. and Thomas, A.L.R. (1992). The aerodynamic and mechanical consequences of elongated tails in the scarlet tufted malachite sunbird: Measuring the cost of a handicap. *Animal Behaviour* **43**, 337–47.

Fagerström, T. and Wiklund, C. (1982). Why do males emerge before females? Protandry as a mating strategy in male and female butterflies. *Oecologia* **52**, 164–66.

Falconer, D.S. (1955). Patterns of response in selection experiments. *Cold Spring Harbor Symposia on Quantitative Biology* **20**, 178–96.

Falconer, D.S. (1981). *Introduction to quantitative genetics.* 2nd edn. Longman, New York.

Felsenstein. J. (1976). The theoretical population genetics of variable selection and migration. *Annual Review of Genetics* **10**, 253–80.

Fischer, J.C.H. (1869). Snestormen den 13de maj 1867 og småfuglene. *Naturhistorisk Tidsskrift* **3** (6), 143–60.

Fisher, R.A. (1915). The evolution of sexual preference. *Eugenics Review* **7**, 115–23.

Fisher, R.A. (1930). *The genetical theory of natural selection.* Clarendon Press, Oxford.

Folstad, I. and Karter, A.J. (1992). Parasites, bright males, and the immunocompetence handicap. *American Naturalist* **139**, 603–22.

Freeland, W.J. (1976). Pathogens and the evolution of primate sociality. *Biotropica* **8**, 12–24.

Fry, C.H. and Smith, D.A. (1985). A new swallow from the Red Sea. *Ibis* **127**, 1–6.

Furman, D.P. (1963). Problems in the control of poultry mites. *Advances in Acarology* **1**, 1–38.

Futuyma, D.J. and Slatkin, M. (ed.). (1983). *Coevolution*. Sinauer, Sunderland.

Gadgil, M. (1972). Male dimorphism as a consequence of sexual selection. *American Naturalist* **106**, 574–80.

Gibson, R.M. and Jewell, P.A. (1982). Semen quality, female choice and multiple mating in domestic sheep: A test of Trivers' sexual competence hypothesis. *Behaviour* **80**, 9–31.

Gillespie, J.H. (1973). Polymorphism in random environments. *Theoretical Population Biology* **4**, 193–5.

Gillespie, J.H. (1992). *The causes of molecular evolution*. Oxford University Press, Oxford.

Gjelstrup, P. and Møller, A.P. (1986). A tropical mite, *Ornithonyssus bursa* (Berlese, 1888) (Macronyssidae, Gamasida) in Danish swallow (*Hirundo rustica*) nests; with a review of mites from Danish birds. *Entomologiske Meddelelser* **53**, 119–25.

Glutz von Blotzheim, U.N. and Bauer, K.M. (ed.). (1985). *Handbuch der Vögel Mitteleuropas*. Vol. **10**. AULA-Verlag, Wiesbaden.

Gomulkiewicz, R.S. and Hastings, A. (1990). Ploidy and evolution by sexual selection: A comparison of haploid and diploid female choice models near fixation equilibria. *Evolution* **44**, 757–70.

Grafen, A. (1990*a*). Sexual selection unhandicapped by the Fisher process. *Journal of theoretical Biology* **144**, 473–516.

Grafen, A. (1990*b*). Biological signals as handicaps. *Journal of theoretical Biology* **144**, 517–46.

Grahn, M. (1992). *Intra- and intersexual selection in the pheasant* Phasianus colchicus. PhD thesis, Lund University, Sweden.

Grant, B.R. (1991). The significance of subadult plumage in Darwin's finches. *Behavioral Ecology* **1**, 161–70.

Grant, P.R. and Grant, B.R. (1992). Hybridization of bird species. *Science* **256**, 193–7.

Grant, P. and Price, T.D. (1981). Population variation in continuously varying traits as an ecological genetics problem. *American Zoologist* **21**, 795–811.

Grossman, C.J. (1985). Interactions between the gonadal steriods and the immune system. *Science* **227**, 257–61.

Gustafsson, L. (1986). Lifetime reproductive success and heritability: Empirical support of Fisher's fundamental theorem. *American Naturalist* **128**, 761–4.

Haigh, J. and Rose, M.R. (1980). Evolutionary game auctions. *Journal of theoretical Biology* **85**, 381–97.

Haldane, J.B.S. (1949). Disease and evolution. *Ricerce Sciencia Supplementum* **19**, 68–76.

Haldane, J.B.S. and Jayakar, S.D. (1963). Polymorphism due to selection of varying direction. *Journal of Genetics* **58**, 237–42.

Halliday, T.R. (1983). The study of mate choice. In *Mate choice* (ed. P. Bateson), pp. 3–32, Cambridge University Press, Cambridge.

Hamilton, W.D. (1964). The genetical theory of social behaviour. *Journal of theoretical Biology* **7**, 1–52.

Hamilton, W.D. (1971). Geometry for the selfish herd. *Journal of theoretical Biology* **31**, 295–311.

Hamilton, W.D. (1980). Sex versus non-sex versus parasite. *Oikos* **35**, 282–90.

Hamilton, W.D. (1982). Pathogens as causes of genetic diversity in their host populations. In *Population biology of infectious disease agents* (ed. R.M. Anderson and R.M. May), pp. 269–96, Verlag-Chemie, Weinheim.

Hamilton, W.D. (1986). Instability and cycling of two competing hosts with two parasites. In *Evolutionary processes and theory* (ed. S. Karlin and A. Nevo), pp. 645–68, Academic Press, New York.

Hamilton, W.D. (1990). Mate choice near or far. *American Zoologist* **30**, 341–52.

Hamilton, W.D. and Zuk, M. (1982). Heritable true fitness and bright birds: A role for parasites? *Science* **218**, 384–7.

Hammerstein, P. and Parker, G. (1987). Sexual selection: Games between the sexes. In *Sexual selection: Testing the alternatives* (ed. J.W. Bradbury and M.B. Andersson), pp. 119–42, Wiley, Chichester.

Harmel, D. (1983). Effects of genetics on antler quality and body size in white-tailed deer (*Odocoileus virginianus*). In *International symposium on antler development in Cervidae* (ed. R.D. Brown), pp. 339–48, Caesar Kleberg Wildlife Research Institute, Kingsville.

Harrison, C.J.O. (1985). Plumage. In *A dictionary of birds* (ed. B. Campbell and E. Lack), pp. 472–4, Poyser, Calton.

Harvey, P.H. and Bennett, P.M. 1985. Sexual dimorphism and reproductive strategies. In *Human sexual dimorphism* (ed. R.D. Martin and F. Newcombe), pp. 43–59, Taylor & Francis, London and Philadelphia.

Harvey, P.H. and Bradbury, J.W. (1991). Sexual selection. In *Behavioural ecology: An evolutionary approach. 3rd ed.* (ed. J.R. Krebs and N.B. Davies), pp. 203–33, Blackwell, Oxford.

Harvey, P.H. and Pagel, M. (1991). *The comparative method in evolutionary biology.* Oxford University Press, Oxford.

Haskins, C.P., Young, P., Hewitt, R.E., and Haskins, E.F. (1970). Stabilized heterozygosis of supergenes mediating Y-linked colour patterns in populations of *Lebistes reticulatus. Heredity* **25**, 575–89.

Hasson, O. (1989). Amplifiers and the handicap principle in sexual selection: A different emphasis. *Proceedings of the Royal Society of London B* **235**, 383–406.

Hasson, O. (1990). The role of amplifiers in sexual selection: An integration of the amplifying and the Fisherian mechanisms. *Evolutionary Ecology* **4**, 277–89.

Hasson, O. (1991). Sexual displays as amplifiers. *Behavioral Ecology* **2**, 189–97.

Hedrick, A.V. (1988). Female choice and the heritability of attractive male traits: An empirical study. *American Naturalist* **132**, 267–76.

Heisler, I.L. (1981). Offspring quality and the polygyny threshold: A new model for the "sexy son" hypothesis. *American Naturalist* **117**, 316–28.

Heisler, I.L. (1984a). A quantitative genetic model for the origin of mating preferences. *Evolution* **38**, 1283–95.

Heisler, I.L. (1984b). Inheritance of female mating propensity for *yellow* locus genotypes in *Drosophila melanogaster. Genetical Research* **44**, 133–49.

Heisler, I.L. (1985). Quantitative genetic models of female choice based on "arbitrary" male characters. *Heredity* **55**, 187–98.

Heisler, L., Andersson, M.B., Arnold, S.J., Boake, C.R., Borgia, G., Hausfater, G., Kirkpatrick, M., Lande, R., Maynard Smith, J., O'Donald, P., Thornhill, A.R., and Weissing, F.J. (1987). The evolution of mating preferences and sexually selected traits. In *Sexual selection: Testing the alternatives* (ed. J.W. Bradbury and M.B. Andersson), pp. 96–118. Wiley, Chichester.

Heisler, I.L. and Curtsinger, J.W. (1990). Dynamics of sexual selection in diploid populations. *Evolution* **44**, 1164–76.

Herroelen, P. (1960). De rui van de boerenzwaluw in Belgisch-Congo. *Gerfaut* **50**, 87–99.

Heywood, J.S. (1989). Sexual selection by the handicap mechanism. *Evolution* **43**, 1387–97.

Hill, G.E. (1991). Plumage coloration is a sexually selected indicator of male quality. *Nature* **350**, 337–9.

Hinde, R.A. (1956). The biological significance of the territories in birds. *Ibis* **98**, 340–69.

Hoelzer, G. A. (1989). The good parent process of sexual selection. *Animal Behaviour* **38**, 1067–78.

Höglund, J. (1989). Size and plumage dimorphism in lek-breeding birds: A comparative analysis. *American Naturalist* **134**, 72–87.

Houde, A.E. (1993). Evolution by sexual selection: What can population comparisons tell us? *American Naturalist* **141**, 796–803.

Houde, A.E. and Endler, J.A. (1990). Correlated evolution of female mating preferences and male color patterns in the guppy *Poecilia reticulata*. *Science* **248**, 1405–8.

Houston, A.I. and Davies, N.B. (1985). The evolution of cooperation and life history in the dunnock, *Prunella modularis*. In *Behavioural ecology: Ecological consequences of adaptive behaviour* (ed. R.M. Sibly and R.H. Smith), pp. 471–87. Blackwell, Oxford.

Hrdy, S.B. (1977). *The langurs of Abu: Female and male strategies of reproduction.* Harvard University Press, Cambridge.

Hrdy, S.B. (1979). Infanticide among animals: A review, classification, and examination of the implications for reproductive strategies of females. *Ethology and Sociobiology* **1**, 13–40.

Hurst, L. (1990). Parasite diversity and the evolution of diploidy, multicellularity and anisogamy. *Journal of theoretical Biology* **144**, 429–43.

Huxley, J.S. (1938a). The present standing of the theory of sexual selection. In *Evolution: Essays on aspects of evolutionary biology* (ed. G.R. de Beer), pp. 11–42, Clarendon Press, Oxford.

Huxley, J.S. (1938b). Darwin's theory of sexual selection and the data subsumed by it, in the light of recent research. *American Naturalist* **72**, 416–33.

Huxley, J.S. (1942). *Evolution: The modern synthesis.* Allen and Unwin, London.

Iwasa, Y., Odendaal, F. J., Murphy, D.D., Ehrlich, P.R., and Launer, A.E. (1983). Emergence patterns in male butterflies: A hypothesis and a test. *Theoretical Population Biology* **23**, 363–79.

Iwasa, Y., Pomiankowski, A., and Nee, S. (1991). The evolution of costly mate preferences. II. The "handicap" principle. *Evolution* **45**, 1431–42.

Jaenike, J. (1978). An hypothesis to account for the maintenance of sex within populations. *Evolutionary Theory* **3**, 191–4.

Janetos, A.C. (1980). Strategies of female mate choice: A theoretical analysis. *Behavioural Ecology and Sociobiology* **7**, 107–12.

Jarry, G. (1980). Dynamique d'une population d'hirondelles rustiques, dans l'est de la région parisienne. *Oiseau* **50**, 277–94.

Jeffreys, A.J., Wilson, V., and Thein, S.L. (1985). Hypervariable "minisatellite" regions in human DNA. *Nature* **314**, 67–73.

Jones, G. (1987). Parental foraging ecology and feeding behaviour during nestling rearing in the swallow. *Ardea* **75**, 169–74.

Karlin, S. and Raper, J. (1990). Evolution of sexual preferences in quantitative characters. *Theoretical Population Biology* **38**, 306–30.

Kasparek, M. (1976). Über Populationsunterschiede im Mauserverhalten der Rauchschwalbe. *Vogelwelt* **97**, 121–32.

Kasparek, M. (1981). *Die Mauser der Singvögel Europas: Ein Feldführer.* Dachverband Deutscher Avifaunisten, Langede.

Kempenaers, B., Verheyen, G.R., van den Broeck, M., Burke, T., van Broeckhoven, C., and Dhondt, A.A. (1992). Extra-pair paternity results from female preference for high-quality males in the blue tit. *Nature* **357**, 494–6.

King, R.C. (ed.) (1975). *Handbook of genetics.* Vol. **4**. Plenum Press, New York.

Kirkpatrick, M. (1982). Sexual selection and the evolution of female choice. *Evolution* **36**, 1–12.

Kirkpatrick, M. (1985). The evolution of female choice and male parental investment in polygynous species: The demise of the "sexy son". *American Naturalist* **125**, 788–810.

Kirkpatrick, M. (1986a). The handicap mechanism of sexual selection does not work. *American Naturalist* **127**, 222–40.

Kirkpatrick, M. (1986b). Sexual selection and cycling parasites: A simulation study of Hamilton's hypothesis. *Journal of theoretical Biology* **119**, 263–71.

Kirkpatrick, M. (1987). The evolutionary forces acting on female mating preferences in polygynous animals. In *Sexual selection: Testing the alternatives* (ed. J.W. Bradbury and M.B. Andersson), pp. 67–82, Wiley, Chichester.

Kirkpatrick, M. (1988). Consistency in genetic models of the sexy son: Reply to Curtsinger and Heisler. *American Naturalist* **132**, 609–10.

Kirkpatrick, M., Price, T., and Arnold, S.J. (1990). The Darwin–Fisher theory of sexual selection in monogamous birds. *Evolution* **44**, 180–93.

Kirkpatrick, M. and Ryan, M.J. (1991). The evolution of mating preferences and the paradox of the lek. *Nature* **350**, 33–8.

Kirpichnikov, V.S. (1981). *Genetic basis of fish selection.* Springer, New York.

Kodric-Brown, A. and Brown, J.H. (1984). Truth in advertising: The kinds of traits favored by sexual selection. *American Naturalist* **124**, 309–23.

Koenig, W.D., Gowaty, P.A., and Dickinson, J.L. (1992). Boxes, barns, and bridges: Confounding factors or exceptional opportunities in ecological studies? *Oikos* **63**, 305–8.

Kondrashov, A. (1988). Deleterious mutations as an evolutionary factor. III. Mating preference and some general remarks. *Journal of theoretical Biology* **131**, 487–96.

Lande, R. (1975). The maintenance of genetic variability by mutation in a polygenic character with linked loci. *Genetical Research* **26**, 221–35.

Lande, R. (1980). Sexual dimorphism, sexual selection and adaptation in polygenic characters. *Evolution* **34**, 292–305.

Lande, R. (1981). Models of speciation by sexual selection on polygenic characters. *Proceedings of the National Academy of Science of USA* **78**, 3721–5.

Lande, R. (1982). Rapid origin of sexual isolation and character divergence in a cline. *Evolution* **36**, 213–23.

Lande, R. (1987). Genetic correlations between the sexes in the evolution of sexual dimorphism and mating preferences. In *Sexual selection: Testing the alternatives* (ed. J.W. Bradbury and M.B. Andersson), pp. 83–94, Wiley, Chichester.

Lande, R. and Arnold, S.J. (1985). Evolution of mating preference and sexual dimorphism. *Journal of theoretical Biology* **117**, 651–64.

Leamy, L. and Atchley, W. (1985). Directional selection and developmental stability: Evidence from fluctuating asymmetry of morphometric characters in rats. *Growth* **49**, 8–18.

Lemel, J. (1993). *Evolutionary and ecological perspectives of status signalling in the Great Tit* (Parus major *L.*). PhD thesis, Univ. of Gothenburg, Sweden.

Lifjeld, J.T. and Slagsvold, T. (1989). How frequent is cuckoldry in pied flycatchers *Ficedula hypoleuca*? Problems with the use of heritability estimates of tarsus length. *Oikos* **54**, 205–10.

Ligon, J.D. (1968). Sexual differences in foraging behavior of two species of *Dendrocopus* woodpeckers. *Auk* **85**, 203–15.

de Lope, F. (1983). La reproduction d'*Hirundo rustica* en Extremadura (España). *Alauda* **51**, 81–91.

de Lope, F. and Møller, A.P. (1993). Female reproductive effort depends on the degree of ornamentation of their mates. *Evolution* (in press).

Low, B.S. (1978). Environmental uncertainty and the parental strategies of marsupials and placentals. *American Naturalist* **112**, 197–213.

Ludwig, W. (1932). *Das Rechts-Links Problem im Tierreich und beim Menschen.* Springer, Berlin.

MacArthur, J.W. (1949). Selection for small and large body size in the house mouse. *Genetics* **34**, 194–209.

Magrath, R.D. (1991). Nestling weight and juvenile survival in the blackbird, *Turdus merula. Journal of Animal Ecology* **60**, 335–51.

Majerus, M.E.N., O'Donald, P., Kearns, P.W.E., and Ireland, H. (1986). Genetics and evolution of female choice. *Nature* **321**, 164–7.

Manly, B.F.J. (1985). *The statistics of natural selection on animal populations.* Chapman and Hall, New York.

Marshall, A.G. (1981). *The ecology of ectoparasitic insects.* Academic Press, London.

Martin, R.F. (1980). Analysis of hybridization between the hirundinid genera *Hirundo* and *Petrochelidon* in Texas. *Auk* **97**, 148–59.

Martin, R.F. and Selander, R.K. (1975). Morphological and biochemical evidence of hybridization between cave and barn swallows. *Condor* **77**, 362–4.

Maynard Smith, J. (1976). Sexual selection and the handicap principle. *Journal of theoretical Biology* **57**, 239–42.

Maynard Smith, J. (1977). Parental investment: A prospective analysis. *Animal Behaviour* **25**, 1–9.

Maynard Smith, J. (1978). The handicap principle: A comment. *Journal of theoretical Biology* **70**, 251–2.

Maynard Smith, J. (1985). Sexual selection, handicaps and true fitness. *Journal of theoretical Biology* **115**, 1–8.

Maynard Smith, J. (1987). Sexual selection: A classification of models. In *Sexual selection: Testing the alternatives* (ed. J.W. Bradbury and M.B. Andersson), pp. 9–20, Wiley, Chichester.

Maynard Smith, J. and Brown, R.L.W. (1986). Competition and body size. *Theoretical Population Biology* **30**, 166–79.

Mayr, E. (1942). *Systematics and the origin of species.* Columbia University Press, New York.

Mayr, E. and Bond, J. (1943). Notes on the generic classification of the swallows. *Ibis* **85**, 334–41.

McKinney, F., Cheng, K.M., and Bruggers, D.J. (1984). Sperm competition in apparently monogamous birds. In *Sperm competition and the evolution of animal mating systems* (ed. R.L. Smith), pp. 523–45, Academic Press, Orlando.

Mendelsohn, J.M. (1973). Some observations on age ratio, weight and moult of the European swallow in Central Transvaal. *Annals of the Transvaal Museum* **28**, 79–89.

Menzel, H. (1984). *Die Mehlschwalbe.* Neue Brehm-Bücherei 548. A. Ziemsen Verlag, Wittenberg-Lutherstadt.

Michener, H. and Michener, J.R. (1938). Bars in flight feathers. *Condor* **40**, 149–60.

Micherdzinski, W. (1980). *Eine Taxonomische Analyse der Familie Macronyssidae Oudemanns, 1936. I. Subfamilie Ornithonyssidae Lange, 1958 (Acarina, Mesostigmata).* Polska Akademia Nauk, Warszawa.

Michod, R.E. and Hasson, O. (1990). On the evolution of reliable indicators of fitness. *American Naturalist* **135**, 788–808.

Milinski, M. and Bakker, T.C.M. (1990). Female sticklebacks use male coloration in mate choice and hence avoid parasitized males. *Nature* **344**, 330–2.

Møller, A.P. (1978). Danske landsvalers *Hirundo rustica* træk i forbindelse med katastrofen i Central Europa i efteråret 1974. *Dansk ornithologisk Forenings Tidsskrift* **72**, 59–60.

Møller, A.P. (1982). Clutch size in relation to nest size in the swallow *Hirundo rustica*. *Ibis* **124**, 339–43.

Møller, A.P. (1983). Changes in Danish farmland habitats and their populations of breeding birds. *Holarctic Ecology* **6**, 95–100.

Møller, A.P. (1984a). Geographical variation in breeding parameters of two hirundines. *Ornis Scandinavica* **15**, 43–54.

Møller, A.P. (1984b). Parental defence of offspring in the barn swallow. *Bird Behaviour* **5**, 110–7.

Møller, A.P. (1985). Mixed reproductive strategy and mate guarding in a semi-colonial passerine, the swallow *Hirundo rustica*. *Behavioural Ecology and Sociobiology* **17**, 401–8.

Møller, A.P. (1986). Mating systems among European passerines. *Ibis* **128**, 234–50.

Møller, A.P. (1987a). Advantages and disadvantages of coloniality in the swallow *Hirundo rustica*. *Animal Behaviour* **35**, 819–32.

Møller, A.P. (1987b). Intraspecific nest parasitism and anti-parasite behaviour in swallows, *Hirundo rustica*. *Animal Behaviour* **35**, 247–54.

Møller, A.P. (1987c). Intruders and defenders on avian breeding territories: The effect of sperm competition. *Oikos* **48**, 47–54.

Møller, A.P. (1987d). Nest lining in relation to the nesting cycle in the swallow *Hirundo rustica*. *Ornis Scandinavica* **18**, 148–9.

Møller, A.P. (1987*e*). Behavioural aspects of sperm competition in swallows *Hirundo rustica*. *Behaviour* **100**, 92–104.

Møller, A.P. (1987*f*). Extent and duration of mate guarding in swallows *Hirundo rustica*. *Ornis Scandinavica* **18**, 95–100.

Møller, A.P. (1987*g*). Mate guarding in the swallow *Hirundo rustica*: An experimental study. *Behavioural Ecology and Sociobiology* **21**, 119–23.

Møller, A.P. (1988*a*). Infanticidal and anti-infanticidal strategies in the swallow *Hirundo rustica*. *Behavioural Ecology and Sociobiology* **22**, 365–71.

Møller, A.P. (1988*b*). Paternity and paternal care in the swallow *Hirundo rustica*. *Animal Behaviour* **36**, 996–1005.

Møller, A.P. (1988*c*). Female choice selects for male sexual tail ornaments in the monogamous swallow. *Nature* **322**, 640–2.

Møller, A.P. (1988*d*). Testes size, ejaculate quality, and sperm competition in birds. *Biological Journal of the Linnean Society* **33**, 273–83.

Møller, A.P. (1988*e*). Ejaculate quality, testes size and sperm competition in primates. *Journal of Human Evolution* **17**, 479–88.

Møller, A.P. (1989*a*). Viability costs of male tail ornaments in a swallow. *Nature* **339**, 132–5.

Møller, A.P. (1989*b*). Natural and sexual selection on a plumage signal of status and on morphology in house sparrows, *Passer domesticus*. *Journal of evolutionary Biology* **2**, 125–40.

Møller, A.P. (1989*c*). Population dynamics of a declining swallow *Hirundo rustica* L. population. *Journal of Animal Ecology* **58**, 1051–63.

Møller, A.P. (1989*d*). Frequency of extra-pair paternity in birds estimated from sex-differential heritability of tarsus length: Reply to Lifjeld and Slagsvold's critique. *Oikos* **56**, 247–9.

Møller, A.P. (1989*e*). Intraspecific nest parasitism in the swallow, *Hirundo rustica*: The importance of neighbours. *Behavioural Ecology and Sociobiology* **25**, 33–8.

Møller, A.P. (1989*f*). Ejaculate quality, testes size and sperm production in mammals. *Functional Ecology* **3**, 91–6.

Møller, A.P. (1990*a*). Male tail length and female mate choice in the monogamous swallow *Hirundo rustica*. *Animal Behaviour* **39**, 458–65.

Møller, A.P. (1990*b*). Changes in the size of avian breeding territories in relation to the nesting cycle and risks of cuckoldry. *Animal Behaviour* **40**, 1070–9.

Møller, A.P. (1990*c*). Effects of parasitism by the haematophagous mite *Ornithonyssus bursa* on reproduction in the barn swallow *Hirundo rustica*. *Ecology* **71**, 2345–57.

Møller, A.P. (1990*d*). Parasites and sexual selection: Current status of the Hamilton and Zuk hypothesis. *Journal of evolutionary Biology* **3**, 319–28.

Møller, A.P. (1990*e*). Effects of an haematophagous mite on the barn swallow (*Hirundo rustica*): A test of the Hamilton and Zuk hypothesis. *Evolution* **44**, 771–84.

Møller, A.P. (1990*f*). Deceptive use of alarm calls by male swallows *Hirundo rustica*: A new paternity guard. *Behavioral Ecology* **1**, 1–6.

Møller, A.P. (1990*g*). Fluctuating asymmetry in male sexual ornaments may reliably reveal male quality. *Animal Behaviour* **40**, 1185–7.

Møller, A.P. (1991*a*). Sexual selection in the monogamous barn swallow (*Hirundo rustica*). I. Determinants of tail ornament size. *Evolution* **45**, 1823–36.

Møller, A.P. (1991*b*). Preferred males acquire mates of higher phenotypic quality. *Proceedings of the Royal Society of London B* **245**, 179–82.

Møller, A.P. (1991*c*). Parasites, sexual ornaments, and mate choice in the barn swallow. In *Bird-parasite interactions: Ecology, evolution and behaviour* (ed. J. Loye and M. Zuk), pp. 328–48, Oxford University Press, Oxford.

Møller, A.P. (1991*d*). The preening activity of swallows, *Hirundo rustica*, in relation to experimentally manipulated loads of haematophagous mites. *Animal Behaviour* **42**, 251–60.

Møller, A.P. (1991*e*). Viability is positively related to degree of ornamentation in male swallows. *Proceedings of the Royal Society of London B* **243**, 145–8.

Møller, A.P. (1991*f*). Parasite load reduces song output in a passerine bird. *Animal Behaviour* **41**, 723–30.

Møller, A.P. (1991*g*). Ectoparasite loads affect optimal clutch size in swallows. *Functional Ecology* **5**, 351–9.

Møller, A.P. (1991*h*). The effect of feather nest lining on reproduction in the swallow. *Ornis Scandinavica* **22**, 396–400.

Møller, A.P. (1991*i*). Defence of offspring by male swallows, *Hirundo rustica*, in relation to participation in extra-pair copulations by their mates. *Animal Behaviour* **42**, 261–7.

Møller, A.P. (1991*j*). Why mated songbirds sing so much: Mate guarding and male announcement of mate fertility status. *American Naturalist* **138**, 994–1014.

Møller, A.P. (1991*k*). Density-dependent extra-pair copulations in the swallow *Hirundo rustica*. *Ethology* **87**, 316–29.

Møller, A.P. (1991*l*). Sexual ornament size and the cost of fluctuating asymmetry. *Proceedings of the Royal Society of London B* **243**, 59–62.

Møller, A.P. (1992*a*). Frequency of female copulations with multiple males and sexual selection. *American Naturalist* **139**, 1089–101.

Møller, A.P. (1992*b*). Nest boxes and the scientific rigour of experimental studies. *Oikos* **63**, 309–12.

Møller, A.P. (1992*c*). Sexual selection in the monogamous swallow (*Hirundo rustica*). II. Mechanisms of intersexual selection. *Journal of evolutionary Biology* **5**, 603–24.

Møller, A.P. (1992*d*). Female swallow preference for symmetrical male sexual ornaments. *Nature* **357**, 238–40.

Møller, A.P. (1992*e*). Patterns of fluctuating asymmetry in weapons: Evidence for reliable signalling of quality in beetle horns and bird spurs. *Proceedings of the Royal Society of London B* **248**, 199–206.

Møller, A.P. (1992*f*). Parasites differentially increase fluctuating asymmetry in secondary sexual characters. *Journal of evolutionary Biology* **5**, 691–9.

Møller, A.P. (1993*a*). Density-dependent conspecific brood parasitism and anti-parasite behavior in the barn swallow. In *The ecology and evolution of brood parasitism* (ed. S.I. Rothstein). Oxford University Press, New York.

Møller, A.P. (1993*b*). Symmetrical male sexual ornaments, paternal care, and offspring quality. *Behavioral Ecology* (in press).

Møller, A.P. (1993*c*). Developmental stability, sexual selection and speciation. *Journal of evolutionary Biology* **6**, 493–509.

Møller, A.P. (1993*d*). Repeatability of female choice in a monogamous swallow. *Animal Behaviour* (in press).

Møller, A.P. (1993*e*). A test of the resource-provisioning model of parasite-mediated sexual selection. MS.

Møller, A.P. (1993*f*). Morphology and sexual selection in the barn swallow *Hirundo rustica* in Chernobyl, Ukraine. *Proceedings of the Royal Society of London B* **252**, 51–7.

Møller, A.P. (1993*g*). Genotype-environment interaction and secondary sexual characters. MS.

Møller, A.P. (1993*h*). Phenotype-dependent arrival time and its consequences in a migratory bird. MS.

Møller, A.P. (1993*i*). Ectoparasites increase the cost of reproduction in their hosts. *Journal of Animal Ecology* **62**, 309–22.

Møller, A.P. (1993*j*). Why are sperm so small? Sperm size, sperm abnormality and selfish DNA. MS.

Møller, A.P. (1993*k*). Sexual selection in the barn swallow *Hirundo rustica*. III. Female tail ornaments. *Evolution* **47**, 417–31.

Møller, A.P. (1993*l*). Fluctuating asymmetry in morphological characters in marginal and central populations. *American Naturalist* (in press).

Møller, A.P. (1993*m*). Sexual selection in the barn swallow *Hirundo rustica*. V. Clinal variation in tail ornaments. MS.

Møller, A.P. (1993*n*). Male ornament size as a reliable cue to enhanced offspring viability in the swallow. MS.

Møller, A.P. (1993*o*). The cost of secondary sexual characters and the evolution of cost-reducing traits. MS.

Møller, A.P. (1993*p*). Female preference for apparently symmetrical male sexual ornaments in the barn swallow *Hirundo rustica*. *Behavioural Ecology and Sociobiology* **32**, 371–6.

Møller, A.P., Allander, K., and Dufva, R. (1990). Fitness effects of parasites on passerine birds: A review. In *Population biology of passerine birds: An integrated approach* (ed. J. Blondel, A. Gosler, J.D. Lebreton, and R. McCleery), pp. 269–80. Springer, Berlin.

Møller, A.P. and Birkhead, T.R. (1992). Validation of the heritability method to estimate extra-pair paternity in birds. *Oikos* **64**, 485–8.

Møller, A.P. and Birkhead, T.R. (1993). Cuckoldry and sociality: A comparative study of birds. *American Naturalist* **142**, 118–40.

Møller, A.P. and Höglund, J. (1991). Patterns of fluctuating asymmetry in avian feather ornaments: Implications for models of sexual selection. *Proceedings of the Royal Society of London B* **245**, 1–5.

Møller, A.P. and de Lope, F. (1993). Differential costs of a secondary sexual character: An experimental test of the handicap principle. *Evolution* (in press).

Møller, A.P., Magnhagen, C., Ulfstrand, A., and Ulfstrand, S. (1993). Phenotypic quality and moult in the barn swallow *Hirundo rustica*. MS.

Møller, A.P. and Pomiankowski, A. (1993*a*). Fluctuating asymmetry and sexual selection. *Genetica* **89**, 267–79.

Møller, A.P. and Pomiankowski, A. (1993*b*). Why have birds got multiple sexual ornaments? *Behavioural Ecology and Sociobiology* **32**, 167–76.

Motro, U. (1982). The courtship handicap: Phenotypic effect. *Journal of theoretical Biology* **97**, 319–24.

Mousseau, T.A. and Roff, D.A. (1987). Natural selection and the heritability of fitness. *Heredity* **59**, 181–97.

Mukai, T. (1964). Polygenic mutation affecting quantitative character of *Drosophila melanogaster*. *Proceedings of the Gamma Field Symposium* **3**, 13–29.

Mukai, Y., Cardellino, R.K., Watanabe, T.K., and Crow, J.F. (1974). The genetic variance for viability and its components in a population of *Drosophila melanogaster. Genetics* **78**, 1195–208.

Muma, K.E. and Weatherhead, P.J. (1989). Male traits expressed in females: Direct or indirect sexual selection? *Behavioural Ecology and Sociobiology* **25**, 23–31.

Nei, M., Maruyama, T., and Wu, C.-I. (1983). Models of evolution of reproductive isolation. *Genetics* **103**, 557–79.

van Noordwijk, A.J. and de Jong, G. (1986). Acquisition and allocation of resources: Their influence on variation in life history tactics. *American Naturalist* **128**, 137–42.

Norberg, U.M. (1990). *Vertebrate flight: Mechanics, physiology, morphology, ecology and evolution.* Springer, Berlin.

Norris, K.J. (1993). Heritable variation in a plumage indicator of viability in male great tits, *Parus major. Nature* **362**, 537–9.

Nur, N. and Hasson, O. (1984). Phenotypic plasticity and the handicap principle. *Journal of theoretical Biology* **110**, 275–97.

O'Donald, P. (1962). The theory of sexual selection. *Heredity* **17**, 541–52.

O'Donald, P. (1967). A general model of natural and sexual selection. *Heredity* **22**, 499–518.

O'Donald, P. (1972). Sexual selection by variations in fitness at breeding time. *Nature* **237**, 349–51.

O'Donald, P. (1980a). *Genetic models of sexual selection.* Cambridge University Press, Cambridge.

O'Donald, P. (1980b). Genetic models of sexual and natural selection in monogamous organisms. *Heredity* **44**, 391–415.

O'Donald, P. (1980c). Sexual selection by female choice in a monogamous bird: Darwin's theory corroborated. *Heredity* **45**, 201–17.

O'Donald, P. (1983). *The Arctic skua.* Cambridge University Press, Cambridge.

O'Donald, P. (1990). Fisher's contributions to the theory of sexual selection as the basis of recent research. *Theoretical Population Biology* **38**, 285–300.

Palmer, R.S. (1972). Patterns of molting. In *Avian biology.* Vol. 2 (ed. D.S. Farner, J.R. King, and K.C. Parkes), pp. 65–155, Academic Press, New York.

Parker, G.A. (1982). Phenotype-limited evolutionarily stable strategies. In *Current problems in sociobiology* (ed. King's College Sociobiology Group), pp. 173–201, Cambridge University Press, Cambridge.

Parker, G.A. (1983a). Arms races in evolution: An ESS to the opponent–independent costs game. *Journal of theoretical Biology* **101**, 619–48.

Parker, G.A. (1983b). Mate quality and mating decisions. In *Mate choice* (ed. P. Bateson), pp. 141–66, Cambridge University Press, Cambridge.

Parker, G.A. (1984). Sperm competition and the evolution of animal mating strategies. In *Sperm competition and the evolution of animal mating systems* (ed. R.L. Smith), pp. 1–60, Academic Press, Orlando.

Parker, G.A. and Courtney, S.P. (1983). Seasonal incidence: Adaptive variation in the timing of life history stages. *Journal of theoretical Biology* **105**, 147–55.

Parsons, P.A. (1990). Fluctuating asymmetry: An epigenetic measure of stress. *Biological Reviews* **65**, 131–45.

Partridge, L. (1980). Mate choice increases a component of fitness in fruit flies. *Nature* **283**, 290–1.

Partridge, L. and Endler, J.A. (1987). Life history constraints on sexual selection. In *Sexual selection: Testing the alternatives* (ed. J.W. Bradbury and M.B. Andersson), pp. 265–77, Wiley, Chichester.

Partridge, L. and Harvey, P.H. (1988). The ecological context of life-history evolution. *Science* **241**, 1449–55.

Petrie, M., Halliday, T.R., and Sanders, C. (1990). Peahens prefer peacocks with elaborate trains. *Animal Behaviour* **41**, 323–31.

Petersen, G.W. (1979). Infestations of *Ornithonyssus bursa*, a haematophagous mite, on starlings over the non-breeding season. *New Zealand Journal of Zoology* **6**, 319–20.

Pomiankowski, A.N. (1987*a*). The "handicap principle" does work — sometimes. *Proceedings of the Royal Society of London B* **127**, 123–45.

Pomiankowski, A.N. (1987*b*). The costs of choice in sexual selection. *Journal of theoretical Biology* **128**, 195–218.

Pomiankowski, A.N. (1988). The evolution of female mate preferences for male genetic quality. *Oxford Surveys in Evolutionary Biology* **5**, 136–84.

Pomiankowski, A., Iwasa, Y., and Nee, S. (1991). The evolution of costly mate preferences. I. Fisher and biased mutation. *Evolution* **45**, 1422–30.

Pomiankowski, A. and Møller, A.P. (1993*a*). The lek paradox resolved: Genetic variation and the strength of sexual selection. MS.

Pomiankowski, A. and Møller, A.P. (1993*b*). Fluctuating asymmetry and the strength of sexual selection. MS.

Poulton, E.B. (1890). *The colour of animals.* Kegan, Paul, Trench and Trubner, London.

Powlesland, R.G. (1978). Behaviour of the haematophagous mite *Ornithonyssus bursa* in starling nest boxes in New Zealand. *New Zealand Journal of Zoology* **5**, 395–9.

Price, P.W. (1980). *Evolutionary biology of parasites.* Princeton University Press, Princeton.

Price, T.D. (1984). The evolution of sexual size dimorphism in Darwin's finches. *American Naturalist* **123**, 500–18.

Price, T. and Schluter, D. (1991). On the low heritability of life-history traits. *Evolution* **45**, 853–61.

Price, T., Schluter, D., and Heckman, N.E. (1993). Sexual selection when the female directly benefits. *Biological Journal of the Linnean Society* **48**, 187–211.

Reeve, E.C.R. 1960. Some genetic tests on asymmetry in sternopleural chetae number in *Drosophila. Genetical Research, Cambridge* **1**, 151–72.

Reeve, E.C.R. and Robertson, F.W. (1953). Studies in quantitative inheritance. II. Analysis of a strain of *Drosophila melanogaster* selected for long wings. *Journal of Genetics* **51**, 276–316.

Reynolds, J.D. and Gross, M.R. (1990). Costs and benefits of female mate choice: Is there a lek paradox? *American Naturalist* **136**, 230–43.

Rice, W.R. (1988). Heritable variation in fitness as a prerequisite for adaptive female choice: The effect of mutation-selection balance. *Evolution* **42**, 817–20.

Ritchie, M.G. (1992). Setbacks in the search for mate-preference genes. *Trends in Ecology and Evolution* **7**, 328–9.

Robertson, F.W. (1955). Selection response and the properties of genetic variation. *Cold Spring Harbor Symposia in Quantitative Biology* **20**, 166–77.

Roelofs, W., Glover, T., Tang, X.-H., Sreng, I., Robbins, P., Eckenrode, C., Löfstedt, C., Hansson, B.S. and Bengtsson, B.O. (1987). Sex pheromone production and perception in European corn borer moths is determined by both autosomal and sex-linked genes. *Proceedings of the National Academy of Science of USA* **84**, 7585–9.

Rohwer, S. (1986). Selection for adoption versus infanticide by replacement "mates" in birds. *Current Ornithology* **3**, 353–95.

Rowley, I. (1983). Re-mating in birds. In *Mate choice* (ed. P. Bateson), pp. 331–60, Cambridge University Press, Cambridge.

Ryan, M.J. (1990). Sexual selection, sensory systems, and sensory exploitation. *Oxford Surveys in Evolutionary Biology* **7**, 157–95.

Ryan, M.J., Fox., J.H., Wilczynski, W., and Rand, A.S. (1990). Sexual selection for sensory exploitation in the frog *Physalaemus pustulosus*. *Nature* **343**, 66–7.

Ryan, M.J. and Rand, A.S. (1990). The sensory basis of sexual selection for complex calls in the túngara frog, *Physalaemus pustulosus* (sexual selection for sensory exploitation). *Evolution* **44**, 305–14.

Sanderson, N. (1989). Can gene flow prevent reinforcement? *Evolution* **43**, 1223–35.

von Schantz, T., Göransson, G., Andersson, G., Fröberg, I,. Grahn, M., Helgée, A., and Witzell, H. (1989). Female choice selects for a viability-based male trait in pheasants. *Nature* **337**, 166–9.

Schluter, D., Price, T. D., and Rowe, L. (1991). Conflicting selection pressures and life history trade-offs. *Proceedings of the Royal Society of London B* **246**,11–7.

Searcy, W.M. (1979). Female choice of mates: A general model for birds and its application to red-winged blackbirds (*Agelaius phoeniceus*). *American Naturalist* **114**, 77–100.

Searcy, W.M. (1982). The evolutionary effects of mate selection. *Annual Review of Ecology and Systematics* **13**, 57–85.

Searcy, W.M. and Andersson, M. (1986). Sexual selection and the evolution of song. *Annual Review of Ecology and Systematics* **17**, 507–33.

Searcy, W.M. and Yasukawa, K. (1981). Does the "sexy son" hypothesis apply to mate choice in red-winged blackbirds? *American Naturalist* **117**, 343–8.

Seger, J. (1985). Unifying genetic models for the evolution of female choice. *Evolution* **39**, 1185–93.

Seger, J. and Trivers, R. (1986). Asymmetry in the evolution of female mating preferences. *Nature* **319**, 771–3.

Selander, R.K. (1966). Sexual dimorphism and differential niche utilization. *Condor* **68**, 113–51.

Selander, R.K. (1972). Sexual selection and dimorphism in birds. In *Sexual selection and the descent of man, 1871–1971* (ed. B. Campbell), pp. 180–230, Aldine, Chicago.

Sharpe, R.B. and Wyatt, C.W. (1885–1894). *A monograph of the Hirundinidae or family of swallows.* Sotheran, London.

Sheldon, B. (1993). Venereal diseases in birds. *Philosophical Transactions of the Royal Society B* **339**, 491–7.

Sheldon, F.H. and Winkler, D.W. (1993). Intergeneric phylogenetic relationships of swallows estimated by DNA–DNA hybridization. *Auk* (in press).

Shields, W.M. (1984). Barn swallow mobbing: Self defence, collateral kin defence, group defence or parental care? *Animal Behaviour* **32**, 132–48.

Sibley, C.G. and Ahlquist, J.E. (1982). The relationships of the swallows (Hirundinidae). *Journal of the Yamashina Institute of Ornithology* **14**, 122–30.

Sibley, C.G. and Ahlquist, J.E. (1990). *Phylogeny and classification of birds.* Yale University Press, New Haven and London.

Siegel, S. and Castellan, Jr., N.J. (1988). *Nonparametric statistics for the behavioral sciences.* 2nd edn. McGraw-Hill, New York.

Sikes, R.K. and Chamberlain, R.W. (1954). Laboratory observations on three species of bird mites. *Journal of Parasitology* **40**, 691–7.

Silberglied, R.E., Shepherd, J.G., and Dickinson, J.L. (1984). Eunuchs: The role of apyrene sperm in Lepidoptera. *American Naturalist* **123**, 255–65.

Slatkin, M. (1984). Ecological causes of sexual dimorphism. *Evolution* **38**, 622–30.

Smirensky, S.M. and Mishchenko, A.L. (1981). Taxonomical status and history of formation of the range of *Hirundo rustica* in the Amur territory. *Zoologicheskij Zhurnal* **60**, 1533–41.

Smith, H.G. and Montgomerie, R. (1991). Sexual selection and the tail ornaments of North American barn swallows. B*ehavioural Ecology and Sociobiology* **28**, 195–201.

Smith, H.G. and Montgomerie, R. (1992). Male incubation in barn swallows: The influence of nest temperature and sexual selection. *Condor* **94**, 750–9.

Smith, H.G., Montgomerie, R., Poldmaa, T., White, B.N., and Boag, P.T. (1991). DNA fingerprinting reveals relation between tail ornaments and cuckoldry in barn swallows, *Hirundo rustica. Behavioral Ecology* **2**, 90–8.

Smith, J.M. and Graves, H.B. (1978). Some factors influencing mobbing behavior in barn swallows (*Hirundo rustica*). *Behavior of Biology* **23**, 355–72.

Smith, J.N.M. and Arcese, P. (1989). How fit are floaters? Consequences of alternative territorial behaviors in a nonmigratory sparrow. *American Naturalist* **133**, 830–45.

Speich, S.M., Jones, H.L., and Benedict, E.M. (1986). Review of the natural nesting of the barn swallow in North America. *American Midland Naturalist* **115**, 248–54.

Stearns, S.C. (1976). Life-history tactics: A review of the ideas. *Quarterly Review of Biology* **51**, 3–47.

Stearns, S.C. (1992). *The evolution of life histories.* Oxford University Press, Oxford.

Stoner, D. (1935). Temperature and growth studies on the Barn Swallow. *Auk* **52**, 400–7.

Stresemann, E. and Stresemann, V. (1966). Die Mauser der Vögel. *Journal für Ornithologie, Sonderheft* **105**.

Sulkin, E.S. and Izumi, E.M. (1947). Isolation of western equine encephalomyelitis virus from tropical fowl mites, *Liponyssus bursa* (Berlese). *Proceedings of the Society of Experimental Biological Medicine* **66**, 249.

Sullivan, M. S., Robertson, P. A., and N. A. Aebischer. (1993). Fluctuating asymmetry measurement. *Nature* **361**, 409–410.

Sutherland, W.J. (1985). Chance can produce a sex difference in variance in mating success and account for Bateman's data. *Animal Behaviour* **33**, 1349–52.

Sutherland, W.J., and M.C.R. de Jong. (1991). The evolutionarily stable strategy for secondary sexual characters. *Behavioral Ecology* **2**, 16–20.

Templeton, J.W., Sharp, R.M., Williams, J., David, D., Harmel, D., Armstrong, B., and Wardroup, W. (1983). Single dominant major gene effect on the expression of antler point number in the white-tailed deer. In *International symposium on antler development in Cervidae* (ed. R.D. Brown), pp. 365–88, Caesar Kleberg Wildlife Research Institute, Kingsville.

Thoday, J.M. (1958). Homeostasis in a selection experiment. *Heredity* **12**, 401–15.

Thomas, A.L.R. (1993). On the aerodynamics of bird tails. *Philosophical Transactions of the Royal Society of London B* **340**, 361–80.

Thornhill, R. and Alcock, J. (1983). *The evolution of insect mating systems.* Harvard University Press, Cambridge and London.

Threlfall, W. and Bennett, G.F. (1989). Avian Haematozoa. *Wildlife Journal* **12**, 3–16.

Tomlinson, I.P.M. (1988). Diploid models of the handicap principle. *Heredity* **60**, 283–93.

Tomlinson, I.P.M. (1989). Models of clines in sexual selection by female choice. *Biological Journal of the Linnean Society* **36**, 331–48.

Trail, P.W. and Adams, E.S. (1989). Active mate choice at cock-of-the-rock leks: Tactics of sampling and comparison. *Behavioural Ecology and Sociobiology* **25**, 283–92.

Trivers, R.L. (1972). Parental investment and sexual selection. In *Sexual selection and the descent of man, 1871–1971* (ed. B. Campbell), pp. 136–79, Aldine, Chicago.

Trivers, R. (1988). Sex differences in rates of recombination and sexual selection. In *The evolution of sex* (ed. R.E. Michod and B.R. Levin), pp. 270–86, Sinauer, Sunderland.

Tuomi, J., Hakala, T., and Haukioja, E. (1983). Alternative concepts of reproductive effort, costs of reproduction, and selection in life-history evolution. *American Zoologist* **23**, 25–34.

Turelli, M. (1984). Heritable genetic variation via mutation-selection balance: Lerch's zeta meets the abdominal bristle. *Theoretical Population Biology* **25**, 138–93.

Turner, A.K. (1980). *The use of time and energy by aerial feeding birds.* PhD thesis. University of Stirling, U. K.

Turner, A.K. (1982). Optimal foraging by the swallow: Prey size selection. *Animal Behaviour* **30**, 862–72.

Turner, A.K. and Rose, C. (1989). *A handbook of the swallows and martins of the world.* Christopher Helm, London.

Vansteenwegen, C. (1982). Longevité et structure des nids d'hirondelles de cheminée (*Hirundo rustica*). *Aves* **19**, 247–51.

Van Valen, L. (1973). A new evolutionary law. *Evolutionary Theory* **1**, 1–30.

von Vietinghoff-Riesch, A. (1955). *Die Rauchschwalbe.* Duncker & Humblot, Berlin.

Wakelin, D. (1978). Genetic control of susceptibility and resistance to parasite infection. *Advances in Parasitology* **16**, 210–308.

Wakelin, D. and Blackwell, J.M. (ed.). (1988). *Genetics of resistance to bacterial and parasitic infection.* Taylor and Francis, London.

Walker, W.F. (1980). Sperm utilization strategies in nonsocial insects. *American Naturalist* **115**, 780–99.

Wallace, A.R. (1889). *Darwinism: An exposition of the theory of natural selection.* Macmillan, London.

Wallace, A.R. (1891). *Natural selection and tropical nature.* Macmillan, London.

Ward, P. and Zahavi, A. (1973). The importance of certain assemblages of birds as "information-centres" for food-finding. *Ibis* **115**, 517–34.

Watt, W.B., Carter, P.A., and Donohue, K. (1986). Females choice of 'good genotypes' as mates is promoted by an insect mating system. *Science* **233**, 1187–90.

Weatherhead, P.J. and Robertson, R.J. (1979). Offspring quality and the polygyny threshold: "The sexy son hypothesis". *American Naturalist* **113**, 201–8.

Weatherhead, P.J. and Robertson, R.J. (1981). In defense of the "sexy son" hypothesis. *American Naturalist* **117**, 349–56.

West-Eberhard, M.J. (1979). Sexual selection, social competition and evolution. *Proceedings of the American Philosophical Society* **123**, 222–34.

West-Eberhard, M.J., Bradbury, J.W., Davies, N.B., Gouyon, P.-H., Hammerstein, P., König, B., Parker, G.A., Queller, D.C., Sachser, N., Slagsvold, T., Trillmich, F., and Vogel, C. (1987). Conflicts between and within the sexes in sexual selection. In *Sexual selection: Testing the alternatives* (ed. J.W. Bradbury and M.B. Andersson), pp. 181–95, Wiley, Chichester.

Wetton, J.H., Carter, R.E., Parkin, D.T., and Walters, D. (1987). Demographic study of a wild house sparrow population by DNA fingerprinting. *Nature* **327**, 147–9.

White, G. (1770). *The natural history of Selborne* (ed. R. Mabey) 1977. Penguin, Harmondsworth.

Whittingham, L.A., Taylor, P.D., and Robertson, R.J. (1992). Confidence of paternity and male parental care. *American Naturalist* **139**, 1115–25.

Wikel, S.K. (1982). Immune responses to arthropods and their products. *Annual Review of Entomology* **27**, 21–48.

Wiklund, C. and Fagerström, T. (1977). Why do males emerge before females? A hypothesis to explain the incidence of protandry in butterflies. *Oecologia* **31**, 153–8.

Williams, G.C. (1966). *Adaptation and natural selection.* Princeton University Press, Princeton.

Winge, Ø. and Ditlevsen, E. (1947). Colour inheritance and sex-determination in *Lebistes*. *Heredity* **1**, 65–83.

Winkler, D.W. and Møller, A.P. (1994). A phylogeny of the Hirundinidae based on morphology. MS.

Winkler, D.W. and Sheldon, F.H. (1993). Evolution of nest-construction in swallows (Hirundinidae): A molecular phylogenetic perspective. *Proceedings of the National Academy of Science of the USA* **90**, 5705–7.

Wirtz, P. (1982). Territory holders, satellite males, and bachelor males in a high density population of waterbuck (*Kobus ellipsiprymnus*) and their associations with conspecifics. *Zeitschrift für Tierpsychologie* **58**, 277–300.

Wittenberger, J.F. (1981). Male quality and polygyny: The "sexy son" hypothesis revisited. *American Naturalist* **117**, 329–42.

Wood, H.B. (1950). Growth bars in feathers. *Auk* **67**, 486–91.

Wright, J. (1992). Certainty of paternity and paternal care. *Animal Behaviour* **44**, 380–1.

Wu, C.-I. (1985). A stochastic simulation study on speciation by sexual selection. *Evolution* **39**, 66–82.

Xia, X. (1992). Uncertainty of paternity can select against paternal care. *American Naturalist* **13**, 1126–9.

Yasukawa, K. (1981). Male quality and female choice of mate in the red-winged blackbird (*Agelaius phoeniceus*). *Ecology* **62**, 922–9.

Zahavi, A. (1975). Mate selection — a selection for a handicap. *Journal of theoretical Biology* **53**, 205–14.

Zahavi, A. (1977). The cost of honesty (further remarks on the handicap principle). *Journal of theoretical Biology* **67**, 603–5.

Zahavi, A. (1987). The theory of signal selection and some of its implications. *Proceedings of the International Symposium on Biological Evolution* pp. 305–27.

Zahavi, A. (1991). On the definition of sexual slection, Fisher's model, and the evolution of waste and of signals in general. *Animal Behaviour* **42**, 501–3.

Zeh, D.W. and Zeh, J.A. (1988). Condition-dependent sex ornaments and field tests of sexual selection theory. *American Naturalist* **132**, 454–9.

Zink, G. (1970). The migrations of European swallows to Africa from data obtained through ringing in Europe. *Ostrich, Supplement* **8**, 211–22.

Zouros, E. (1981). The chromosomal basis of sexual isolation in 2 sibling species of *Drosophila*: *Drosophila arizonensis* and *Drosophila mojavensis*. *Genetics* **97**, 703–18.

Zuk, M. (1991). The role of parasites in sexual selection: Current evidence and future directions. *Advances in the Study of Behavior* **21**, 39–68.

Author index

Subject index